2/99

Food
Safety

SECOND PRINTING

Julie Miller Jones

eagan press
St. Paul, Minnesota, USA

The second printing of this book contains revised material on pages 20 and 22–29 to reflect changes in federal food labeling laws that were announced in early 1993 after the first printing.

Library of Congress Catalog Card Number: 91-73975
International Standard Book Number: 0-9624407-3-6

©1992 by the American Association of Cereal Chemists, Inc.
Reprinted 1993, 1995, 1998

Printed in the United States of America on acid-free paper.

American Association of Cereal Chemists, Inc.
3340 Pilot Knob Road
St. Paul, MN 55121-2097, USA

Dedication

I would like to dedicate this book to my parents, who instilled in me the love of learning and the idea that, as a woman, I needed a career, and to my husband David and my sons Christopher and Nicholas Jones.

Preface

We are bombarded daily with seemingly conflicting information about the safety of our food and the food supply. Newspapers, talk shows, books, news magazines, and even fashion magazines discuss food safety. Interpreting the information is frequently very difficult, even for professionals. Often it is difficult to determine the true picture when aspersions are cast on a substance used in the food supply, as the information may be in a research journal held by only one or two libraries in a state. In some cases, the information may be only partially presented, providing a distorted view of the situation. In other cases, the information about a substance may be accurately presented but neither the consumer nor the scientist is able to extrapolate from the reactions to large doses fed under experimental conditions to what will happen when extremely small amounts are consumed under normal dietary conditions.

It is my hope that this book will prove useful to students and professionals alike, especially those just entering or contemplating entering upon a career in food or nutrition. While the information in each chapter may or may not be news to experienced researchers in the field of its particular subject matter, all 15 chapters taken as a whole should provide a perspective on food safety issues not usually available in a single volume. For each issue, the risks and benefits to consumers are considered, as well as environmental, agricultural, and economic concerns. Widely available, inexpensive food is among the benefits considered important for the whole of society.

I hope that food scientists and other food professionals, dietitians and professional nutritionists, food editors and other journalists, home economists, and students will all find this a ready reference for their food safety questions.

I would like to thank all of the staff at Eagan Press for their tireless work in helping to get the manuscript into book form and also Elwood Caldwell for his suggestions on the content of the chapters and his editorial expertise.

Julie Miller Jones

Contents

An Overview of Food Safety

CHAPTER 1

For many things in life, what we want is not necessarily the same as what actually is. Food clearly fits this truism. On an emotional level, we think of food as that which sustains life. This endows food with an almost spiritual quality. On a factual level, we are bewildered to learn that food is comprised of an array of chemicals put there by Mother Nature herself. This undermines our notion that food is an uncomplicated, pure source of essential nutrients. Despite being told that only a small percentage of dietary chemicals possess nutritional significance, we want desperately to cling to the myth that natural foods and combinations of them are inherently pure and salutary.

We don't mind learning that some of the constituents of food enhance nutritive value; in fact, we pay little attention to these discoveries because it is only what we expect. On the other hand, we feel betrayed when we learn that some constituents or combinations of natural foods contain things that decrease nutritional value or, worse yet, are toxic (Daniel, 1991; Kada, 1983; Rosin, 1982). This violates a basic sense of order that we feel ought to occur, and as a result we either remain uninformed or disbelieve the reports.

Since it fits much better with the way we believe things ought to be, we shift our attention from natural toxicity to toxicity from chemicals added to food either intentionally or unintentionally during production and processing. We also choose not to acknowledge that cooking, storing, and processing create new components and different interactions, which may either create toxic factors or render them completely innocuous.

Complexity is increased still more because of the interactions among body fluids and components of the diet. Such interactions may prove either beneficial or harmful. For example, human saliva has the inherent capacity to inactivate a wide spectrum of cancer-causing agents (carcinogens) that enter the body through food. This is true even though the inactivation capacity may be overpowered by excessive use of known carcinogens such as tobacco (Rosin and Stich, 1983). On the other hand, this very same fluid, saliva, is a rich source of nitrate, which can be

1

reduced to nitrite. Under certain conditions, nitrites react with proteins in the stomach to become carcinogenic nitrosamines. Such possible interactions make foods and food safety a complicated business—rarely a realization we want to face.

IS THERE A FOOD SAFETY CRISIS?

In Western industrialized countries, concerns about safe food have replaced those about adequate food. Many feel that such concerns are a product of our convenience society. However, as early as just after the Civil War, Dr. W. O. Atwater, a scientist in the U.S. Department of Agriculture (USDA) and director of the first nationwide program of human nutrition research, warned in *Harper's Weekly* that city people were in constant danger of buying unwholesome meat—dealers were unscrupulous and consumers uneducated. It was common knowledge to New Yorkers at that time that their milk was diluted with water. Not uncommonly, coffee was adulterated with charcoal, cocoa with sawdust, olive oil with coconut oil, and butter with margarine. Milk was preserved with formaldehyde, meat with sulfurous acid, and butter with borax (Foster, 1982). Upton Sinclair's book *The Jungle* warned of unwholesome meat and unsanitary conditions.

Thus, concerns about safe food are definitely not new in the United States. As in Atwater's day, we seem daily to be faced with reports of a food or food constituent whose safety has come under scrutiny. In the face of these frequent allegations, it is hard not to succumb to the belief that a food safety crisis exists.

Conflicting Views

Articles in the popular press claim that we are overfed and that our food has been overprocessed, oversalted, oversaturated, and over-sugared, in the end leaving us undernourished. As if such charges weren't bad enough, many articles claim that our food is devitalized; colored; filled with chemicals, drugs, and synthetic ingredients; polluted by agricultural and environmental chemicals; and grown on impoverished land puffed up by the use of chemical fertilizers and other aids. Moreover, the chemicals used in the growing or processing of food are frequently alleged to cause adverse effects, even cancer, in humans or laboratory animals.

Articles with opposing views, frequently in the medical and scientific press rather than the popular press, state that these claims are either totally untrue or blown out of proportion. In fact, most bona fide nutritionists and academic, corporate, and government food scientists, including those from the Food and Drug Administration (FDA), claim that the United States has the safest, least expensive, and most varied food supply in the world (Foster, 1982). Unfortunately safety suffers from having a pretty boring newsbite and tends to be discussed only in stories that shock and scare.

The success of delivering an enormous variety of food for diversified lifestyles has created some of the problem. Consumers are now faced with the complex task of selecting the right balance from a vast array of food choices. It is my strong contention that malnourishment that exists in the United States results in the majority of cases from poor personal food choices rather than from a deficient food supply or even inadequate access to food.

DEFINITIONS

In any discussion of food safety, some agreed-upon definitions of *safety, hazard*, and *toxicity* are crucial, since these are basic concepts.

Box 1.1

A Passage from Upton Sinclair's 1906 novel, *The Jungle*

With one member trimming beef in a cannery, and another working in a sausage factory, the family had a first-hand knowledge of the great majority of Packingtown swindles. For it was the custom, as they found, whenever meat was so spoiled that it could not be used for anything else, either to can it or else to chop it up into sausage. With what had been told them by Jonas, who had worked in the pickle rooms, they could now study the whole of the spoiled-meat industry on the inside. . . . Jonas had told them how the meat that was taken out of pickle would often be found sour, and how they would rub it up with soda to take away the smell, and sell it to be eaten on free-lunch counters; also of all the miracles of chemistry which they performed, giving to any sort of meat, fresh or salted, whole or chopped, any color and any flavor and any odor they chose. . . . Also, after the hams had been smoked, there would be found some that had gone to the bad. Formerly these had been sold as "Number Three Grade," but later on some ingenious person had hit upon a new device, and now they would extract the bone, about which the bad part generally lay, and insert in the hole a white-hot iron. After this invention there was no longer Number One, Two, and Three Grade—there was only Number One Grade.

(From Chapter 14)

Absolute safety is the assurance that damage or injury from use of a substance is impossible. Any ethical scientist must concede that some risk exists with any food, any chemical or, indeed, any human endeavor. Thus, absolute safety is unattainable. This is extremely frustrating to the average consumer, who would like to see a good eating "Seal of Approval" guaranteeing 100% safety before consuming any food.

Foods safe under normal conditions will never qualify for a seal of approval if they are consumed in excessive quantities or used in an unusual manner. Relative food safety can then be defined as the practical certainty that injury or damage will not result from a food or ingredient used in a reasonable and customary manner and quantity (Hall, 1988; Hall, 1991).

Relative Safety

The two following examples illustrate the difference between absolute safety and relative safety. Most would agree that peanut butter is safe to eat, yet peanut butter causes an average of seven asphyxiations in young children each year. All would agree that water is not only safe but essential. However, even pure water is "poisonous" if it is

Figure 1.1. Consumers may be confused by conflicting reports in the media.

not used in a reasonable and customary manner and quantity. A Massachusetts woman died from drinking too much water. In this case, the woman's mother had recently died of cancer, and the young woman believed that drinking lots of water would purify her internal organs so that she would not die of cancer. She didn't! She died of renal shutdown caused by drinking many gallons of water in a short period of time (Minneapolis Tribune, 1977). Thus, any substance, no matter how beneficial or how low its toxicity, has some level or condition of use that may be injurious.

Furthermore, food safety refers not only to the food itself, but also to the person ingesting it. Foods considered safe for most people when used in a reasonable and customary manner and quantity can be extremely toxic, even lethal, to certain sensitive or allergic individuals. For example, properly cooked fish is considered both a safe and nutritious food. As a low-fat source of protein that is often rich in certain types of fatty acids (omega-3) believed to be beneficial, its consumption is frequently encouraged. However, for a person with a severe fish allergy, one bite of fish could prove fatal. Thus, the answer to the question of whether fish is safe to eat is a qualified yes. It depends upon who is ingesting it.

Toxicity vs. Hazard

An understanding of food safety is improved by defining two other basic concepts, toxicity and hazard. *Toxicity* is the capacity of a substance to produce harm or injury of any kind (chronic or acute) under any conditions. This might include the capacity to damage the developing fetus (teratogenicity), to alter the genetic code (mutagenicity), or to induce cancer (carcinogenicity). Furthermore, any deviation from normal is viewed as a possible negative effect, even though the change may seem to be positive, such as increased growth rates or enhanced nutrient absorption. The change is assumed to be negative until proven beneficial. *Hazard* is the relative probability that harm or injury will result when the substance is used in a proposed manner and quantity. Assessments of whether a food or ingredient is safe should not be based on its inherent toxicity but on whether or not a hazard is created.

Unfortunately, the public does not perceive the difference between hazard and toxicity. The press often scoops findings from journals or scientific meetings that describe the toxic effects resulting from ingestion of high levels of a food constituent. Upon reading or hearing of these press reports, many consumers assume that a constituent is hazardous at current consumption levels. This assumption occurs because of an inability to distinguish between toxicity and hazard.

Consumers are not alone. All too frequently, scientists themselves are unable to extrapolate from studies of high-dose toxicity to assess the actual human hazard from low-dose ingestion of the food or food constituent.

The inability to give an accurate assessment of hazard leaves a void that is filled in one of two ways: 1) self-appointed experts pontificate their assessment of the hazard with alacrity in books or on talk shows or 2) bewildered consumers, muddled by questions of safety, decide to believe the worst and stop eating the food anyway.

A recent example of the inability to differentiate between toxicity and hazard resulted from studies on methylene chloride, the chemical used to decaffeinate coffee. This chemical, also used to dry-clean clothes, was shown to be carcinogenic when inhaled by laboratory rats. Interestingly, it was not carcinogenic when rats ingested it. For the purpose of this example, let's assume that the effects of inhaling the compound are the same as those of ingesting it, which they are not. The amount of methylene chloride residue that remains in the decaffeinated coffee would necessitate drinking 50,000 cups of decaf coffee per day to reach the dose level that caused cancer in rats. Coffee itself contains over 350 compounds that have been identified. Many of these natural compounds at 50,000 times their customary dose have toxicities much greater than the toxicity of methylene chloride. Even if all the naturally occurring chemicals didn't exert adverse effects, excessive water consumption would clearly cause renal shutdown long before the hazardous dose of residue was reached. The FDA calculated the increased cancer risk from drinking two and a half cups of decaffeinated coffee per day to be, at most, one in a million—which is pretty close to zero!

Now, admittedly, two alternative, ostensibly natural, processes are available for the decaffeination of coffee. One uses ethyl acetate, a compound produced naturally during the ripening of fruit, the other, carbon dioxide and water. Decaffeinated coffee processed by these

---

Box 1.2

Basic Concepts

Toxicity = Capacity to harm

Hazard = Likelihood of harm

Safety = No likelihood of harm when used normally

methods has the allure of being safer because it doesn't employ any chemicals that are carcinogenic. However, the degree to which it is safer is miniscule, and the difference in the cost of the methods is not.

FOOD AND SCIENCE IN THE INFORMATION AGE

The decaffeinated coffee story is just one example of scientific advances creating more data than can be rationally managed or fully understood. Analytical chemists can now measure levels that toxicologists are unable to evaluate for biological significance. The zealous attempt to make food safer through microscopic scrutiny of food constituents has made researchers and consumers aware of small risks that were unknown in the past. Analytical chemistry has become so sophisticated that we may soon have enough data to ban the whole food supply!

Scientists and academicians believe that the only food safety crisis is that which exists in people's minds as a result of incomplete reporting of scientific information coupled with distortion of the facts by fearmongers, hucksters, and misdirected "consumer interest groups." Dr. Sanford Miller (1984), former director of the FDA's Bureau of Foods, said, "We live in neo-muckraking times, and since muckrakers need to have muck to rake, each report of a possible problem of food is treated as if it represented a major threat to individual life and to the survival of the civilization." In addition, financial support of many of these consumer interest groups is dependent on keeping their names before the public. They thus seek muckraking issues to garner donations and support.

The food industry itself is frequently guilty of delivering mixed messages. The scientific and technical affairs staff of one company may spend time trying to assure the consumer that a certain product, additive, or ingredient is safe. At the very same time, the marketing department of the same company may be advertising another product in their line as not having any artificial additives or preservatives. No wonder consumers are confused and unsure where to place their trust!

The word *crisis* implies a turning point or decisive stage in the progress of anything. This book looks at many food safety issues. With facts presented from all sides of these issues, a decision can be made about whether real crises exist or whether they have been created. If the crisis is real, we can make recommendations for managing it in such a way that will prevent a turn for the worse and enable a change for the better. If the crisis is created, a massive public education program will need to be launched.

WHAT ARE THE FOOD SAFETY ISSUES?

Specific food safety concerns differ markedly depending on the segment of the population involved. Consumers are most concerned about pesticides and additives. Since both of these are linked in the consumer's mind to cancer, they create great fear. A 1986 study by the Food Marketing Institute showed that 76% of consumers considered pesticides a serious hazard. Additives still concern consumers, although the percentage who are very concerned has decreased in the past few years (Bruhn, 1991; FMI, 1986, 1990; Lecos, 1984). Irradiation concerns a specific group of consumers. Many consumers stated that they would prefer irradiation if it had the capability to eliminate additives and pesticides (Bruhn et al, 1986).

The level of concern about any one of the issues is dependent on the level of attention it receives from the media. However, several issues have been noted for which both interest and concern have risen steadily in the past decade. One such issue is nutrition; another is fear of product tampering. Ironically, fear of product tampering is on the rise at the same time as increased use of self-serve barrels in grocery stores and co-ops—avenues for all kinds of contamination and tampering.

Microbiological Contamination, the Top Priority

It is also interesting, perhaps even alarming, that most consumers are not concerned about microbiological contamination, despite solid

---

Box 1.3

Food Safety Issues

Additives, colors, and flavors
Antibiotics and other feed additives
Fertilizers and other growing aids
Irradiation
Microbiological contamination
Naturally occurring food toxicants
Nutrition
Pesticides
Pollutants (e.g., PCBs)
Processing, packaging, and labeling
Tampering

(Expanded from Hall, 1988)

evidence that, of all the hazards, it is the one most likely to occur. Many homes have unsafe food storage and preparation practices. These may cause flulike and other illnesses that are never traced to improper handling of food. Consumers rarely consider their own food practices a hazard (Gravani and Williamson, 1991; Rucker et al, 1977).

Microbiological contamination of food is the FDA's top priority, as measured by both dollar outlay and staff activity. Although only 6% of the reported cases of foodborne disease in 1981 (Bryan, 1982) involved microbial contamination of processed food sold at grocery stores, it was still considered the most likely hazard to occur from manufactured food and therefore the top priority. (The other 94% of the cases were from incidents in a home or restaurant.)

Extraneous Matter, GMP, HACCP

Like the FDA, the food industry is most concerned about the microbiological safety of its products. In addition, many quality control checks are in place that help assure that foods are free of extraneous matter such as glass, machine filings, and insect parts. Interest in these two areas (microbiological and physical hazards) stems from experience that marks them as the hazards most likely to occur during manufacturing. In addition to being regulated by the FDA, large food companies in the United States adhere to a code of manufacturing practice known as Good Manufacturing Practice (GMP). This code helps to assure that products are manufactured under conditions of proper storage and sanitation. Many also employ an elaborate system known as HACCP (hazard analysis and critical control points) to make sure that there is no chance for contamination or error during processing.

Product Tampering

Industry has recently increased its activity in the area of product tampering. Since a fatal incident in 1984 involving malicious addition of poison to the analgesic Tylenol, over 100 incidents of food tampering have occurred. Even though the current risk of ingesting food from a package that has been tampered with is far less than the risk of being struck by lightning, the industry realizes that it must be proactive with this issue. As a result, tamper-resistant packaging, such as seals around bottle tops, and tamper-evident packaging are rapidly taking over.

Pesticide Residues

In the later part of the 1980s, the FDA raised pesticides to a higher level of concern. The elevation was not caused by an increase in adverse effects known to be caused by pesticides, but rather by reports that many pesticide chemicals and inert ingredients in pesticides had not been adequately tested using the most up-to-date methods. Because of this, Congress mandated the FDA to undertake a testing program for these substances (Sun, 1984).

Nutrition

Historically, the food industry was not very concerned about nutrition. Recently, however, industry activity in this area has increased dramatically (Sloan and Curley-Leone, 1984). Pronounced consumer interest has forced the industry to provide nutrition information, and consumer demand for products with reduced calories, fat, cholesterol, and sodium has initiated a whole array of products to fill this need (Fig. 1.2). The connection between diet and chronic disease has prompted the industry to make extensive use of health claims.

Box 1.4

Tamper-Evident Packaging

Film wrappers: A transparent film is wrapped securely around a product container. Usually the film bears a message such as "Do not use product if the film is broken."

Sealed cartons: All flaps of a carton are securely sealed, and the carton must be visibly damaged when opened.

Bubble packs: The product and the container are sealed in plastic and mounted in or on a display card. The plastic must be torn or broken to get at the product.

Shrink seals and bands: Bands or wrappers with designs or writing are shrunk by heat or drying to seal the union of the cap and container. The seal must be cut or torn to use the product.

Pouches: The product is enclosed in a pouch that must be torn to reach the product.

Seals: Paper or foil with a distinctive design is sealed over the container under the cap or over the carton flaps and bottle top. The seal must be broken to get to the product.

Breakable caps: This container has a cap that breaks away partially or completely.

(Data from Brody, 1983)

Activity and interest in nutrition have increased not just for the food industry but for the consumer and the FDA as well. Nutritional imbalance is clearly a hazard that does occur. Deficiencies that nutrition surveys always flag for certain population subgroups include vitamins A and B-6, iron, magnesium, and calcium. Nutritional excesses are an even greater problem than deficiencies in the United States. Excess calories, fat, sodium, and even protein are thought to put the population

Figure 1.2. Food processors have created many new products in response to consumer demand. (Courtesy AVEBE America, Inc.)

at risk for degenerative diseases such as cancer, hypertension, and coronary heart disease (DHHS/PHS, 1988; NAS/NRC, 1989).

Along with increased activity and interest has come an attempt to fix the blame. Some people blame the food industry for poor nutrition. Occasionally the charge is justified because some segments of the industry offer foods that are high in salt and fat (especially saturated fat) and low in fiber. On the other hand, no consumer is forced to purchase any of these products. Unfortunately, consumer food choices are mostly based not on nutritive value but on taste. Even when the industry offers nutritious items that consumers have requested, consumers may not actually buy them. In other words, we desire to eat one thing with our minds and another with our mouths.

Natural Toxicants

Natural toxicants, including carcinogens, often go unnoticed as a consumer concern and are of only moderate concern to the industry and the FDA. One reason why these remain unnoticed is that the media do not focus on this area, and the laws deal only with hazards created by food additives, not with hazards naturally associated with foods. The other is the consumer belief mentioned earlier, that natural foods couldn't possibly cause any problems. They are risks that appear to consumers as both old and familiar and therefore without any shock value. In actuality, natural toxicants can be hazardous, especially if the diet is distorted by excessive consumption of certain foods or the omission of foods that provide essential nutrients. The 1988 passage of a voter initiative in California (known as Proposition 65), which requires the labeling of all carcinogens in food, may have awakened consumers to the area of natural carcinogens and toxicants.

Even though government, industry, consumers, and consumer activists have different degrees of concern for the various issues that comprise food safety, these various segments can and must work together to ensure that our food supply is and remains safe. The fact that these various and often antagonistic segments have different perspectives creates the crucially needed checks and balances that can maintain our food supply as the safest possible.

REFERENCES

Bruhn, C. M. 1991. Consumer perceptions: Safety means more than microbiology. Presented at the annual meeting of the Institute of Food Technologists, Dallas, June 2.

Bruhn, C. M., Schultz, H. G., and Sommer, R. 1986. Attitude change toward food irradiation and conventional and alternative consumers. Food Technol. (Chicago) 40(1):86-91.

Brody, A. L. 1983. Don't tamper with packaging. Cereal Foods World 28:520.

Bryan, F. L. 1982. Microbiological Safety of Foods in Feeding Systems. Am. Board of Military Suppliers Rep. 125. National Academy Press, Washington, DC.

Daniel, J. W. 1991. Meeting report: Naturally occurring toxins. Food Safe. Notebook 2(5):47.

DHHS/PHS. 1988. The Surgeon General's Report on Diet and Health. U.S. Dep. Health Hum. Serv., Pub. Health Serv., Washington, DC.

FMI. 1986. Consumer Attitudes and Trends. The Research Division, Food Marketing Institute, Washington, DC.

FMI. 1990. Consumer Attitudes and Trends. The Research Division, Food Marketing Institute, Washington, DC.

Foster, E. M. 1982. Is there a food safety crisis? Food Technol. (Chicago) 36(8):82-83, ff.

Gravani, R. B., amd Williamson, D. M. 1991. Consumer food preparation practices: Report of a national survey. Presented at the annual meeting of the Institute of Food Technologists, Dallas, June 5.

Hall, R. L. 1988. Food Safety in the Year 2000. Speech at the IFT/AMA Conference, Washington, DC, March 1988.

Hall, R. L. 1991. Toxicological burdens, and the shifting burden of toxicology. Presented at the annual meeting of the Institute of Food Technologists, Dallas, June 2.

Kada, T. 1983. Desmutagens: An overview. Pages 63-67 in: Carcinogens and Mutagens in the Environment, Vol. I. H. F. Stich, ed. CRC Press, Boca Raton, FL.

Lecos, C. 1984. Pesticides and food: Public worry no. 1. FDA Consumer 18:12-13, 15.

Miller, S. A. 1984. Food safety legislation—The uneasy relationship between science and politics. Concepts Toxicol. 1:159-169.

Minneapolis Tribune. 1977. Woman dies from too much water. Feb. 25.

National Academy of Sciences/National Research Council (NAS/NRC). 1989. Diet and Health: Implications for Reducing Chronic Disease Risk. National Academy Press, Washington, DC.

Rosin, M. P. 1982. Inhibition of genotoxic activities of complex mixtures. Pages 259-270 in: Carcinogens and Mutagens in the Environment, Vol. VI. H. F. Stich, ed. CRC Press, Boca Raton, FL.

Rosin, M. P., and Stich, H. F. 1983. The identification of antigenotoxic/anticarcinogenic agents in food. Pages 141-154 in: Diet, Nutrition, and Cancer. D. Roe, ed. Alan R. Liss, New York.

Rucker, M. H., Tom, P. Y., and York, G. K. 1977. Food safety—What do the experts say? J. Nutr. Ed. 9(4):158-161.

Sloan, A. E., and Curley-Leone, L. 1984. Food products in the 1980s. Cereal Foods World 29:360-363.

Sun, M. 1984. Better living through chemistry? Science 223(4641):1154-1155.

Regulating Food Safety

CHAPTER 2

HISTORY OF FOOD LAWS

Food laws and regulations existed in some form in most ancient cultures, as in modern cultures, to deal with food safety and consumer concerns of the day. Many scholars and theologians believe that the Mosaic dietary laws were written partially because of food safety concerns (Regenstein and Regenstein, 1979), although bacteria and parasites were not discovered until several thousand years later. For instance, prescriptions against eating meat that did not come from a cud-chewing animal with a cloven hoof (Exodus 23:19) are thought to be due to concerns about the disease since identified as trichinosis. (Pork, meat from a non-cud-chewing animal without a cloven hoof, can carry *Trichinella spiralis*, an organism that can form cysts in human muscle.) Thus, neither Jews nor Moslems eat pork to this day if they still follow dietary laws set out in the Old Testament. Another dietary law from the Old Testament bars the eating of milk and meat in the same meal—it says one should "not seethe a kid in its mother's milk" (Leviticus 21:9-12). This food safety concern may have stemmed from the possibility of spoilage in protein foods held in a hot climate with inadequate refrigeration, or it may have simply represented maintenance of an ethnic identity.

Just as biblical writings contained food laws dealing with food safety, so did the writings of other early cultures. Very old Chinese documents contain food laws. In addition, certain government officials were responsible for preventing the sale of adulterated food. Sanskrit law imposed fines on persons selling adulterated oils or spices. Records from Greece and Rome also indicate that unwholesome and adulterated food was a problem for them. The philosopher Pliny recommended the value of a kitchen garden for unadulterated market supplies. Galen, a physician who followed the tradition of Hippocrates, warned against use of adulterated products. Roman civil law tried to regulate fraudulent food practices by stating that anyone who substituted one substance for another or sold spoiled goods was liable for the offense of *stellionatus*.

Punishment for such an offense would result in temporary exile or condemnation to the mines.

During the Middle Ages, fraudulent practices were common and legislation was needed. Spices were in high demand and were often highly diluted. Flour often had chalk in it. Weights were also tampered with so that customers did not receive a fair weight. The first recorded food law known to be in the English language was proclaimed by King John in 1202 prohibiting the adulteration of bread. In 1266 a law was passed to protect against short weight and unsound meat.

Adulteration became more sophisticated with each passing century. According to Drummond and Wilbraham in *The Englishman's Food* (1940), gin in the 1750s included the following: oil of vitriol (sulfuric acid!), oil of almonds, oil of turpentine, lump sugar, lime water, rose

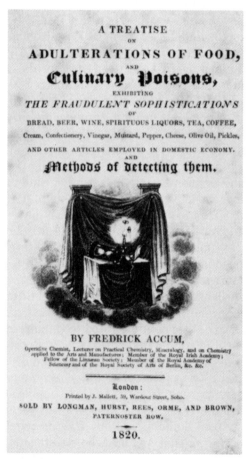

Figure 2.1. Accum's frontispiece page.

water, alum, and salt of tartar. While the lump sugar, lime water, rose water, and tartaric acid would not have been harmful, the other substances were indeed questionable. The word *intoxication* would take on a whole new meaning. In the American colonies, Massachusetts became the "Nutmeg State" because it had a big industry in wooden nutmegs. In the South, coffee was regularly adulterated with the bitter ground root of chicory (French endive). Coffee sold in Louisiana still may contain chicory; the difference is that coffee containing chicory is now labeled as such.

In 1820 a German chemist living in London, Frederick C. Marcus, whose pen name was Accum, published *A Treatise on Adulteration of Food and Culinary Poisons* (Figure 2.1). A pirated version was also published in the United States. He pointed out that many common foodstuffs were adulterated. As a result of Accum's exposé and other investigations, most notably that of the Lancet Commission, the British Parliament in 1860 passed a food law that remained Britain's basic food law for many years (Kinder and Green, 1978).

FOOD LAWS AND REGULATIONS IN THE UNITED STATES

Concern about food contamination also resulted in the passage of food laws in the United States, the first of which was adopted in the 1780s in Massachusetts. This law penalized the seller of diseased, unwholesome, and corrupted product. Another law penalized the seller of wooden nutmegs. Other states also had a variety of statutes dealing with food (Janssen, 1975).

An Act againſt ſelling unwholeſome Proviſions.

WHEREAS ſome evilly diſpoſed perſons, from motives of avarice and fil-
thy lucre, have been induced to ſell diſeaſed, corrupted, contagious or un-
wholeſome proviſions, to the great nuiſance of public health and peace :
    Be it therefore enacted by the Senate and Houſe of Repreſentatives, in Ge-
neral Court aſſembled, and by the authority of the ſame, That if any perſon
ſhall ſell any ſuch diſeaſed, corrupted, contagious or unwholeſome proviſions, whether for meat or drink, knowing the ſame without making
it known to the buyer, and being thereof convicted before the Juſtices
of the General Seſſions of the Peace, in the county where ſuch offence
ſhall be committed, or the Juſtices of the Supreme Judicial Court, he
ſhall be puniſhed by fine, impriſonment, ſtanding in the pillory, and
binding to the good behaviour, or one or more of theſe puniſhments,
to be inflicted according to the degree and aggravation of the offence.

    [This act paſſed *March* 8, 1785.]

Figure 2.2. The first comprehensive food adulteration law passed in the United States, by Massachusetts in 1785. (From Janssen, 1975)

At the federal level, no food legislation was passed until late in the 19th century, although the U.S. Department of Agriculture (USDA) began to be interested in the problem of adulterated feed and food earlier than that. People became increasingly aware that protection was beyond what each state could do alone.

Concern about adulterated tea brought about the first federal food law, the Tea Act of 1883. This law attempted to prevent the sale of adulterated teas, but it was useless because it provided no standards for tea and no method of enforcement. Fourteen years later a new Tea Act was passed that provided for a body of experts to set minimum quality standards and to inspect tea for adulteration. During this same period, several other federal food laws were passed. One taxed and

Figure 2.3. Harvey Wiley. (Courtesy FDA)

regulated the sale and manufacture of oleomargarine, and another dealt with "filled" dairy products (in which dairy constituents such as milk fat were replaced by cheaper materials). By 1906, when the first Food and Drug Act was passed, nearly 200 measures had been introduced into the U.S. Congress in various attempts to deal with adulterated and misbranded foods and feeds.

In 1883 Dr. Harvey Wiley became chief chemist of the Chemical Division of the USDA. He proceeded to direct a department study of food adulteration and published several bulletins between 1887 and 1893, showing that nearly all classes of foods were adulterated. While some of the adulterants were found to be simply fraudulent, others were poisonous. The ultimate passage of a comprehensive federal food law, the Food and Drug Act of 1906, was a triumph for Dr. Wiley, who was committed to increasing safety and decreasing fraud in the food marketplace (Janssen, 1975; Levine et al, 1985).

In that same year, Congress passed the Meat Inspection Act, which prevented the sale in interstate commerce of any unwholesome meat

Table 2.1. Descriptions of Adulterated and Misbranded Foods

Adulterated Food	Misbranded Food
Is or contains a poisonous or deleterious substance that may render it injurious to health. (If the substance is present in the food naturally, it is not considered adulterated if the quantity contained is not injurious.)	Has labeling that is false or misleading.
	Is offered for sale under the name of another food.
Is or contains any food or color additive deemed unsafe.	Imitates another food without the word *imitation*.
Is or contains any filthy, putrid, or decomposed substance.	Has a container that is made, formed, or filled such that it is misleading.
Is or contains any residue in excess of the established tolerances.	Has a label that does not contain the name of the manufacturer, packer, or distributor.
Is or contains any substance that increases bulk or weight.	Has a label that does not contain an accurate statement of the quantity or the ingredients.
Is prepared or packed under unsanitary conditions.	Has a label that makes a nutritional claim and does not contain nutritional labeling.
Is taken from any part of a diseased animal.	Has a label that does not contain the required information on the correct display panel or does not use the proper format for the nutritional label (see Table 2.2).
Is packed in material that contains poisonous or injurious material.	
Is damaged or has inferiority concealed.	
Is prepared with a valuable constituent omitted.	

product. This act made mandatory the inspection of animals and carcasses for interstate commerce and set up the inspection service.

Although the Food and Drug Act of 1906 was a major breakthrough, it was weak and ineffectual despite several attempts to strengthen it by amendments. A new law, the Food, Drug, and Cosmetic Act of 1938, included all the provisions of the 1906 law but also fixed the problems that made the former law unenforceable. This law ensured enforcement by assigning it to a federal agency, the Food and Drug Administration (FDA). It defined food, food preservatives, and artificial colors. Conditions under which a food would be regarded as adulterated or misbranded (Table 2.1) were outlined. The law barred the sale of food prepared under unsanitary conditions or containing any deleterious or unsafe substances. Standards of quality and fill were specified.

Standards of Identity

The Standard of Identity program was devised to protect the consumer against fraudulent practices. Before its enactment, products were sometimes labeled falsely. For instance, there are records of a product labeled as strawberry jam that contained no strawberries; it was pectin, colored and flavored to look like strawberry jam. From the beginning of the use of standards of identity until the 1970s, foods continued to be added to the list, until it included over 200 foods.

Testing to see whether a food complies with a standard of identity or with other tenets of the food law is done by the FDA. In addition, the FDA periodically tests food from all regions of the United States for levels of pesticides, radionuclides, and other contaminants in what is called the FDA Market Basket Survey. Food manufacturing facilities are periodically inspected to ensure that they meet standards of cleanliness and adhere to the guidelines of Good Manufacturing Practice as set out by industry and the FDA together (Bauman and Taubert, 1984).

The Nutrition Labeling and Education Act of 1990 requires that all standard-of-identity foods have complete nutritional labeling.

Box 2.1

Standards of Identity for Foods

Establish
   Minimum quality standards
   Required ingredients and the allowed quantities
   Permitted ingredients
   Standards for color, form, and packing medium, where appropriate
Describe specific processing requirements, if any

Amendments to the Food, Drug, and Cosmetic Act

The Food, Drug, and Cosmetic Act of 1938 has had five subsequent major amendments: the Pesticide Chemical Amendment (1954), the Miller Food Additive Amendment (1958), the Color Additive Amendment (1960), the Fair Packaging and Labeling Act (1966), and the Nutrition Labeling and Education Act (1990).

The Pesticide Chemical Amendment prohibited the marketing of raw agricultural products with pesticide residues above certain allowed tolerances. This is discussed further in Chapter 13.

The Miller Food Additive Amendment required proof of safety before the use of any new additives in the food supply. The cost of proving the additive safe was made the responsibility of the manufacturer who wished to market or use it or foods containing it. Passage of this amendment marked a substantial shift, in that food additives would need to be tested before being used in food, whereas previously, additives had to be proven unsafe by the FDA or other government body.

Substances in use at the time the amendment was passed could continue to be used if they were "generally recognized as safe" (GRAS) by experts qualified by training and experience to make such a judgment or if their safety could be assumed from prior use in the food supply. An important part of this amendment was its so-called Delaney clause (named after the congressman chiefly responsible for including it), which stated that no additive would be deemed safe if any quantity of it was found to induce cancer in humans or any animal species. Food additives are discussed further in Chapters 9 and 10.

The Color Additive Amendment was added in 1960 to establish rules for the use of color additives that were somewhat similar to the rules for other additives, but it allowed color and flavor substances found to be carcinogenic to be referred to a committee for evaluation of their safety. (Color additives are discussed in Chapter 10.)

FOOD LABELS

Labels perform several functions. In addition to their obvious function of attracting the consumer's attention, food labels are legal documents. According to the Fair Packaging and Labeling Act of 1966 (CFR 16:500-503), all labels must contain the same basic information, outlined in Table 2.2. Each piece of information that is required has a very important function. Obviously, the name, style, and form tell the buyer or consumer what the product is. The net weight allows price and value comparisons of different sized and competing products. The address enables the consumer to contact the manufacturer for additional information, product complaints, and other matters. Probably the piece of information that helps the consumer most is the ingredient statement. This gives the consumer some idea of the relative amounts of the various components in a product, since they must be listed in order by weight from the most to the least prevalent in the final product. Many consumers carefully scrutinize the ingredient statement for foods or additives that must be avoided, either by choice (as in the case of a vegetarian) or for medical reasons (as in the case of a person with an allergy).

For many years, foods subject to a federal Standard of Identity were exempt from having an ingredient statement. After 1967, optional ingredients were required to be listed. Since 1991, all ingredients, whether optional or mandatory under the Standard, must be listed. The original exemption was granted because the ingredients allowed and their proportions were set by law. In the 1930s when the Food, Drug, and Cosmetic Act was passed, most food shoppers were familiar enough with cooking skills to know what ingredients went into such foods, so the lack of an ingredient statement did not cause any problems. Now, however, some consumers are so far from the basics of food prep-

Table 2.2. Food Label Musts

Required Item	Extra Information Needed	Exemptions
Common or usual name of the product	Style and form, if that is important	...
Net weight	All packages must be labeled with both English and metric units.[a] If package contains between 1 and 4 pounds, its contents must be stated in terms of total weight in ounces and also weight in ounces and pounds.	
Name, address, and zip code of manufacturer, distributor, or packer	...	...
Ingredients in order by weight	...	Standard of Identity foods Spices, colors (except FD&C #2), and flavors
Statement that product contains artificial color or flavor (if any)	...	Butter, cheese, and ice cream
"USDA Inspected and Passed" on all processed meat and poultry products	...	...

[a]Takes effect in February 1994.

aration they would not know, for example, that creamed corn contains starch or modified food starch.

The Nutrition Label

In 1973 the FDA promulgated regulations about nutritional labeling. These were revised by the Nutrition Labeling and Education Act (NLEA), which was passed November 8, 1990. Regulations required by the act were published in November 1991. These regulations were subject to a comment period, and final regulations were published in January 1993. The required nutrition label formats are given in Box 2.2. Under NLEA, all packaged food, from pickles to caviar, must be labeled. The USDA has promulgated similar labeling regulations for items containing meat so that all foods in the United States will be labeled in the same format and with nearly identical rules. The 20 most common fresh fruits and vegetables and the 20 most common fresh fish items must be voluntarily labeled in over 60% of grocery stores, or this will also become a mandatory provision. The only exempted foods are those served by restaurants and delis, those with very small packages (less than 12 square inches of available label space), those produced by companies with very small annual sales, and those that are not a significant source of nutrients, such as spices.

The nutrition label is required to include information on total calories and calories from fat and on amounts of total fat, saturated fat, cholesterol, sodium, total carbohydates, dietary fiber, sugars, protein, vitamin A, vitamin C, calcium, and iron. Manufacturers may voluntarily declare information on calories from saturated fat and amounts of polyunsaturated and monounsaturated fat, soluble and insoluble fiber, sugar alcohols, other carbohydrate, potassium, additional vitamins and minerals for which reference daily intakes have been established, and the percent of vitamin A present as β-carotene. The information reflects the nutrient content of the food as packaged. The manufacturer may also choose to give nutrition information of the food as prepared. All food labels must bear this information by May 1994.

Serving sizes have been determined in the regulations. Therefore, a single 12-ounce can of pop can no longer be called two servings, and serving sizes cannot be made abnormally small so that foods look like they are lower in fat and sodium than they really are. Serving sizes must be given in common household measures as well as by weight.

The law requires that the nutrient values printed on each package should be obtained from laboratory analysis or from approved databases. Values for fresh fish, meat, and produce may be derived from handbooks. Added vitamins and minerals must be at 100% of the stated amount at the end of the anticipated shelf life of the product. The inherent

Box 2.2

Examples of Nutritional Labels[a]

Shortened format:
label for vegetable soup

Simplified format:
label for soft drink

Nutrition Facts

Serving Size 1 cup (245g)
Servings Per Container 2

Amount per Serving

Calories 55   Calories from Fat 20

	% Daily Value*
Total Fat 1 g	**2%**
Sodium 800mg	**33%**
Total Carbohydrate 31g	**11%**
Dietary Fiber 4 g	**16%**
Sugars 0g	
Protein 2g	

Vitamin A 20% • Vitamin C 4% • Iron 2%

Not a significant source of saturated fat, cholesterol, and calcium.

* Percent Daily Values are based on a 2,000 calorie diet. Your daily values may be higher or lower depending on your calorie needs:

	Calories:	2,000	2,500
Total Fat	Less than	65g	80g
Sat Fat	Less than	20g	25g
Cholesterol	Less than	300mg	300mg
Sodium	Less than	2,400mg	2,400mg
Total Carbohydrate		300g	375g
Dietary Fiber		25g	30g

Calories per gram:
Fat 9 • Carbohydrate 4 • Protein 4

Nutrition Facts

Serving Size 1 can (240 ml)

Amount per Serving

Calories 145

	% Daily Value*
Total Fat 0 g	**0%**
Sodium 20mg	**1%**
Total Carbohydrate 36g	**12%**
Sugars 36g	
Protein 0g	**0%**

* Percent Daily Values are based on a 2,000 calorie diet.

[a]Federal Register (1993).

vitamin and mineral content may deviate as much as 20% from the labeled value in an individual sample. To be in compliance with the law, some food manufacturers understate the vitamin and mineral value in the product to be sure it will be in compliance even at the end of its shelf life. On the other hand, calories, fat, cholesterol, and sodium must not be labeled at less than they are in the product.

The FDA does not accept responsibility for monitoring the accuracy and format of every food label, although there are very specific rules about the format and the placement of the nutrition label, and any variance in the allowed format could cause a product to be judged misbranded. The FDA depends primarily on voluntary compliance. The burden of assuring accuracy and legibility rests with the food industry, although the FDA will take action if label information appears to constitute misbranding or otherwise violates the Fair Packaging and Labeling Act. The USDA, in contrast, requires premarketing approval of labels for all foods (meat- and poultry-based) under its jurisdiction.

Label Descriptors

In addition to nutritional labels, many products also have what are known as nutritional descriptors such as "light," "low-calorie," or "very low sodium." Until the NLEA, many of these terms had no legal definition and were therefore used in many different ways by companies, leading to a lot of consumer confusion and controversy.

Box 2.3 gives descriptors allowed by the NLEA. Low-calorie foods can be so labeled if they contain no more than 40 Calories per reference amount. A reduced-calorie food must have calories reduced by at least one third and must be labeled to compare it to the food it is replacing. Foods normally low in calories must be labeled so as not to confuse the consumer. For instance, celery may not be labeled as low-calorie celery, as some might think that labeled celery has fewer calories than unlabeled celery. For these foods, the label may read: celery, naturally low in calories.

Products making any cholesterol claim are required either to have less than 2 grams of saturated fat per serving or to clearly state very near the cholesterol claim the level of fat in the product. The FDA

Figure 2.4. Meat and poultry inspection mark.

is concerned that cholesterol labeling not be used in a misleading manner. Therefore, the labeling rules do not allow labeling of products that are always cholesterol-free as such unless the label clearly states that all foods of this category are cholesterol-free, not just the brand in question. Examples of products on which this has been abused are canned pineapple and peanut butter. Since these products are from plants, all items in the group are cholesterol-free.

Special fat-labeling rules are allowed on meat and poultry. These are allowed to have the descriptor "lean" if they contain no more than 10% fat and "extra lean" if they have less than 5% fat. Some consumer activists object to the break given to meat and poultry; others think

---

Box 2.3

New Terminology[a]

Free: Contains no more than an amount that is "nutritionally trivial" and unlikely to have a physiological consequence.

Fresh: Can refer only to raw food that hasn't been processed, frozen, or preserved or to freshly baked bread.

High: A serving provides 20% or more of the recommended daily value.

Less: Term may be used to describe nutrients if the reduction is at least 25%.

Light: Term may be used on foods that have one third fewer calories than a comparable product. Any other use of "light" must specify whether it refers to the look, taste or smell; for example, "light in color."

More: Term may be used to show that a food contains at least 10% more of a desirable nutrient, such as fiber or potassium, than a comparable food.

Source of: A serving has 10–19% of the recommended daily intake of the nutrient.

Calorie Free: Has less than 5 Calories per reference amount.

Low Calorie: Has less than 40 Calories per reference amount and per 50 grams of food if reference amount is small.

Reduced Calories: Has 25% fewer Calories per reference amount than the comparison food.

Cholesterol Free: Has less than 2 milligrams of cholesterol per reference amount.

Low in Cholesterol: Has 20 milligrams or less cholesterol per serving and per 100 grams of food.

Reduced Cholesterol: Has 25% less cholesterol per reference amount than appropriate comparison food.

it is necessary to encourage and maintain fat reduction in these foods, which are important nutritionally but also can be a significant source of dietary fat.

Reference Values

The recommended daily intakes (RDIs) used on the nutrition label are a renaming of the U.S. RDAs, which were used on food labels starting in 1973. The values are calculated from the recommended dietary allowances (RDAs), as set by the National Research Council of the National Academy of Sciences. The actual RDIs for some nutrients were scheduled to be changed by the proposed NLEA regulations.

Fat Free: Has less than 0.5 gram of fat per reference amount and no added fat or oil.

Low Fat: Has 3 grams or less of fat per reference amount and per 50 grams of food if reference amount is small.

Low in Saturated Fat: Has 1 gram or less of saturated fat per serving and not more than 15% of the food's calories come from saturated fat.

(Percent) Fat Free: Term may be used only in describing foods that qualify as low fat.

Reduced Fat: Has at least 25% less fat per reference amount than appropriate comparison food.

Sodium Free and Salt Free: Has less than 5 milligrams of sodium per reference amount.

Very Low Sodium: Has less than 35 milligrams per reference amount and per 50 grams of food if reference amount is small.

Low Sodium: Has less than 140 milligrams of sodium per reference amount and per 50 grams of food if reference amount is small.

Reduced Sodium: Has no more than half the sodium of appropriate comparison food.

Sugar Free: Has less than 0.5 gram of sugar per reference amount.

Reduced Sugar: Has 25% less sugar per reference amount than appropriate comparison food.

[a]From Federal Register (1993).

However, a low passed in October 1992 placed a one-year moratorium on any changes. NLEA added daily reference values (DRVs) for food components for which an RDA does not exist. However the National Academy of Sciences has recommended that the intakes of these nutrients be limited, as excess intake of them is thought to be partly responsible for the onset of chronic disease. On the standard nutrition label, DRVs are calculated for both 2,000 and 2,500 Calories (Box 2.2).

Health Claims

Under NLEA, only approved health claims are allowed. The seven currently permitted health claims are those that make connections between: 1) fiber-containing grain or fruit and vegetable products and cancer, 2) fiber-containing (particularly soluble-fiber-containing) grain or fruit and vegetable products and coronary heart disease, 3) fruits and vegetables high in antioxidants and cancer, 4) calcium and osteoporosis, 5) dietary saturated fat and cholesterol and risk of coronary heart disease, 6) dietary fat and cancer, and 7) sodium and hypertension. However, health claims are not allowed if a food contains one or more of four substances in an amount above a disqualifying (disclosure) level. The substances and disclosure levels are: fat, 11.5 grams; saturated fat, 4 grams; cholesterol, 60 milligrams; and sodium, 480 milligrams. Nutrient content claims are allowed provided they are factual and use the terminology defined in Box 2.3. If a positive claim is made for one nutrient but the food contains another factor such as fat at above the disclosure level, then a statement about the level of fat must be made in the same place as the nutrient content claim.

REGULATION BY GOVERNMENT AGENCIES OTHER THAN FDA

In addition to the labeling and other aspects of foods covered by the Food, Drug, and Cosmetic Act, other government agencies also regulate aspects of food safety and fraudulent practices. The USDA inspects meat and poultry to ensure the wholesomeness of animals that are slaughtered under the provisions of the Wholesome Meat Act of 1967 and the Wholesome Poultry Act of 1968. Egg processing plants undergo continuous inspection as a result of the Egg Products Inspection Act of 1970.

The Agricultural Marketing Service of the USDA gives some raw commodities quality grades. These are usually unrelated to nutritional quality but refer primarily to texture, taste, and appearance. For instance, with beef grading, the highest grades will be the juiciest, the most tender, and the most flavorful, but they are not the most nutritious choice as they actually carry the most fat. A lower quality grade is the best nutritional buy.

The Federal Insecticide, Fungicide, and Rodenticide Act (FIFRA), passed in 1960s and strengthened in 1972 and 1988, requires the registration of economic poisons (as they are legally called) with the USDA, but pesticide tolerances in food are set by the FDA. The Environmental Protection Agency (EPA) evaluates their environmental impact. Thus, three agencies are involved in the question of pesticides in foods.

Other government bodies that don't seem so closely aligned with food or agriculture also regulate food. For instance, the National Marine and Fisheries Service of National Oceanic and Atmospheric Administration at the Department of Commerce conducts fish inspection (Sackett, 1982). The Department of Commerce also defines weights and measures, and shares in administration of the Fair Packaging and Labeling Act. This act requires that the contents of the package must be fairly represented both by the package and by the advertising claims made for the product. The Federal Trade Commission governs what is said about the product in advertising.

Federal laws and regulations are found in the *Code of Federal Regulations* (CFR), Title 21. In addition to the U.S. laws and agencies that monitor food federally, every state has a Department of Health or Department of Agriculture, or both, that work with the federal agencies and regulate intrastate activities. Some state agencies are very active. Food service, retail food stores, and fluid milk operations are regulated on the local level. In addition, there are laws in other countries as well as international agreements about regulations. The Codex Alimentarius Commission of the Food and Agricultural Organization in Rome administers an international food law code (Codex) (FAO, 1990; Kimbrell, 1982).

REFERENCES

Bauman, H. E., and Taubert, C. 1984. Why quality assurance is necessary and important to plant management. Food Technol. (Chicago) 38(4):101-102.

Code of Federal Regulations. 1990. Title 21, Foods and Drugs, Parts 1–199. Superintendent of Documents, U.S. Government Printing Office, Washington DC.

Drummond, J. C., and Wilbraham, A. 1940. The Englishman's Food. Jonathan Cape, London.

FAO. Codex Alimentarius, Abridged Edition. 1990. Secretariat of the Joint FAO/WHO Food Standards Programme. Food and Agriculture Organization, Rome.

Federal Register. 1973. V 38(13) pt. III:2124-2164, Jan. 19.

Federal Register. 1993. V 58(3), pp. 632-691, 2066-2964, Feb. 6.

Janssen, W. J. 1975. America's first food and drug laws. Consumer reprint. DHEW publication (FDA) (US) 76-1005.

Kimbrell, E. F. 1982. Codex Alimentarius food standards and their relevance to the U.S. standards. Food Technol. (Chicago) 36(6):93-96.

Kinder, F., and Green, N. R. 1978. Meal Management, 5th ed. Macmillan, New York. pp. 30-118.

Levine, A. S., Labuza, T. P., and Morley, J. E. 1985. Food technology. A primer for physicians. New Engl. J. Med. 312:628-634.

Regenstein, J. M., and Regenstein, C. E. 1979. An introduction to the Kosher dietary laws for food scientists and food processors. Food Technol. (Chicago) 33(1):89-99.

Sackett, I. D. 1982. Quality inspection activities of the National Marine Fisheries Service. Food Technol. (Chicago) 34(6):91-92.

Establishing the Safety of Food Components

CHAPTER 3

Sola dosis facit venenum.

Only the dose makes the poison.
—Paracelsus, 16th Century

WHO DECIDES WHAT IS SAFE FOOD?

Ancient rulers and kings determined food safety by using a food taster. The judgment was easy: if the taster lived, then the food was proclaimed "fit victuals."

As food tasters no longer find steady employment, we may assume that eating may be less risky now than in the days of Lucretia Borgia, when adding poison to food was commonly practiced to murder political enemies. Although we currently face little risk of being purposefully poisoned through our food, as in Renaissance Italy, some people fear that we may be poisoned inadvertently by chemicals in the food whose harmful effects have yet to be discovered.

Unlike the royal food tasters of the past, toxicologists today are primarily concerned with chronic rather than acute exposure to poisons. The tasters have been replaced by scientists from the FDA, academia, and industry. All three are armed with sophisticated analytical equipment capable of identifying the presence of a poison or a toxicant at the level of parts per million (ppm), parts per billion (ppb), or even parts per trillion (ppt)! Table 3.1 gives some examples of these amounts. In most cases, our ability to detect these poisons has far outstripped our ability to grasp the meaning of their presence.

Frequently, we are left wondering whether the 1538 dictum of Paracelsus that "only the dose makes the poison" still applies. Although there is no scientific dispute that this dictum does apply to classic toxic responses, many consumers and journalists fail to understand the concept. As a case in point, a network television program focusing on safety and quality of food in the United States included a report on the modern broiler industry. The report accurately stated that arsenic was added to the rations of commercially produced chickens. However, the report failed to state that arsenic at low levels is an essential nutrient for chickens. For poultry (and many other species, including humans) too little arsenic is just as noxious as too much (Nielsen, 1984). Arsenic, like other trace elements, is a clear-cut example of a case in which only the dose makes the poison (Table 3.2). However, most consumers recognize arsenic only as a poison, so they infer that poisons are being added to rations of commercially raised chickens.

While this controversy is fairly easy to quell with facts, another debate brews about whether the dictum of Paracelsus can be applied in all circumstances, particularly with respect to carcinogens. For some carcinogens, a clear dose response can be demonstrated, while for others it cannot. In the latter case, it is feared that a single molecule has the potential to initiate or promote a cancer growth.

WHO IS RESPONSIBLE FOR SAFE FOOD?

Food safety in the United States is regulated by both the food industry and the government. The food industry is very interested in ensuring that the food it sells is safe. In doing so, its members are protecting their own self-interest as well as that of consumers. The reputation of a company as one that sells safe food is crucial for consumer confidence. Only one death from a can of food with botulism can wipe out a several million dollar corporation, even one with a long history.

To ensure safety, companies employ quality control experts, food

Table 3.1. Examples of Small and Large Quantities

Size	Unit	Example
Small[a]	1 ppm	1 gram in a ton 1 drop of vermouth in 1,000 quarts of gin 1 mouthful of food in a lifetime
	1 ppb	1 drop of vermouth in 10,000 gal of gin 1 second in 32 years 1 pinch of salt in 2,000 pounds of potato chips 1,000 times less than 1 ppm
	1 ppt	1 grain of sugar in an Olympic pool 1 needle in a 100,000-ton haystack
Large[b]	A billion seconds ago A billion minutes ago A billion hours ago A trillion seconds ago	World War II just ended St. Paul was writing his epistles No known civilization 30,000 B.C.

[a] After Klaassen (1986).
[b] After Elton (1982).

chemists, and food microbiologists to continuously assess food safety. Despite the industry's emphasis on food safety, only 8% of consumers, according to a 1986 survey by the Food Marketing Institute (FMI, 1986), believed that food safety was a responsibility of the food industry.

By statute, the government has been charged with ensuring the wholesomeness of the food supply. Yet only 29% of consumers said that food safety was a responsibility of the government. The government has recently requested a more thorough look at food safety through a variety of its agencies (Benbrook, 1990; Scroggins, 1990). These renewed efforts may increase consumer confidence.

The survey also showed that 48% of the consumers surveyed felt personally responsible for ensuring their own food safety. Clearly, consumers should be responsible for inspecting packages for visible evidence of tampering, microbial growth, or other contamination. Further, food should be stored and used, and the product should be served, in such a way as to prevent microbiological growth and maintain a high-quality, safe product. However, it is ironic that we as consumers see ourselves as having greater responsibility for food safety than either the government or the food industry since we frequently lack the tools and the skills to assess many crucial aspects of food safety. For instance, most of us are unable to determine that our food is free of 1) harmful bacteria that don't affect the color, flavor, or appearance of the food, such as the very dangerous *Listeria*, 2) radioactive elements such as strontium-90, or 3) excessive levels of pesticides or pollutants such as polychlorinated biphenyls (PCBs). Another irony is that many food safety problems are caused by unsafe food practices in the home (Gravani, 1991).

The government monitors the food supply to try to protect us from many of the hazards we cannot detect. Furthermore, both international and national scientific commissions try to evaluate the toxicity of chronic and acute doses of substances that might exist naturally in food or might be added intentionally or inadvertently.

Table 3.2. Deficient, Safe, and Toxic Daily Levels[a] of Certain Trace Elements[b]

Element	Deficiency Level	Safe and Adequate Level for Humans[c]	Toxic Level	Fatal Level
Arsenic	<15–25 ng	Not established	...	0.76–1.95 mg/kg
Fluorine	<2 mg	2–10 mg	10–20 mg	>20 mg
Selenium	<50 μg	50–200 μg	200–1,000 μg	>1 mg

[a] ng = nanogram, μg = microgram, mg = milligram, kg = kilogram.
[b] Data from Freiden (1984), NRC/NAS (1989), and Spallholz (1989).
[c] According to The National Research Council (NRC/NAS, 1989).

TOXICITY TESTING OF FOODS

The FDA has a protocol for the testing of substances to be added to the food supply. The testing procedures are revised frequently to keep up with the ever-advancing field of toxicology (Squire, 1984). FAO/WHO has an expert committee of scientists from many countries of the world that meets regularly to evaluate any new data on the

Figure 3.1. FDA inspectors check a tank in an ice cream plant for any conditions that can promote the growth and spread of bacteria. (Courtesy FDA)

toxicity of additives, colorings, and pesticides used anywhere in the world (FAO/WHO, 1974).

How Are Toxicity Tests Conducted?

Toxicity testing usually involves feeding the chemical in question to laboratory animals. Ideally, the metabolic fate of the chemical in the body is determined early in the test procedure. To do this, the compound can be labeled with an isotope before it is administered to the animal. Metabolic studies then determine whether the substance is completely excreted, rapidly or slowly excreted, stored, or metabolized into a compound that is more or less toxic than the compound that was fed. If the ingested compound is metabolized into one or more other compounds, their metabolic fate and safety must also be determined (Kimbrough, 1984).

A major objective of these studies is to establish that the same patterns of absorption, distribution, metabolism, and elimination occur in the experimental animals used for the toxicity tests as in humans (Vettorazzi, 1980). Obviously, further testing should be done on animals with pathways for metabolism and disposition of the chemical that are most similar to the human pathways (Parke, 1979). Unfortunately, this is not always the case. Sometimes species are selected that are low-cost, convenient, and have low food consumption and short life spans. Apart from these considerations, it is important that the pharmacokinetics in experimental animals and humans be comparable when determining appropriate dosages for experimentation and when extrapolating animal data to predict human safety (Cordle and Miller, 1982; Farber and Guest, 1984).

Levels and Types of Exposure

Definitions of various types of exposure appear in Table 3.3. Toxic effects from a single exposure may be different from those produced by chronic exposure. In general, dilution of the dose reduces the effect. For instance, a single dose that produces an immediate and severe effect might produce less than half the effect when given in two doses and no effect when given in 10 doses over a period of a day or more. Chronic effects may result from repeated exposures if the chemical accumulates before being metabolized or excreted, if it produces an irreversible toxic effect, or if repeated exposures don't allow the system time to recover from damage.

After the metabolic studies, the effects of the various types of exposure are tested in experimental animals. Acute toxicity of the chemical is measured first. Federal regulations require that an acute toxicity test of the chemicals be made on at least two species of animals, with one species being a nonrodent. The chemicals are fed at different dose levels. Detailed records of growth and behavior patterns, feed consumption, external characteristics, and clinical values (such as those from blood and urinary analysis) are kept. After death, all animals are autopsied to evaluate internal organs for any gross or microscopic changes, to check for tumors, and to certify the cause of death. The acute tests help to predict which systems are most likely to be affected in lower-dose chronic studies (Turnball, 1984; Vettorazzi, 1980).

The usual way to report acute toxicity is with the LD_{50}, which stands for lethal dose 50. This is the dosage of a substance required to kill half the animals in the experimental group. Table 3.4 gives the relative degrees of toxicity of any substance, using numbers derived from the LD_{50} test. Table 3.5 shows lethal dose comparisons for some common products in household amounts and for some products in the more commonly reported scientific units, milligrams per kilogram. Box 3.1 gives an example of supertoxicity.

After the acute toxicity tests, subacute toxicity tests are made on two or three animal species—usually rats, mice, and dogs. The chemical in question is placed in the feed and drinking water for two to three months. During the feeding period, the animals are carefully observed. A dose response curve is established (Figure 3.2). The highest dose that produces no harm, in what appears to be the most sensitive species, is then used to estimate a safe dose for human consumption.

Table 3.3. Definition of Types of Exposure to Toxic Substances[a]

Level of Effect	Dose	Time of Exposure
Acute	High	A few days
Subacute	Moderate	A month
Subchronic	Low	One to three months
Chronic	Very low	More than three months

[a] Adapted from Klaassen (1986).

Table 3.4. Lethal Dose 50 Comparisons[a]

How Toxic	Approximate Amount Needed for LD_{50}[b]	Dosage Needed for LD_{50} (mg/kg of body weight of the animal)
Practically nontoxic	>1 qt	>15,000
Slightly	1 pt to 1 qt	5,000–15,000
Moderately	1 oz to 1 pt	50–500
Very	1 tsp to 1 oz	20–50
Extremely	7 drops to 1 taste	<5–20
Super	<7 drops	<5

[a] After Kimbrough (1984) and Hassall (1982).
[b] The dosage that will kill 50% of the test population.

This lowest level that produces no effect is called the no observable effect level (NOEL). The NOEL terminology was adopted in 1983 by the WHO/FAO committee on food safety. NOEL is used in setting allowed levels of food additives and pesticides (Kimbrough, 1984; Kolbye, 1984; Parke, 1979). In most cases, the NOEL value for the most sensitive experimental animal is *divided by at least 100,* and that becomes the maximum value that should be ingested by a human.

Chronic toxicity tests follow the acute and subacute toxicity studies. Animals are fed the compound for their complete normal life span. Since three to four years is the normal life span of a rodent, it is also the customary length of a chronic toxicity test. The dose levels in these studies range from a fairly low dose to dosages 100 to 1,000 times the amount that might normally be consumed by humans.

Such studies yield information on both cumulative toxicity and any chronic effects of low doses. Animals are monitored as described for acute tests. Two rodent and one nonrodent species are required by the protocol for chronic feeding trials. Both sexes must be used for these tests.

Table 3.5. Estimated Toxicity (Lethal Dose 50) of Common Products

	Amount	
Substance	In Pounds[a]	In Milligrams per Kilogram[b]
Common Household Products[c]		
Cloves, ground	6.5	...
Horseradish, grated	6.5	...
Whiskey, 86 proof	5.0	...
Sugar, granulated	4.2	...
Vanilla extract	4.0	...
Window cleaner	4.0	...
Gasoline	2.9	...
Nail polish	2.9	...
Cream of tartar	2.3	...
Soap, toilet	1.5	...
Herbicide, Tordon 22-K	1.5	...
Salt (NaCl)	0.7	...
Baking soda	0.5	...
Aspirin	0.3	...
Some Chemical Agents[d]		
Ethyl alcohol	...	10,000
Sodium chloride (salt)	...	4,000
Iron tablets (ferrous sulfate)	...	1,500
Nicotine	...	1
Dioxin	...	0.001
Botulinum toxin	...	0.00001

[a] For a 150-pound human.
[b] These units allow comparison with values commonly found in rat experiments.
[c] After Hodge and Downs (1961). Based on extrapolations from rat experiments.
[d] After Klaassen (1986).

The amount of material to be fed in toxicological studies is a source of some debate. For the chronic feeding studies, the levels fed are greater than the NOEL but less than the LD_{50}. For carcinogenicity testing, levels of less than 5% of the diet are suggested (Cabral, 1984). Some believe that even 5% of the diet represents a dose that would never occur in normal eating patterns. Others feel that since the experimental animal's life span is short, high doses are the only alternative.

Generational tests are undertaken to ensure that the animals reproduce normally and that the offspring show no teratogenic or mutagenic effects from the test substance. Tests last over three generations (Ehling et al, 1983).

Cost of the Testing Procedure

The testing of a food additive or pesticide is costly and time-consuming. The outlined test procedures take a minimum of five years and 130,000 separate measurements (Turnball, 1984). The costs of acute and chronic tests for each substance range from $500,000 to $1,000,000 in 1984 dollars. The teratogenic and mutagenic tests range between $250,000 and $500,000. Several volumes of data are generated for each substance.

Epidemiological Tests

Data on humans come from records available on persons who are

---

Box 3.1

A Supertoxic Substance

An example of a supertoxic substance is the heart drug digoxin. The dose level is 0.25 to 1 milligram (mg) per person. That is a dose of 0.0036–0.014 mg per kilogram (kg) for a 70-kg person. The lethal dose for humans is in the extremely toxic range, 10–20 mg per person. Thus, for this drug, only a 10-fold margin exists between the therapeutic dose and the lethal dose (Kimbrough, 1984). Obviously, the margin of safety for constituents of the food supply needs to be much larger.

The amounts of chemicals in our water and our food are usually measured in parts per million (milligrams of the substance per kilogram of food or liter of water) or parts per billion (micrograms of the substance per kilogram of food or liter of water). The average adult consumes about 2 liters (quarts) of water and, at most, 200 grams (1/2 pound) of a given food item. If the chemical were present in both food and water at a concentration of 1 ppm, the person would consume about 2.2 mg. For a 70-kg (150-pound) person, that would amount to 0.3 micrograms per kilogram per day, a very small amount.

occupationally, accidentally, or voluntarily exposed to the compound in question. For example, to asses the bladder carcinogenicity of saccharin, death certificates of diabetics were analyzed to ascertain whether they had a higher than normal rate of bladder cancer. Such epidemiological data must be compared with those from animal experiments. One of the factors that is compared is the route of exposure to the toxic substance. Exposure may occur by several routes, listed here in order from most toxic to least toxic: intravenous, inhalation, intraperitoneal, subcutaneous, intramuscular, intradermal, oral, topical (Klaassen, 1986). Routes of exposure must be compared with ways the population might be exposed to the substance in food, and test doses must be compared with amounts normally consumed (Ballantyne, 1985; Weil, 1984).

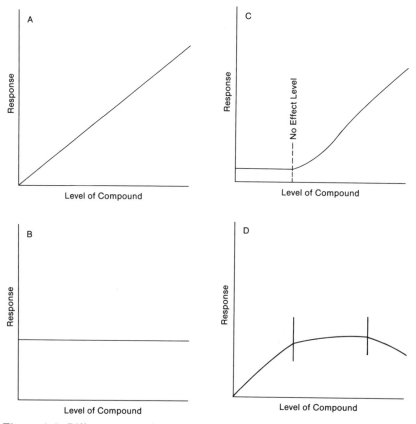

Figure 3.2. Different types of dose response curves. **A,** linear; **B,** no dose response; **C,** curvilinear; **D** dose response showing enhanced, optimal, and finally adverse effects of an increasing dose.

ESTABLISHING AND MONITORING SAFE LEVELS OF USE IN THE FOOD SUPPLY

Experts from the FDA and the FAO/WHO evaluate all the data from the tests and from human exposure. If a substance is deemed safe for use in food, allowable levels are determined. NOEL is used as the basis for establishing an acceptable daily intake (ADI) and a virtually safe level (VSL) for a compound. The ADI is the amount of substance that can be consumed every day for an entire lifetime with the practical certainty, on the basis of known facts, that no harm will result (Turnball, 1984). ADIs may be expressed for a group of related compounds. The VSL has been suggested for carcinogens for which a dose-response relationship has been established from linear extrapolation (Farber and Guest, 1984).

To set the ADI, NOEL is divided by at least 100 to ensure a margin of safety. Such a 100-fold margin of safety is necessary to allow for the fact that the number of animals tested is small compared to the size of the human population that may be exposed. Further, the most sensitive individual is assumed to be 10 times more sensitive than the most susceptible animal (Lowy and Manchon, 1984). For carcinogens, a safety factor of 1,000 has been suggested (Squire, 1984).

Food intake data must then be analyzed to make sure that no population group will exceed the ADI for the substance in question. A variety of methods is used, including dietary recall, food diaries, weighed food records, food frequency assessments, food disappearance data, and the FDA Market Basket Survey (Box 3.2). Each method of assessing food intake has its own inherent errors that may cause estimates of over- or underconsumption of the substance in question. Results from all these methods are pooled to increase the possibility of accurately predicting intake.

Many unanswered questions remain despite these attempts to ensure that the intake of various food constituents does not exceed the ADI. For instance, most methods fail to consider seasonal, regional, cultural, and socioeconomic differences in food consumption as well as effects on persons restricted to a few foods, such as infants (Lindsay, 1986).

Evaluation of all food substances for toxic hazards is a Herculean task. One food alone may have 500 minor chemical constituents, in addition to its seven to 10 major ones. Years could be spent evaluating the safety of all the substances in just one food.

Since testing of all the constituents is neither possible nor sensible, some rational approach must be considered. One suggestion is the use of a decision tree to categorize substances into low, moderate, or high risk (Cramer et al, 1978; Wodika, 1977). Factors such as the food's chemical structure and its similarity to other known hazardous materials, metabolism and storage in the body, frequency of use, and the like are used to determine the risk category. Testing efforts can then be concentrated on the high-risk group. High-risk substances typically have low LD_{50} values and appear to have immediate or prolonged irreversible effects. Frequency of exposure is assessed for the decision tree using the questions in Box 3.3.

WHAT ABOUT CARCINOGENS IN FOOD?

With respect to carcinogens in foods, the dictum of Paracelsus that "only the dose makes the poison" is considered by some to be as outdated as royal food tasters. Safe levels for carcinogens have not been established. One theory proposes that a single molecule of certain compounds is all that is necessary to initiate damage to the DNA—damage that in 20–30 years may cause fulminating cancer. The other theory is that

---

Box 3.2

The Market Basket

The Market Basket program was begun in 1961. The program assumes that the eater is a teen-age boy and that he consumes 4,000 calories. Programs have been added to include typical diets of six-month-olds and two-year-olds. Different foods have been added to account for our changing dietary habits, so that the foods now analyzed contain fast-food items and convenience foods in addition to staples and unprocessed foods. Foods are bought at 30 retail stores around the country, frozen, and shipped to an FDA laboratory for analysis. The foods are tested for several hundred chemicals, including a wide variety of pesticides; industrial chemicals, including polychlorinated biphenyls (PCBs); and metals such as cadmium, zinc, lead, and mercury. Of these chemicals, only 70–80 are found regularly, and in all but a few cases the levels found have been well below the acceptable standards. In those few instances where they were not, corrective actions were taken. One example was too much iodine in the diet from the extensive use of iodine sanitizing solutions. Practices in the industry were changed, and the FDA continues to monitor for high concentrations of this element. Another example was PCBs found in cereal boxes made from recycled office paper. Office paper had always been considered a clean source of paper for use in food packaging material. The advent of carbonless paper containing PCBs changed that. Luckily, the monitoring system was in place so that food packages could be manufactured to eliminate PCB contamination (Worthy, 1983).

a threshold level exists that must be exceeded before a substance has carcinogenic effects—this is an extension of the dictum of Paracelsus: only the dose makes the carcinogen. In either case, it is important to know the stages of carcinogenesis, as outlined in Box 3.4.

The One-Molecule Hypothesis

The first theory mentioned above is known as the one-molecule hypothesis. Believers in this hypothesis (Baldwin, 1979; Haseman, 1984; Pim, 1981) hold the following tenets.

1. Cancer is induced by the reaction of a single molecule with a single site on the DNA.

2. The reaction is irreversible and probably results in a neoplasm (tumorous growth).

3. The reaction is two-stage, and the stages may occur months— even years—apart.

Box 3.3

Questions to Determine Exposure Frequency

Who is exposed?
    Age
    Sex
    Race
    Health status
    How many?

By what route?
    Oral
    Dermal
    Inhalation
    Multiple routes

Through what
   chain of events?
    Production
    Manufacturing
    Storage
    Bioaccumulation
      waste or discharge

How are they exposed?
    Occupationally
    Environmentally
    As consumers

How frequently?
    Continuously
    Regularly, periodically
    Irregularly, but repeatedly
    Single incidence

In what amount?
    Grams per year
    Grams per incident
    Milligrams, per liter
      or square meter

(Adapted from Cramer et al, 1978)

4. A substance shown to be carcinogenic to any animal is carcinogenic to humans.

5. The effects of carcinogens are cumulative.

6. Since testing must be done with animals as human surrogates, doses tested must be higher than humans experience in a lifetime.

7. Well-nourished, otherwise pampered, experimental animals understate the risk of a substance.

8. It is impossible to set a safe level for a molecule known to be carcinogenic.

Support for the one-molecule hypothesis comes from the so-called megamouse study in which 2-acetylaminofluorene (AAF), a known bladder carcinogen, was fed for 33 months to 24,000 rats. The AAF was fed at seven dose levels from 150 to 30 ppm. As the dose level decreased, the number of animals in the group was increased to enable

---

Box 3.4

Stages of Carcinogenesis

1. Initiation

During this stage, some event or chemical begins the transformation of a single cell into a multicellular, malignant tumor. In most cases, this is believed to involve a permanent, irreversible change, probably consisting of a genetic mutation.

Once initiation has begun, the cell has the potential to develop into a tumor, provided that the other essential steps of carcinogenesis take place. If these other steps do not take place, the initiated cell may remain dormant indefinitely with no apparent effect on the organism.

2. Promotion

During this stage, the initiated cell clones itself and expands within a given tissue. As the number of these target cells increases, there are many sites where changes can produce fully malignant cells.

3. Progression

As the target cells multiply further, other changes such as chromosomal rearrangements, point mutations, and additions or deletions to genetic material, as well as nongenetic changes, may take place until the tumor consists of a heterogeneous population of cells. Eventually, one or more of these cells emerges with all the characteristics of a neoplastic cell.

(Adapted from Stott et al, 1989)

detection of the effects of very small doses. Results in this case show a molecule, AAF, for which no dose level appears to be safe (Klaassen, 1986). What this study may say is that for some carcinogens no threshold can be established; however, it is also clear that a strong carcinogen such as AAF would never be allowed as a pesticide or a food additive.

The Threshold Hypothesis

The other theory is called the threshold hypothesis. Believers in the threshold hypothesis (Baldwin, 1979; Francis, 1986a, 1986b; Frayssinet, 1984; Haseman, 1984; Kolbye, 1984; Lowy and Manchon, 1984; Slaga, 1989; Squire, 1984; Stott et al, 1989) believe that carcinogens, like other toxic compounds, have a no-effect level and a threshold dose. This hypothesis has the following tenets.

1. Animal systems have a variety of defense mechanisms able to detoxify and dispose of xenobiotic chemicals. Many factors determine whether a particular chemical is activated to become a carcinogen or is rendered innocuous.

2. The high doses used in many experiments for carcinogenicity can change normal processes of cellular activation and deactivation and can overwhelm natural defense mechanisms. Tumors may be due to secondary effects of toxicity and tissue damage from excessive amounts of the fed compound.

3. Even in cases where the DNA is actually damaged, repair of the damage has been documented. Only under conditions in which the repair mechanisms are exhausted will the damage become irreversible.

4. Not all species show the same susceptibility.

5. Dose should accurately reflect what reaches the tissue, not simply what is administered. For instance, some substances are quite toxic, but as they are absorbed with difficulty, they never reach the target tissue.

6. Accurate epidemiological data should have precedence over animal studies.

7. It is possible to establish a no-effect level.

Both theories are subject to criticism. Those skeptical of the threshold hypothesis cite evidence that some agents can directly alter DNA and conceivably could damage DNA with one molecule. Counterarguments show evidence that even tissues suffering extensive DNA alkylation did not necessarily become cancerous. In fact, they show that a chronic low dose of a chemical may force the DNA repair mechanisms to function more efficiently (Fournier and Thomas, 1984).

While it is conceivable that a single molecule might irreversibly damage the DNA, many believe it is statistically improbable (Borzelleca, 1984). The improbability of this happening is described in Box 3.5.

Use of High Dosages

Extrapolation from high doses to low doses is fraught with problems (Portier, 1989). At high doses, normal defenses and metabolism may be disordered. In some experiments, doses of noncancerous substances are so high that they are toxic in other ways. In these cases, tumors may result that are not related to any direct carcinogenicity of the material but simply to the fact that a toxic material was given at high doses over extended periods of time. Further, several scientists argue that single chronic tests overestimate the probability of a false positive result, i.e., finding that the molecule is a carcinogen when it is not (Ames and Gold, 1990; Kirshman, 1981). Also the model used for extrapolation is of critical importance. If the wrong one is chosen, it can under- or overestimate the risk by many magnitudes (Portier, 1989).

It is obviously unethical to experiment with humans, and accordingly, experimental animals must be used. Since the lifetime of most experimental animals is much shorter than that of humans, a large dose must be administered, which would be similar to the low, long-term chronic dose that a human might receive over an 80–90 year life span. Use of a human lifetime dose in the life span of an experimental animal

---

Box 3.5

The Improbability of a Single Molecule as a Carcinogen

Saccharin, a molecule considered to be mildly carcinogenic, is put forward to illustrate this point. An epidemiological study was done to assess the relationship between steady use of low doses of saccharin by 5 million people and cancer incidence. The expected increase in cancers among steady users of low doses of saccharin was not seen in this study. The average daily intake of saccharin was 0.1 gram or 3×10^{20} molecules of saccharin. Even with this large number of molecules, an increase in the cancer risk was not shown.

Another criticism of the single-molecule hypothesis is that all scientific experiments to determine carcinogenicity would be invalid, since it would be impossible to eliminate contamination by single molecules. For instance, a picogram (10^{-12} gram) of saccharin, an amount less than can be measured by sophisticated analytical techniques, contains more than two billion molecules. Thus, in any experiment, many contaminants may be present below analytical levels. If all that is needed is a single molecule, then no experiment could truly establish that increased cancer risk was actually due to the agent under study, as a few contaminating molecules left on the glassware or animal's cage from a prior experiment could dramatically alter the results.

may create as many questions as it attempts to answer (Crouch, 1983, 1989; Hallenbeck and Cunningham, 1986; Kolbye, 1984).

To further expand on the problem of high dosing, consider an example using vitamins or minerals. A human lifetime dose of vitamin A, iron, or selenium would kill or severely impair an experimental animal if it were given to a rodent during its three-year life span.

Besides the problem of high dosages, there is also the problem of compounds that behave differently in different species (Yang et al, 1989). For instance, benzene is carcinogenic to humans but not to rats; thalidomide is teratogenic to humans but not mice; and aspirin is teratogenic to mice but not humans (Baldwin, 1979; Reiches, 1981; Weil, 1984). A diet of 10% potato starch is toxic to other species but not to humans. Green beans cause pulmonary lesions in dogs and rats (Elias, 1982). The list can go on but certainly shows that species react very differently even to normal foods (Hallenbeck and Cunningham, 1986).

Different responses occur not only between different species (Byrd et al, 1990), but also between different strains. Even animals of the same strain but of different ages may have different responses (Clayson, 1985).

In addition to the different responses by animals, the problem is even more complex because many foods decrease or increase the toxicity or carcinogenicity of a particular constituent. For example, cabbage inactivates mutagens; fatty acids and ascorbic acid inhibit nitrosamines; phenols increase nitrosamines; and sulfite inhibits coffee mutagens. The preceding are just a few examples from an extensive and growing list (Kada, 1983; Rosin, 1982). Adding to the confusion is the fact that chemical carcinogens have additive, synergistic, or inhibitory effects on the activity of other carcinogens. Agents that are inhibitory either counteract the metabolism of the carcinogen, increase the breakdown of the carcinogen, or favor detoxification mechanisms (Slaga, 1989).

Sorting Out Inconclusive Data

If high doses and species differences create many questions, it could logically be asked why the experiments aren't more appropriate. The regrettable fact is that the best available methods are being used. Until a major breakthrough occurs in this area, we will be stuck in the muddle that we are in, trying to make both sense and sane food policy from data that severely overestimate or underestimate the risks.

Because science can't give a definitive answer, some feel that we must follow a conservative approach. In other words, we must consider that any molecule shown to cause cancer in animals when administered at any level and under any conditions is potentially carcinogenic in humans. In the other camp are those who think that animal data must be tempered with knowledge from experience with human cancer. This experience tells us that food carcinogens would most likely increase cancers in areas where they have direct contact, that is cancers of the liver, esophagus, stomach, pancreas, bladder, colon, and rectum. Over the past 15 years in the United States, age-adjusted incidence of cancers at all sites has increased nearly 15% (NCI, 1990). There has been decreased incidence of stomach and pancreatic cancer and increased incidence of cancer of the colon or rectum, urinary tract, bladder, liver, and kidney (Fig. 3.3). The question that must be posed in the face of these data is whether the increases in these cancers are due to contaminants in food, to other factors in our environment, or to some combination of both. If these increases are due to food contaminants, then perhaps a more conservative policy is warranted.

The HERP Index

In 1987, Dr. Bruce Ames, a biochemist from University of California at Berkeley, set up an index to help assess the cancer hazard from a variety of substances (Ames, 1987). He called it the HERP index. The acronym comes from Human Exposure/Rodent Potency. The

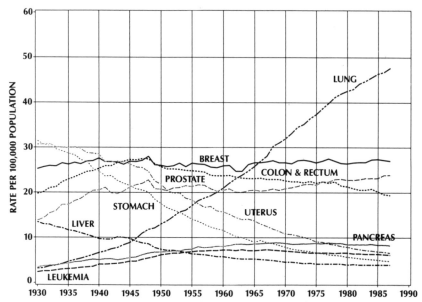

Figure 3.3. Cancer deaths, by site, in the United States, 1930–1987. Rates are for the population standardized for age on the 1970 U.S. population. Rates are for both sexes combined except for breast and uterus (female population only) and prostate (male population only). (Reprinted, with permission, from American Cancer Society, 1991)

number takes into consideration both the potency of the carcinogen and the degree of exposure. The higher the HERP number, the greater the risk. The calculations should not be viewed as an absolute determination of risk, but rather as a guide to rational (as opposed to hysterical) thinking and to priority setting. In an editorial in *Business Week* (Carey and Hamilton, 1990), Ames was quoted as saying that there is more carcinogenic potential in a peanut butter sandwich than in a muffin highly contaminated with ethylene dibromide (EDB). EDB is a pesticide whose registration was terminated by EPA because of presumptive carcinogenicity. Ames also mentioned that the risk from eating a peanut butter sandwich is so low that he doesn't worry about eating one (Lipkin, 1988). Table 3.6 gives some of the HERP numbers.

WHAT, THEN, IS SAFE FOOD?

Scientists are identifying toxic compounds in food faster than we can process the information, creating a mounting conundrum. Despite careful testing of food additives and pesticides, the procedures leave many unanswered questions. Experimental animals are healthy animals fed optimum diets and grown with ideal conditions. However, the substances being tested may ultimately be used by humans on suboptimal or deficient diets, by those on drugs both legal and illicit, by those with a variety of diseases, by those subjected to environmental hazards, and by young and old alike. In addition, the toxicity of a compound is rarely assessed in combination with another compound.

Table 3.6. HERP[a] Index for Possible Carcinogenic Potency[b]

HERP	Average or Reasonable Daily Human Exposure		Carcinogen and Dose per 150-Pound Person	
	Substance	Amount	Carcinogen	Dose
4.7	Wine	8 ounces	Alcohol	30 milliliters
2.8	Beer	12 ounces	Alcohol	18 milliliters
0.1	Basil	1-gram dry leaf	Estragole	3.8 milligrams
0.1	Mushroom	½ ounce	Hydrazines	
0.06	Saccharin diet cola	One	Saccharin	95 milligrams
0.03	Comfrey tea	1 cup	Symphytine	38 micrograms
0.03	Peanut butter	One sandwich	Aflatoxin	2 parts per billion
0.003	Bacon, cooked	100 grams	Nitrosamine	0.04 micrograms
0.001	Chlorinated tap water	1 liter	Chloroform	83 micrograms
0.0004	Grain products	Daily dietary intake	EDB[c]	0.42 microgram
0.0002	PCBs	Daily dietary intake	PCBs[d]	0.2 microgram

[a] HERP = human exposure/rodent potential.
[b] Adapted from Ames (1987).
[c] Ethylene dibromide.
[d] Polychlorinated biphenyls.

Thus, uncertainties about food safety seem infinite.

However, the greatest uncertainty lies in the area of food carcinogens, as consumers are especially fearful about carcinogens in the environment and no less concerned about them in food. Passage of state statutes similar to Proposition 65 in California are testaments to fear of carcinogens in all aspects of life.

The great burden is that we must rationally deal with this fear and uncertainty. Since many important (natural) foods contain small amounts of toxins and carcinogens, diets designed to be free of these create nutritional problems far riskier than the small and unknown risks we are trying to avoid.

Choosing to include foods in our diets that will enhance health is a much more important coping strategy than excluding foods. One of the benefits of a nutritionally sound diet is that it fortifies the body's natural detoxifying mechanisms. Further, low-fat and calorie-restricted diets appear to minimize the effects of carcinogens (CAST, 1987; Kritchevsky, 1987).

Undue fear or excessive worrying is perhaps as deleterious as the rather small risks that may arise from the use of many of the compounds in question. If legislation like Proposition 65 were passed generally, the required warnings on carcinogen-containing foods would be so numerous that they would neither advance public health nor make an informed consumer.

It is interesting that we seem to have created an information paradox: for each quantum of uncertainty we have attempted to reduce, our uncertainty has multiplied!

REFERENCES

American Cancer Society. 1991. Cancer Facts and Figures—1991. The Society, Atlanta, GA.

Ames, B. 1987. Carcinogens in food and the environment. Science 236:271-281.

Baldwin, R. H. 1979. Regulating carcinogens—The bitter and the sweet. CHEMTECH 9(3):156-162.

Ballantyne, B. 1985. Evaluation of hazards from mixtures of chemicals in the occupational environment. J. Occup. Med. 27(2):85-94.

Benbrook, C. M. 1990. What we know, don't know, and need to know about pesticide residues in food. Pages 140-150 in: Pesticide Residues and Food Safety: A Harvest of Viewpoints. B. G. Tweedy, H. J. Dishburger, L. G. Ballantine, J. McCarthy, and J. Murphy, eds. American Chemical Society, Washington, DC.

Borzelleca, J. F. 1984. Extrapolation of animal data to man. Concepts Toxicol. 1:294-304.

Byrd, D. M., Crouch, E. A., and Wilson, R. 1990. Do mouse liver tumors predict rat tumors? A concordance between tumors induced at different sites in rats and mice. Prog. Clin. Biol. Res. 331:19-31.

Cabral, J. R. P. 1984. Critical analysis of long-term tests for carcinogenicity. Food Addit. Contam. 1(2):101-107.

Carey, J., and Hamilton, J. O. 1990. Heresy in the cancer lab. Bus. Week 3182 (Oct. 15):58.

CAST. 1987. Nutritional epidemiology of cancer. Pages 34-40 in: Diet and Health. Report 111. Council for Agricultural Science and Technology, Ames, IA.

Clayson, D. B. 1985. Problems in interspecies extrapolation. Pages 105-124 in: Toxicological Risk Assessment, Vol. 1. D. B. Clayson, D. Krewski, and I. Munro, eds. CRC Press, Boca Raton, FL.

Cordle, F., and Miller, S. A. 1982. The use and limitation of epidemiological and animal data to describe current cancer risk in the U.S. Pages 3-12 in: Carcinogens and Mutagens in the Environment, Vol. I. H. F. Stich, ed. CRC Press, Boca Raton, FL.

Cramer, G. M., Ford, R. A., and Hall, R. L. 1978. Estimation of toxic hazard—A decision tree approach. Food Cosmetic. Toxicol. 16:255-276.

Crouch, E. A. 1983. Uncertainties in interspecies extrapolations of carcinogenicity. Basic Life Sci. 24:658-665.

Crouch, E. A. 1989. TOX-RISK: A program for fitting dose-response formulae and extrapolating between species. Risk Anal. 9:599-603.

Ehling, U. H., Averbeck, D., Cerutti, P. A., Friedman, J., Greim, H., Kolbye, A. C., and Mendelsohn, M. L. 1983. Review of the evidence for the presence or absence of thresholds in the induction of the genetic effects by genotoxic chemicals. Mutat. Res. 123:281-341.

Elias, P. S. 1982. Methods for the detection of carcinogens and mutagens in food. Pages 201-209 in: Carcinogens and Mutagens in the Environment, Vol. I. H. F. Stich, ed. CRC Press, Boca Raton, FL.

Elton, G. A. H. 1982. Sources of carcinogens and mutagens in food. Pages 67-73 in: Carcinogens and Mutagens in the Environment, Vol. I. H. F. Stich, ed. CRC Press, Boca Raton, FL.

FAO/WHO. 1974. Toxicological evaluation of certain food additives with a review of general principles and of specifications. WHO Report 539. Joint FAO/WHO Expert Committee on Food Additives. World Health Organization, Geneva.

Farber, T. M., and Guest, G. B. 1984. A strategy for the assessment of the carcinogenic potency of veterinary drug residues: Recommendations of the FDA. Food Addit. Contam. 1(2):213-219.

FMI. 1986. Consumer Attitudes and Trends. The Research Division, Food Marketing Institute, Washington, DC.

Fournier, P. E., and Thomas, G. 1984. Mechanisms of chemical carcinogenesis: Recent advances. Food Addit. Contam. 1(2):73-80.

Francis, F. J. 1986a. Testing for toxicity. Sci. Food Agric. (CAST). 4(2):10-13.

Francis, F. J. 1986b. Testing for carcinogens. Sci. Food Agric. (CAST). 4(2):13-15.

Frayssinet, C. 1984. The principle of a threshold dose in chemical carcinogenesis. Food Addit. Contam. 1(2):89-94.

Frieden, E. 1984. Biochemistry of the Essential Ultratrace Elements. Plenum Press, New York.

Gravani, R. B., and Williamson, D. M. 1991. Consumer food preparation practices: Report of a national survey. Presented at the Annual Meeting of the Institute of Food Technologists, Dallas, TX, June 5.

Hallenbeck, W. H., and Cunningham, K. M. 1986. Quantitative Risk Assessment for Environmental and Occupational Health. Lewis Publishers Inc., Chelsea, U. K.

Haseman, J. K. 1984. Statistical issues in the design, analyses, and interpretation of animal carcinogenicity studies. EHP, Environ. Health Perspect. 58:385-392.

Hassall, K. A. 1982. The Chemistry of Pesticides: Their Metabolism, Mode of Action and Uses in Crop Protection. Verlag Chemie, Deerfield Beach, FL.

Hodge, H. C., and Downs, W. L. 1961. The approximate oral toxicity in rats of selected household products. Toxicology. Appl. Pharmacol. 3:689-695.

Kada, T. 1983. Desmutagens: An overview. Pages 63-67 in: Carcinogens and Mutagens in the Environment, Vol. I. H. F. Stich, ed. CRC Press, Boca Raton, FL.

Kimbrough, R. D. 1984. Relationship between dose and health effects. Clin. Lab. Med. 4:507-519.

Kirshman, J. C. 1981. Definition of toxicological and physiological effects. Pages 1-3 in: Impact of Toxicology on Food Processing. J. C. Ayres and J. C. Kirshman, eds. AVI, Westport, CT.

Klaassen, C. D. 1986. Principles of toxicology. Pages 11-32 in: Casarett and Doull's Toxicology. C. D. Klaassen, M. O. Amdur, and J. Doull, eds. Macmillan, New York.

Kolbye, A. 1984. Interaction of safety evaluation and government guidelines. Concept Toxicol. 1:72-76.

Kritchevsky, D. 1987. The role of calories and energy expenditure in carcinogenesis. Food Nutr. News 59(3):63-64.

Lindsay, D. G. 1986. Estimation of the dietary intake of chemicals in food. Food Addit. Contam. 3(1):71-88.

Lipkin, R. 1988. Risky business of assessing danger. Insight (on the News) 4(21):8-11.

Lowy, R., and Manchon, P. 1984. Determination of a virtually non-carcinogenic dose. Food Addit. Contam. 1(2):109-120.

National Research Council/National Academy of Sciences (NRC/NAS). 1989. Recommended Dietary Allowances. National Academy Press, Washington, DC.

NCI. 1990. Cancer Statistics Review 1973–1987. Natl. Cancer Inst., Bethesda, MD.

Nielsen, F. H. 1984. Ultratrace elements in human nutrition. Annu. Rev. Nutr. 4:21-41.

Parke, D. V. 1979. Toxicological evaluation of food additives and contaminants. Proc. Nutr. Soc. Aust. 4:61-71.

Pim, L. R. 1981. The Invisible Additives. Doubleday, Garden City, NY.

Portier, C. J. 1989. Quantitative risk assessment. Pages 164-173 in: Carcinogenicity and Pesticides: Principles, Issues, and Relationships. N. N. Ragsdale and R. E. Menzer, eds. American Chemical Society, Washington, DC.

Reiches, N. A. 1981. Environmental carcinogens: The human perspective. Pages 19-42 in: Management of Toxic Substances in the Environment. B. W. Cornaby, ed. Ann Arbor Science, Ann Arbor, MI.

Rosin, M. P. 1982. Inhibition of genotoxic activities of complex mixtures. Pages 259-270 in: Carcinogens and Mutagens in the Environment, Vol. VI. H. F. Stich, ed. CRC Press, Boca Raton, FL.

Scroggins, C. D. 1990. Consumer attitudes toward the use of pesticides and food safety. Pages 182-191 in: Pesticide Residues and Food Safety: A Harvest of Viewpoints. B. G. Tweedy, H. J. Dishburger, L. G. Ballantine, J. McCarthy, and J. Murphy, eds. American Chemical Society, Washington, DC.

Slaga, T. J. 1989. Critical events and determinants in multistage skin carcinogenesis. Pages 78-93 in: Carcinogenicity and Pesticides: Principles, Issues, and Relationships. N. N. Ragsdale and R. E. Menzer, eds. American Chemical Society, Washington, DC.

Spallholz, J. E. 1989. Nutrition: Chemistry and Biology. Prentice Hall, Englewood Cliffs, NJ.

Squire, R. A. 1984. Carcinogenic potency and risk assessment. Food Addit. Contam. 1(2):221-231.

Stott, W. T., Fox, T. R., Reotz, R. H., and Watanabe, P. G. 1989. Mechanisms of chemical carcinogenicity. Pages 43-74 in: Carcinogenicity and Pesticides: Principles, Issues, and Relationships. N. N. Ragsdale and R. E. Menzer, eds. American Chemical Society, Washington, DC.

Turnball, G. J. 1984. Pesticide residues in food—A toxicological view. J. R. Soc. Med. 77:932-935.

Vettorazzi, G. 1980. Handbook of International Food Regulatory Toxicology, Vol. 1. Evaluations. SP Medical and Scientific Books, New York.

Weil, C. S. 1984. Some questions and opinions on issues in toxicology and risk assessment. Am. Ind. Hyg. Assoc. 45:663-670.

Wodicka, V. O. 1977. Food ingredient safety criteria. Food Technol. (Chicago) 31(1):84-88.

Worthy, W. 1983. Federal food analysis program lowers detection limits. Chem. Eng. News 61(Mar. 7):23-24.

Yang, R. S. H., Huff, J., Germolec, D. R., Luster, M. I., Simmons, J. E., and Seely, J. C. 1989. Biological issues in extrapolation. Pages 142-163 in: Carcinogenicity and Pesticides: Principles, Issues, and Relationships. N. N. Ragsdale and R. E. Menzer, eds. American Chemical Society, Washington, DC.

Risk-Benefit

CHAPTER 4

WHAT IS RISK, AND WHAT DOES IT HAVE TO DO WITH FOOD, ANYWAY?

The word *risk* implies uncertainty along with possible danger, injury, or loss. Risk can be physical, psychological, or monetary. It is not the same as *hazard*, but rather a way of judging the degree of hazard. Two concepts are embodied in the term—both the *magnitude of loss* due to the occurrence of an undesirable event and the *probability of its occurrence* (Munro and Krewski, 1983).

We tolerate varying degrees of risk depending on the situation. However, many of us believe that some areas in life ought to be risk-free, and for many, food is clearly one of these areas. It is disillusioning to face the reality that nothing is without risk and that—like it or not—food is included (Wilson and Crouch, 1987). One documented risk associated with eating is the risk of choking. Each year in the United States, nearly 3,000 deaths occur from choking (Koop, 1985), making this the sixth most frequent cause of accidental death (Heimlich and Uhley, 1979; Winter and Brown, 1982).

Although consumers rarely contemplate their own voluntary starvation, most are subconsciously aware of the risk associated with not eating anything, and are thus (perhaps unconsciously) forced into a risk-benefit decision between the risks of eating nothing and the risk of eating and then choking. Obviously, the decision is in favor of eating, as the benefit far outweighs the risk of possible choking.

As with eating, every aspect of our lives involves risk-benefit decisions, conscious or not (Lee, 1989; Roberts, 1978; Slovic, 1987; Wildavsky, 1979). Care must be exercised so that we have a clear understanding of how a certain action or position may put us at risk. On the stock market, the safest investment may actually be quite risky in the sense that yields are so low that they don't keep up with inflation. In like manner with foods and other life decisions, the course of action appearing to be the least risky may indeed not be so. People who avoid vegetables because of the possibility of certain pesticide residues may not get

53

adequate numbers of fresh fruits and vegetables and, as a result, may increase their cancer risk for lack of cancer-preventing agents.

We must be cognizant that our attempts to avoid certain risks always introduce other ones. For instance, we may read about the risk incurred each time we ride in a car and therefore decide to stay at home. While there are probably fewer risks encountered at home, it is clearly not risk-free. In the United States, 200 electrocutions involving appliances and wiring occur annually in the home (Wilson and Crouch, 1987), and falls in the home result in 6,000 deaths annually. Further, staying home prevents earning an income or going somewhere to meet friends and family to have fun. Table 4.1 gives some common life risks.

Reality vs. Perception: Is Life Riskier Today Than in Grandma's Day?

Most people seem to believe that life is becoming riskier, even though most objective measures show the contrary (Crouch and Wilson, 1981; Doll, 1979). Cancer rates, except for lung cancer, have decreased or remained stable, as was shown in Figure 3.7. Lung cancer's increase is attributed to increased smoking. Also, life expectancy continues to increase.

Real differences in risk may be less important than people's perception of what is riskier. One survey of these perceptions compared the amount

Table 4.1. Common Risks in the United States[a]

Cause of Death	Risk of Death per 1,000,000 Persons per year
Travel	
Walking	40
Motorcycling	20,000
Car	20–30
Airplane	9
Sports	
Canoeing	400
Motor boating	30
Skiing	170
Car racing	1,200
Food	
Wine, one bottle per day	75
Beer, one bottle per day	20
Aflatoxin, 4 tbsp peanut butter per day	40
Charcoal broiled steak, ½ lb per week	0.4
Other	
Smoking, 20 cigarettes per day	2,000–5,000
Falls	70–90
Lightning	0.05–0.1

[a] Adapted from Oser (1978).

of risk experienced in day-to-day living with that experienced 20 years before. Only 6% of 1,500 people said there was less risk than there had been 20 years earlier, whereas 78% said there was currently more risk. Of 401 top corporate executives, 36% said there was less risk, and 38% said more; 26% of the 47 members of Congress who responded said the risk was less, and 55% said more (Crouch and Wilson, 1981). While these data on risk do not make any specific reference to food, food certainly can be assumed to be one of the factors included in the public's perception of increased risk.

It is easy to get caught up in the myth of the wholesome, risk-free way Grandma ate, as described in the *Whole Earth Catalog* (The Point Foundation, 1986). The nostalgic vision of food in Grandma's day considers neither the effort required to preserve and prepare the food nor the limited quality, variety, and availability of food from the root cellar and the preserving pantry, especially in late winter.

Was Grandma's diet actually as risk free as we tend to visualize it? No, Grandma and her household suffered some very real risks, including botulism from improperly canned vegetables. Even properly canned food had the problem of low nutrient content, since foods were boiled for 6 hours in a conserver. Root-cellared vegetables in the late winter also suffered significant nutrient losses.

Lack of variety and availability plagued the meal planners of Grandma's day. Few choices often translated into limited availability of nutrients, especially vitamins provided by fresh fruits and vegetables. Scurvy was a real risk, as vegetables canned in a conserver or held in a root-cellar lost considerable vitamin C—this is why old cookbooks contained recipes for spring tonics. In addition to the loss of nutrients, the vegetables could be withered and moldy. The mold could possibly contain aflatoxin. Just because Grandma had never heard of it and didn't worry about it didn't mean that it wasn't there!

Grains in Grandma's day may have been free of pesticides, but crop losses at times were catastrophic. Histories of American settlers on the Great Plains recount crop failures due to plagues of locusts and other pests. Nor were crop failures the only food loss, when losses of stored grain to rodents, insects, birds, and molds are considered. The child's story *The House That Jack Built* shows the inevitability of rodents where there is flour. Accounts in literature such as in *The Mayor of Casterbridge* by Thomas Hardy allude to problems with moldy grains in bread. Food quality, as well as food availability, was affected. So much for the good old days!

Perceptions of risk have changed. We have long forgotten the risks of Grandma's larder and instead have romanticized life in those times. In place of these old risks, we have found new risks in substances occurring in parts per million as opposed to locusts in swarms of millions.

Increased media attention has heightened our awareness of risks that were not only unknown to Grandma but frequently unknown even last year (Munro and Krewski, 1983). Some of these risks include the fear of radioactivity caused by the Chernobyl nuclear incident in the Soviet Union in 1986 and the fears of contamination by EDB in 1984 and of Alar in 1989 in the United States.

The media attention may also make some believe that it is riskier to live in highly industrialized societies (with risks similar to those in the United States) than in more agrarian or developing societies.

Figure 4.1. Ad for a conserver from an old cookbook.

Clearly, different risk profiles do occur in these societies. Predominant health risks in developing countries remain those associated with the prevalence of communicable disease, other natural hazards, and the lack of a safe and nutritious food supply (Munro and Krewski, 1983). Crop losses to drought, pests, poor soil quality, and overfarming of marginal land are common in many developing countries. Risks from inadequate sanitation and refrigeration and from unsafe water supplies are very real in Third World countries and are responsible for infection, parasites, and high rates of infant mortality.

Thus, life in both industrialized countries and developing countries is risky, but the risks are clearly different. Before we rush headlong to reduce risks so that they are like those of another culture, we should contemplate the real nature and extent of both sets of risks. Table 4.2 is a good case in point. It tells what the cancer risks are for a variety of countries as compared with those for the United States. For instance, if (as some suggest) Americans change their dietary habits to eat like the Japanese, will this reduce or increase their risks? Clearly the risk for certain cancers is much lower in the United States while the risk for other cancers is much higher.

Is Risk Taking Rational?

When it comes to health risks, the answer is emphatically, no! True rationality about these risks would imply changed behavior, in which no one would smoke or use cocaine, and everyone would wear seat belts and control fat and calorie intakes. Each stress-managed day would include adequate exercise. We would never waste time or money achieving

Table 4.2. Cancer Death Rates per 100,000 Population for Selected Countries[a]

Subjects	Country	All sites	Colo-rectal	Esopha-geal	Lung	Stomach	Oral	Prostate or Breast
Males	U.S.	213.6	26.2	5.4	68.1	9.2	5.8	22.3
	Chile	107.7	9.9	12.8	25.3	64.9	3.3	17.0
	Germany	244.8	32.4	4.9	64.0	34.4	3.5	24.1
	Hong Kong	229.1	16.9	14.3	65.6	5.8	21.2	2.9
	Japan	186.7	15.0	9.5	28.3	70.2	2.2	3.6
	Scotland	269.8	32.7	9.2	108.5	25.4	4.4	17.2
	Sweden	197.5	25.0	3.8	33.4	20.9	3.4	32.3
Females	U.S.	136.3	20.2	1.5	17.2	4.4	2.0	27.3
	Chile	153.8	10.0	6.0	6.9	30.4	0.9	14.1
	Germany	154.8	24.8	1.0	6.8	18.3	0.9	25.1
	Hong Kong	125.3	11.9	4.4	30.4	9.2	7.1	10.7
	Japan	108.7	11.1	2.2	8.2	34.9	0.8	5.7
	Scotland	165.8	26.5	5.3	23.1	13.4	1.9	31.3
	Sweden	140.9	18.6	3.8	7.7	10.8	1.4	23.5

[a] Data from the American Cancer Society (1985).

a glamorous tan. We'd be careful not to let our hand slip during a cocktail party discussion of artificial sweeteners and unconsciously grab a handful of fat-laden peanuts that could contain carcinogenic aflatoxins. (Ironically, the sweeteners that were the subject of discussion are statistically less dangerous than drugs, fat, aflatoxin, or suntanning.)

Many people choose not to stop smoking, not to exercise, and not to reduce blood lipid levels even though the risk of coronary heart disease is then six times higher than when none of these risk factors is present (82 per 1,000 compared to an average risk of 13 per 1,000). Our choices are frequently made with full awareness of the risks. However, we choose to ignore the risks, either because the personal benefit outweighs the risk or because the cost of changing habits is too great.

In addition to our inherent inertia in changing our habits, we also have the problem of being unable to comprehend real differences in risk when the risk probabilities are smaller than 1 in 100 (Fischhoff et al, 1977). Furthermore, studies indicate that we overestimate some risks, such as homicides or shark bite, and underestimate others, especially diseases such as diabetes (Press, 1986). To make matters worse, we are falsely overconfident about what we actually know with regard to risks. In one study, even those who were certain that they had only a one-in-1,000 chance of being wrong in their risk estimate were actually wrong as much as 20% of the time (Fischhoff et al, 1977; Press, 1986).

ACCEPTABILITY OF RISKS

If a risk is immediate, direct, or involves an element of fear or major catastrophe, it creates much greater outrage and is much less acceptable than if the risk is delayed, indirect, or commonplace. Thus, the results of a major plane crash or catastrophic nuclear power plant accident are perceived as more serious than automobile accidents, even though the number of traffic injuries and fatalities in the same time span may be substantially greater.

People who study risk have placed risk behavior in a framework showing when risks are considered acceptable and when they are not (Crouch and Wilson, 1981; Johnson, 1981; Litai et al, 1983; Munro and Krewski, 1983; Oser, 1978; Rowe, 1977; Slovic, 1987). Table 4.3 lists a number of risk acceptability factors.

A risk that is voluntary is much more acceptable than one that is involuntary. Obviously, the involuntary risks of drinking water, breathing air, and eating food should be far less than the voluntary risks of hang-gliding or piloting small aircraft.

The voluntary nature of risk is sometimes debatable. Using a car to get to work may be the only viable alternative; thus, we might argue that the risk was necessitated and not totally voluntary. Whether voluntary or involuntary, the annual risk is one in 50,000 of being killed in an automobile on U.S. roads.

Risks are also less acceptable if people feel that they are foisted upon them rather than personally chosen. For instance, a person will not fear the risk of coronary heart disease from the inactivity, smoking, or high-fat diet that was chosen willingly, but will greatly fear possible cancer from traces of pesticide or a food additive that there was no choice but to ingest. This fear occurs despite the fact that the risks from pesticides and additives are both smaller and less well documented than the risks associated with inactivity, smoking, or high-fat diets.

A risk that is a necessity—for which no less-risky or other alternative exists—is more acceptable than one associated with luxuries or less risky alternatives. For instance, pesticides may be more readily accepted when the survival of a group is dependent on an adequate harvest. The possible long-term health risks of certain pesticides in this instance pale beside concerns about starvation.

A risk that is catastrophic is less acceptable than one that has a less severe outcome. For instance, a plane crash with 200 people on board is headline news, while the 200 auto fatalities that occur daily in the United States are rarely even thought of unless one is personally affected.

Natural risks are more acceptable than those that are created. The risks from smoke and dust that emanate from a volcano are far more tolerable than the risks of smoke and dust from industrial burning,

Table 4.3. Risk Factor[a] Acceptability

Greater Acceptability	Lesser Acceptability
Voluntary	Involuntary
Chosen	Foisted
Necessity	Luxury
No alternative	Alternatives available
Natural	Created
Controllable	Uncontrollable
Personal error	Corporate error
Occupational exposure	Nonoccupational exposure
Occasional exposure	Continuous exposure
Immediate consequences	Delayed consequences
Common or old hazard	Uncommon hazard or dread disease
Ordinary	Catastrophic
Clear benefit	Unclear benefit
Nonvital	Vital

[a] Sources: Litai (1983); Oser (1978).

although the net result may be the same. Since chemicals and technology are perceived as created by humans, the risks they create are less acceptable. Both are mistrusted and not understood by the general public and therefore feared.

Risks are less acceptable if they are the result of corporate activity rather than the result of a person's own careless or improper use of a carefully and clearly labeled substance.

Risks that have delayed consequences are less acceptable than those with immediate effects. The delay period often creates a nagging anxiety about whether the consequences will appear at some time in the future.

In like manner, risks that have clear outcomes are easier to deal with than those for which the outcomes are unclear. For instance, the risk of eating food from bulging cans is great, and the outcome is clear, so this food is discarded. However, the fact that the cancer risk from the ingestion of large amounts of saccharin-containing foods is extremely small and unquantifiable makes it unclear what risk, if any, is associated with the daily ingestion of a single serving of saccharin-containing diet beverage.

Risks that are old are more acceptable than new risks. For instance, the risk of botulism has been with us for years, and most of us don't think about it unless it has been in a recent news story. Even the mention of botulism doesn't create the same kind of fear that the mention of nitrate does. Despite the fact that botulinum toxin is very lethal, more deadly than cobra venom, the possible role of nitrate in causing cancer creates greater consumer fear. This scenario is especially interesting as botulism is much more likely to occur in canned or vacuum-packed meats that do not contain nitrate than in those that do.

Risks are less acceptable if they result in a disease that creates an extraordinary amount of fear and for which the cure is often not successful. In the United States, cancer and AIDS are probably feared as much as leprosy was in biblical times. Thus, reports that something in the food supply may increase the risk of cancer create great fear. Controversy over the presence or potential presence of cancer-causing agents in the food supply has increased dramatically over the past decade (Ames and Gold, 1991; Lorentzen, 1984).

Risks that are vital are more acceptable than those that are nonvital. Thus, drug therapy may be chosen even though the therapy itself creates another vital risk, whereas the chance of decreased quality in food due to increased shelf life is less acceptable because it is viewed in the United States to be a nonvital risk.

In some cases, it is rational not to be rational about risks. With respect to foods, it is probably not healthy to live in an age of food

McCarthyism where we are unduly frightened to eat and the consumption of any single food item brings on extraordinary guilt. Nor is it healthy to live in a state of psychological distress due to media-generated fear about food. In some cases, extreme fear, especially about food carcinogens, can limit food selection so dramatically that the diet becomes nutritionally unbalanced.

HOW DO WE MEASURE RISKS AND BENEFITS?

Risks and benefits are assessed by the public, by scientists, by legislators, and by regulators. Neither risk nor benefit can be easily quantified.

Opinion polls are used to gauge public concern about risk. A 1979 Canadian survey of nearly 25,000 adult shoppers revealed that the majority were very concerned about the possible health effects of food additives. Only a small fraction of the respondents knew any of their benefits. Thus, consumers apparently overestimate some risks and are unaware of many benefits (Carpenter, 1991; Kroger and Smith, 1984; Munro and Krewski, 1983; Oser, 1978). This has been likened to "shearing an accountant's ledger in half and only showing the liabilities" (Garrand, 1976). Under these conditions, it is very difficult for consumers to make informed risk-benefit decisions.

Even without the additional biases ascribed to risk, the risk-benefit equation is a very difficult one to solve. Both sides of the equation usually require value judgments. In addition, there is always the problem of incomplete scientific data and assumption-laden extrapolations that must be plugged into the equation. Furthermore, incompletely understood science generates both fear and conflict when risks and benefits are being debated (Fischhoff et al, 1983; Hattis and Kennedy, 1986; Lave, 1987; Miller, 1979).

Answers to the question of how risk should be measured frequently have the most agreement. The best available scientific assessment should be used, even though there is often no agreement as to what best is. The FDA and the food industry use a procedure called risk assessment. This uses a base of scientific research to establish possible harm that may result to an individual or population from exposure to a substance or process (Weil, 1984; Wilson and Crouch, 1987).

Risk assessment must be based on exposure level and frequency together with inherent toxicity (Ames et al, 1987; Scheuplein, 1986). Further, it must consider all available data, including historical and epidemiological data. Prediction of carcinogenic chemicals is done by comparing the structure in question with that of known carcinogens. Those with similar structures are more likely to be carcinogenic. Thus, highly probable risks are separated from improbable ones, which enables scientists and regulatory agencies to concentrate their limited resources on those chemicals most likely to be a problem to humans (Clausi, 1982; Phillips et al 1987).

Correlations derived from epidemiological data also can identify risks. However, extreme care must be used in evaluating correlations, and one must be constantly aware that the relationships found through correlations may not be causal. A correlation can erroneously suggest a risk. For example, research has found an association between TV watching and coronary heart disease. Yet, TV watching does not cause coronary heart disease; it may indicate a sedentary lifestyle, which is a risk factor for that disease. A humorous example of this kind of erroneous deduction concerns a physician who took a series of patient histories and learned that patients with cirrhosis had consumed gin and soda, scotch and soda, and whiskey and soda, from which he deduced that cirrhosis is caused by the consumption of large amounts of soda.

Results of Risk Assessment

Care must be taken that data for risk assessment are not misunderstood. For instance, in studies that do not show increased risk due to the ingestion of a certain food or additive, the risk is often reported as zero, when in actuality the risk was simply smaller than could be detected by the measuring technique (Wilson and Crouch, 1987).

The methodologies for risk assessment must produce results that are consistent with human experience. Valid methodologies will be developed only if logical, defensible, and reasoned answers are sought rather than simply expedient ones. We must take great care with some of the current methodologies such as extrapolation (Lorentzen, 1984). Linear extrapolation must not be chosen simply because it is the most conservative (Sugarman, 1990). An article by the vice-president of research and development of agricultural products for Dow Chemical Company pointed out the fallacy of extrapolation, using the time record for running the mile. Between 1875 and 1985, the time needed to run a mile dropped in a linear fashion from four minutes 25 seconds to three minutes 40 seconds. Using linear extrapolation, in 654 years, the record time to run one mile would be zero minutes and zero seconds. The writer believes that by using linear extrapolation, the risks proposed for EDB and other chemicals have been overstated (Gehring, 1985). Using the wrong assumptions for extrapolation also can lead to the wrong conclusions. For instance, a philosopher in the 1800s predicted that London would perish in its own horse manure by the turn of the century. He did not foresee the advent of the internal combustion engine with its own (but different) form of pollution.

Another problem with extrapolation is that different approaches can give values for risk that differ by several orders of magnitude (Lorentzen, 1984). Depending on the model used, risk calculations for annual increases in human cancers due to saccharin ingestion vary from 0.22 to 1,444,000. Widely varying predictions of risk have been made for many other chemicals.

Finally, we must realize the purpose of risk assessment. Knowing about risk is valuable only in a relative sense. A documentable risk of one in 1,000 of contracting cancer from exposure to a chemical is meaningful, but it is of little operational value unless this chemical may suitably replace a needed chemical having a risk of, for instance, one in 100.

FOCUSING RISK ELIMINATION EFFORTS

A question that must be answered is: Where should we focus the attention for risk elimination? An approach that some consumer activists suggest is that the goal should be *zero risk*, implying that we must address all risks (Pim, 1981). As previously discussed, zero risk may well be desirable but, unfortunately, is unattainable.

A second approach, like that of lawyer James Turner, author of *The Chemical Feast*, suggests that we should not focus on all known or even all imaginable risks, but rather on *manageable* risks. He then suggests that additives, pesticides, and animal drugs are the risks most easily managed, therefore the easiest to eliminate (Turner, 1979), and should be immediately removed from the food supply unless an overwhelming case of need can be made. If the case for the need is strong, he suggests that the risk information be vigorously sought. The problem with this approach is that if the only focus is on what is manageable, we could eliminate foods or useful chemicals without justification and create other hazards through their loss. We could also miss real dangers that are not easily managed and, by ignoring them, leave them to fester.

The National Academy of Sciences, after a two-year study, suggested that focus should be directed toward compounds of *high toxicity and high rates of use*. Their report suggested that raw commodities, ingredients, and additives, intentional and unintentional, should receive the same scrutiny (Clausi, 1979; Hutt and Sloan, 1979). Toxicity would be assessed using the food safety decision tree method developed by the Scientific Committee of the Food Safety Council (FSC, 1978, 1982). This third approach is currently being used by the FDA (Hutt and Sloan, 1979, Miller, 1979). Existence of three such disparate approaches clearly shows the lack of agreement about where to focus our efforts with respect to risk.

Purpose of Risk Acceptance

The answer to the question of for what purpose we will accept risk is dependent on who is giving the answer. Most food scientists feel that small risks are necessary to ensure an abundant food supply that has the minimum number of pests and the maximum amount of nutrients and is inexpensive enough so that nutritious diets are available to everyone.

For some consumers with busy lifestyles, some risks are tolerated to minimize the amount of time they spend in the kitchen while still providing adequate diets for their families. These consumers do accept the risks from foods that may be processed, contain additives and pesticides and high levels of sugar and fat, and may not contain the same ingredients (hence nutrients) that would have been there if the foods were prepared from scratch. In contrast, other consumers may find these risks unacceptable and are willing to pay substantially higher prices and, at times, to accept lower quality produce while trying to minimize exposure to chemicals.

Ironically, the pesticide load in "organic" produce has been measured at the same level as that in regular grocery store produce (Newsome, 1990).[1] Levels of heavy metals (mercury, cadmium, and lead) and other pollutants were found to be high in some organic produce since the sewage used as fertilizer can be a source of these pollutants. Thus,

[1]The reason that the government has been unwilling to have an organic category for vegetables and fruits is that it has been unable to analytically determine a difference in the pesticide levels of organically grown produce and thus cannot assure consumers that they are getting what they are paying for.

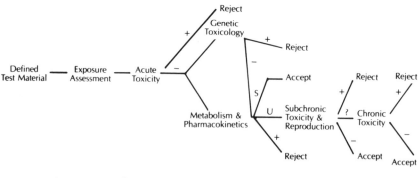

+ = presents socially unacceptable risk
− = does not present a socially unacceptable risk
S = metabolites known and safe
U = metabolites unknown or of doubtful safety
? = decision requires more evidence

Figure 4.2. The safety decision tree. (Reprinted, with permission, from Clausi, 1982)

in an attempt to minimize one risk, consumers may exchange it for another risk of equal or even greater magnitude. The question about the purpose for which the risk will be accepted thus has no agreed-upon answer.

The questions of who is at risk, who benefits, and whether these risks are shared equally are also difficult to answer (Campt et al, 1991). If pesticides are used on crops, everyone can benefit from lower food prices and food containing a minimum of insect parts and rodent hairs, but is the risk equally distributed? The pesticide sprayer, the farmer, the migrant worker, and their respective families may take the brunt of the risk, although one could argue that they may also stand to reap the greatest benefit in profits from higher yields and improved crop quality. However, the poorly paid migrant worker may be exposed to high pesticide levels with neither the knowledge that the crop was sprayed nor the benefit in dollars from the increased crop yield. Thus, the controversy about who is at risk and how equally the risks are shared remains unresolved (Darby, 1985).

Acceptance of Risk-Benefit Determination

Public agreement of the expected risks and benefits will be improved if certain things take place (Crouch and Wilson, 1981).

1) The experts reach general agreement about the basic facts. (Distinguished dissidents, as there frequently are in science or in the press, reduce agreement.) Risks will be better accepted if a good experimental and theoretical basis for the facts has been published.

2) One of the affected parties has had direct experience with a similar product or system.

3) Both the risks and the benefits have been clearly stated and fairly publicized.

4) The concerned groups know that all available alternatives and the risks and benefits of each course of action were seriously considered.

At some point, society will need to decide the amount of risk it is willing to assume to achieve a certain benefit. Further, it must decide whether the risks are known with certainty or are conjectural, hypothetical, unsubstantiated, or indeterminate. Finally, it should have a good way to distinguish between acceptable and unacceptable risks. For instance, cancer risks of one in a million due to chlorination of water may be acceptable, while cancer risks of one in a thousand from regular inhalation of gasoline vapors are clearly not acceptable, according to the U.S. Occupational Health and Safety Administration (Hallenbeck and Cunningham, 1986). Table 4.4 shows what standards are used currently by the FDA to regulate carcinogenic substances.

If risk is to be legislated, the return in health and consumer benefits for each proposed alternative or for each incremental step in risk reduction should be made available to consumers. Costs should also be made available, so that society can best allocate its limited resources and maintain a high-quality, abundant food supply available to all segments of society at a reasonable price (Affleck, 1979; Johnson, 1981). If a substance is to be banned, the losses from banning should be considered (Johnson, 1981; Lave, 1987).

We must seek to have the very best answer to what it means if, for instance, a substance causes 2.4 cancers per 1,000 rats fed over one milligram per day for over six months, and be able to translate its meaning into terms that the population will understand (Omang, 1983). Further, we must vigorously continue to improve our information with respect to risks and benefits, and all sides should be flexible enough to change our thinking to account for new discoveries and changed assumptions.

As Harry Truman once said, "Some questions cannot be answered, but they must be decided." Ultimate decisions are not in the domain of science. Neither do they belong to soothsayers reading entrails. They must be the informed decisions of an enlightened and well-educated government and populace.

Table 4.4. Legal Standards for Food Constituents That May Be Carcinogenic[a]

Food Constituent or Contaminant	Legal Standard	Section of Act
Natural toxicant	"Does not ordinarily render the food injurious"	402
Added poisonous or deleterious substance	Adulterated, subject to seizure	402
Added poisonous or deleterious substance that cannot be avoided by Good Manufacturing Practices	Tolerance set on what is necessary to "protect the public health," considering "the extent to which the use cannot be avoided"	406
Food additives Color additives	"Reasonable certainty of no harm" and Delaney amendment	409,706
Pesticides	Tolerance based on whether it is "safe for use" considering the benefits	408
Animal drugs	"Reasonable certainty of no harm" and Delaney amendment	512

[a] As regulated in the U.S. Federal Food, Drug, and Cosmetic Act.

REFERENCES

Affleck, J. G. 1979. Toward realistic risk/benefit decisions. The New York Times. Sunday April 29.

Ames, B. N., and Gold, L. S. 1991. Cancer prevention strategies greatly exaggerate risks. Chem. Eng. News 69(1):28-32.

Ames, B. N., Magau, R., and Gold, L. S. 1987. Ranking possible carcinogenic hazards. Science 236:271-280.

Campt, D. D., Roelofs, J. V., and Richards, J. 1991. Pesticide regulation is sound, protective, and steadily improving. Chem. Eng. News 69(1):44-47.

Carpenter, W. D. 1991. Insignificant risks must be balanced against great benefits. Chem. Eng. News 69(1):37-39.

Clausi, A. S. 1979. Revising the U.S. food safety policy. Food Technol. (Chicago) 33(11):65-67.

Clausi, A. S. 1982. A proposed food safety evaluation process. Food Technol. (Chicago) 36(11):42-67.

Crouch, E. A. C., and Wilson, R. 1981. Risk/Benefit Analysis. Ballinger, Cambridge, MA.

Darby, W. J. 1985. The significance of benefits in regulatory decision making. Pages 89-96 in: Toxicological Risk Assessment. Vol. 2. D. B. Clayson, D. Krewski, and I. C. Munro, eds. CRC Press, Boca Raton, FL.

Doll, R. 1979. The pattern of disease in the post infection era. Proc. R. Soc. London, B 205:47.

Fischhoff, B., Slovic, P., and Lichtenstein, S. 1977. Knowing with certainty: The appropriateness of extreme confidence. J. Exp. Psych., Human Percept. Perform. 3:552-564.

Fischhoff, B., Slovic, P., and Lichtenstein, S. 1983. The public vs. expert. Pages 234-240 in: The Analysis of Actual Versus Perceived Risk. V. T. Covello, W. G. Flamm, J. V. Rodricks, and R. G. Tardiff, eds. Plenum Press, New York.

Food Safety Council (FSC). 1978. Proposed system for food safety assessment. Food Cosmet. Toxicol. 16 (Suppl. 2):1-139.

Food Safety Council (FSC). 1982. A proposed food safety evaluation process. Food Technol. (Chicago) 36:163-170.

Garrand, M. 1976. Guest comment. Chem. Eng. News Oct 25.

Gehring, P. J. 1985. On chemicals and risk. CHEMTECH 15:522-525.

Hallenbeck, W. H., and Cunningham, K. M. 1986. Quantitative Risk Assessment for Environmental and Occupational Health. Lewis Publishers, Inc., Chelsea, MI.

Hattis, D., and Kennedy, D. 1986. Assessing risks from health hazards: An imperfect science. Technol. Rev. 89:60-71.

Heimlich, H. J., and Uhley, M. H. 1979. The Heimlich maneuver. Clin. Symp. 31(3):3-32.

Hutt, P. B., and Sloan, A. E. 1979. Food Safety Report. Nutr. Policy Issues, 6. National Academy of Sciences, Washington, DC.

Johnson, E. L. 1981. Pesticide regulation. The Pesticide Chemist and Modern Technology. ACS Symposium 160. American Chemical Society, Washington, DC.

Koop, C. E. 1985. The Heimlich maneuver. Public Health Rep. 100:557.

Kroger, M., and Smith, J. S. 1984. An overview of chemical aspects of food safety. Food Technol. (Chicago) 38(7):62-64.

Lave, L. B. 1987. Health and safety risk analyses: Information for better decisions. Science 236:291-295.

Lee, K. 1989. Food neophobia: Major causes and treatments. Food Technol. (Chicago) 43(12):62-73.

Litai, D., Lanning, D. D., and Rasmussen, N. C. 1983. The public perception of risk.

Pages 213-222 in: The Analysis of Actual Versus Perceived Risks. V. T. Covello, W. G. Flamm, J. V. Rodricks, and R. G. Tardiff, eds. Plenum Press, New York.

Lorentzen, R. J. 1984. FDA procedures for carcinogenic risk assessment. Food Technol. (Chicago) 38(10):108-111.

Miller, S. A. 1979. Achieving food safety through regulation. Food Technol. (Chicago) 33(11):57-60.

Munro, I. C., and Krewski, D. R. 1983. Regulatory considerations in risk management. Pages 156-166 in: Carcinogens and Mutagens in the Environment, Vol. II. Naturally Occurring Compounds. H. F. Stich, ed. CRC Press, Boca Raton, FL.

Newsone, R. 1990. Organically grown foods, scientific status summary. Food Technol. (Chicago) 44(12):123-130.

Omang, J. 1983. Perception of risk: A journalist's perception. Pages 267-271 in: The Analysis of Actual Versus Perceived Risks. V. T. Covello, W. G. Flamm, J. V. Rodricks, and R. G. Tardiff, eds. Plenum Press, New York.

Oser, B. L. 1978. Benefit/Risk: Whose? What? How much? Food Technol. (Chicago) 32(8):55-58.

Phillips, J. C., Purchase, R., Watts, P., and Gangolli, S. D. 1987. An evaluation of the decision tree approach. Food Addit. Contam. 4(2):109-123.

Pim, L. 1981. The Invisible Additives. Doubleday, Garden City, NY.

The Point Foundation. 1986. The Essential Whole Earth Catalog. Doubleday, Garden City, NY.

Press, F. 1986. Speaking about risk. CHEMTECH 16:336-338.

Roberts, H. R. 1978. Regulatory aspects of the food additive risk/benefit problem. Food Technol. (Chicago) 32(8):59-61.

Rowe, W. D. 1977. An Anatomy of Risk. Wiley-Interscience, New York.

Scheuplein, R. J. 1986. The role of risk assessment in food safety policy. Basic Life Sci. 38:563-571.

Slovic, P. 1987. Perception of risk. Science 236:280-285.

Sugarman, C. 1990. Assessing risk: A risky business. Food Technol. (Chicago) 44(8):60-64.

Turner, J. M. 1979. Putting food safety policy into proper focus. Food Technol. (Chicago) 33(11):63-64 ff.

Weil, C. S. 1984. Some questions and opinions on issues in toxicology and risk assessment. Am. Ind. Hygiene Assoc. J. 45:663-670.

Wildavsky, A. 1979. No risk is the highest risk of all. Am. Sci. 67:32.

Wilson, R., and Crouch, E. A. C. 1987. Risk assessment comparisons. Science 236:267-270.

Winter, F. W., and Brown, R. E. 1982. Food accidents. Pages 353-366 in: Adverse Effects of Foods. E. F. P. Jelliffe and D. B. Jelliffe, eds. Plenum Press, New York.

Naturally Occurring
Food Toxicants

CHAPTER 5

Raw, natural foods contain a bewildering array of chemicals, none of which was added during processing or growing; all were put there by Mother Nature herself. Many people are surprised to learn that, of the thousands of chemicals in the diet, only a few possess nutritional significance.

The chemicals in food may be divided into four groups. The first includes those items that we expect to find in our food—the nutrients, e.g., chemicals that are essential to life and must be obtained from the diet. These are proteins, fats, carbohydrates, vitamins, and minerals. The second group describes a fairly large number of chemicals that neither enhance nor detract from the wholesomeness and nutrient quality of the food. These chemicals behave almost as if encapsulated, moving through the gastrointestinal tract unchanged and exerting no effect on surrounding tissues or other nutrients as they pass. The third group includes chemicals that have beneficial functions but cannot be defined as true nutrients. For example, these may be chemicals that quench potential carcinogens (i.e., anticarcinogens) or increase the rates of cholesterol excretion or nutrient absorption. The fourth group includes chemicals that impair absorption, inhibit or destroy nutrients (i.e., antinutrients), or, even worse, are toxins or carcinogens.

In this chapter the focus is on the chemicals from group four—naturally occurring toxicants, mutagens, teratogens, antinutrients, or carcinogens—and some from group three—anticarcinogens. The realization that many foods naturally contain minute amounts of toxicants gives perspective for dealing with all of the toxic components, natural or added, that occur in food in small quantities. Provided that the diet is nutritionally adequate and widely varied, these small amounts of otherwise toxic compounds can be tolerated or rendered harmless.

Certain health food enthusiasts contend that telling consumers about natural food toxicants is a smoke screen used by the food industry to shift the consumer's focus away from what they feel are the "real issues," such as pesticides and additives. Ironically, much less is known about the toxicity and quantity of natural toxicants ingested or about the interaction of naturally occurring toxins and nutrients than is known about most additives and pesticides.

DESCRIBING A NATURAL FOOD TOXICANT

Toxic compounds in nature have a striking diversity of chemical structures. They are like vitamins in that they are comprised of many dissimilar compounds. For both vitamins and toxicants, it would be nice to be able to predict biological function from chemical structure, in the same way that the classification of chemicals into risk categories relies on the principle of chemical similarity to a compound that has known toxicity. Although classification by structure is helpful in identifying which molecules are more likely suspects than others, it has only limited usefulness. For instance, modification of chemical structure by changing only one group on the molecule can strip a vitamin of its activity as a vitamin or, worse still, make it a vitamin antagonist. An analogous mechanism occurs for toxicants, whereby the change of one attached chemical substituent can render the compound either substantially more toxic or completely innocuous. Looking at the chemical structure does not always enable the prediction of the molecule's biological effect. The only way to ascertain whether any chemical exerts a biological effect is through a bioassay in which an animal (or other test medium derived from a living system) is actually administered the compound in question.

The occurrence of toxic compounds in a food may have come about because the animal or plant from which it is derived was forced to evolve a means of protection from insects, microorganisms, nematodes, grazing animals, and humans. Frequently, that means of protection was a chemical one.

Plants did not evolve to be food for humans. In fact, far fewer species of plants are used for food than are not so used because many of the latter contain poisonous levels of one or more compounds. Toxic compounds in plant materials certainly may have limited plants as food for early humans. Archaeological studies by Leakey in Africa indicate that primitive people were primarily meat eaters until they learned to use fire for cooking plant material to remove or inactivate some natural toxic factor (Leopold and Ardrey, 1972).

Many plants that we use as food contain small amounts of toxic chemicals. Consider, for example, two commonly eaten foods, the potato and the tomato. Both of these come from plants that are members of the nightshade family, a fact which makes them immediately suspect because the nightshade family is known to contain some quite toxic compounds called alkaloids. The alkaloid in the potato is solanine. In the United States, healthy potatoes contain 1-5 milligrams of solanine per small potato (100 grams); in the United Kingdom, the amount is 2.5-15 milligrams per 100 gram. The most solanine occurs in the eyes (which are, of course, cut away) and in the peel, which is, ironically, the portion with the most varied selection of nutrients. If the potato is sunburned (green under the skin) or blighted, the solanine levels can increase sevenfold or more.

Under current FDA regulations, 20 milligrams of solanine per 100 grams of potato can render it unfit to eat. A survey of potatoes in the United Kingdom showed that two out of 133 samples exceeded the maximum recommended level (Davies and Blincow, 1984). Unfortunately, this plant toxin is neither washed away nor decreased by cooking. Cooked potatoes with high concentrations of solanine may have a bitter taste and cause a burning sensation in the throat.

Well-documented cases show illness resulting from the consumption of sunburned potatoes and of potato shoots, teratogenic effects from consumption of potato sprouts, and even death from consumption of blighted potatoes (Farrer, 1983; Keeler, 1980; Morgan and Coxon, 1987; Sharma and Salunkhe, 1989). In addition, we do not know the long-term effects of consumption of potatoes at the maximum safe level of solanine as currently specified (Gormley et al, 1987).

Why does the nutritious potato contain such a noxious chemical in this fine source of vitamins and complex carbohydrates? The answer is simple: solanine is an excellent natural pesticide that perpetuates the survival of the potato plant by providing protection from the Colorado beetle and the leaf hopper, common potato pests.

Another alkaloid, tomatine, is found in the tomato. The ripe red fruit may contain 36 milligrams of tomatine per small tomato (100 grams); the green fruit has much higher levels. Tomatine protects the plant from infections and is very effective against the Colorado beetle (Silverstone, 1985). Thus, we can see from looking at only one of the 150 compounds in the potato, one of our simpler foods, and one compound in the tomato that each of these plants developed at least one naturally occurring pesticide chemical to promote the survival of the species. Dr. Bruce Ames (he of the Ames mutagenicity test) has calculated that our intake of natural pesticides swamps our intake of manufactured pesticides 10,000 to one (Ames, 1983).

Another reason that naturally occurring toxins are in plants can be that they are the metabolic or waste products of normal plant metabolism. These compounds may be very beneficial to the plant but less than useful in mammalian metabolism. It is ironic that we currently ascribe high value to that which is natural, yet natural compounds from a variety of plant species are among the most lethal poisons known. Even D-Con advertises its rat poison as natural, with total accuracy!

CONDITIONS UNDER WHICH NATURAL TOXICANTS CREATE A PROBLEM

Toxicants found in normal food may cause a problem in the following four scenarios:

1) a normal food constituent eaten in normal amounts is toxic due to some inborn error of metabolism or some pharmacological reaction;

2) a normal food constituent is eaten in normal amounts by an individual with abnormal sensitivity due to allergy or other disease state;

3) a normal food constituent is eaten in abnormal amounts so that toxicity results; and/or

---

Box 5.1

Inborn Errors of Metabolism

These are examples of normal constituents eaten in normal amounts by persons with a genetic inability to metabolize them.

Inborn Error	**Offending Food or Food Constituent**
Celiac sprue	Glutens in wheat, rye, oats, and barley
Lactase deficiency or intolerance	Lactose in milk products
Sucrase deficiency	Sugar (sucrose) from all sources
Fructose intolerance	Fructose from sucrose and fruits and vegetables
Galactosemia	Milk products
Phenylketonuria	Phenalanine from protein and aspartame
Other disorders of amino acid metabolism	Proteins with the offending amino acid(s)
Hemochromatosis	Iron; excess levels are stored
Wilson's disease	Copper; excess is stored

4) an abnormal food constituent is eaten in normal amounts.

In the first case, a substance that is nutritious for most people is deleterious to others. An example is the inborn error of metabolism of the sugar in milk known as lactose intolerance. Lactose must be broken into its two simple sugar constituents, glucose and galactose, in order to be absorbed. Some individuals lack the enzyme lactase, which is necessary to split lactose, and must rely on products in which lactose is fermented, such as aged cheese or yogurt. A second example is people who are taking drugs that inhibit the monoamine oxidase enzyme. People taking these drugs must not eat foods that have been fermented or ripened (cheeses, smoked or pickled meat or fish, wines or beer, sauerkraut, or yeast extracts). Such foods contain tyramine, which requires normally functioning monoamine oxidase to prevent a reaction that includes severe headaches, rapid heart rate, increased blood pressure, and even cardiac failure and/or intercranial hemorrhages (McCabe, 1986). This is discussed further in the section on vasoactive amines later in this chapter.

The second scenario occurs when an individual is hypersensitive or allergic to a particular substance or group of substances. Again, the degree of sensitivity may cause reactions that vary dramatically— from momentary sneeze to anaphylactic shock.

The third scenario occurs when a person tends to misuse a food or consumes it in large amounts. An example is the eating of a large amount of rhubarb. The lowest recorded fatal dose of rhubarb stalks is nine pounds (although the leaves are very toxic and fatalities have occurred from eating small amounts of them) (Leiner, 1980). A 1982 report told of renal damage and hematuria (blood in the urine) from the consumption of two quarts of Worcestershire sauce per month (Rivera and Salcedo, 1982). These examples show that exaggerated consumption is required before untoward effects are seen. In some cases, an irrational desire to eat "natural" or to follow bizarre diet patterns may cause the selection of too much of a certain food and hence of the toxic compounds it may contain.

An example of a food misused is the bay leaf, which can physically damage the mucosa of the gastrointestinal tract if ingested whole (Johns, 1980). (Bay leaf should be crushed or removed before the food is served.)

An example of the fourth scenario occurs in honey if the bee collects nectar from plants containing poisonous (even carcinogenic) alkaloids and these alkaloids are transferred to the honey (Deinzer et al, 1977). Another is the eating of puffer fish (*fugu*) in Japan. Puffer fish is considered a great delicacy, but if the toxic portions are not excised with extreme care, consumption of the fish can be lethal. This scenario can also occur when the consumer either willingly or unknowingly eats normal amounts of a compound that is toxic. This occurred in 1984 and 1985 when cheap imitation vanilla was brought back from Mexico. It had been made from tonka bean extract banned from use in the United States (Concon, 1988) and contained high levels of the banned food constituent coumarin (Hogan, 1983), which has been shown to damage various body organs, including the liver (Havender and Meister, 1986; Hopkins, 1984).

Some people are concerned that the effects of several low doses of

---

Box 5.2

Allergies

Food allergies result when foods eaten in normal, or in some cases minute, amounts cause problems for individuals with abnormal sensitivity. An allergic reaction takes place when a specific immunoglobulin in the body reacts with the offending food protein to cause an antibody-antigen reaction. Some foods are more likely than others to do this. The 10 most common food allergens are:

Milk	Fish and/or shellfish
Wheat	Egg
Nuts	Peanuts
Citrus	Melons
Strawberries	Soy

The symptoms vary widely and can affect nearly all body systems. Some of the systems and symptoms observed are listed here.

Gastrointestinal tract: Canker sores, burping, burning, nausea, vomiting, diarrhea, constipation, gas, abdominal bloating and pain, ulceration

Skin: Rashes, acne, hives

Ear and eye: Ear infections, dizziness, ringing of the ears, tearing, photophobia, blurring

Muscle and joint: Pain in the muscles and joints, swelling, feelings of sluggishness

Nervous system: Headache, drowsiness, learning disorders, restlessness, hyperactivity, allergic epilepsy

Cardiovascular system: Increased heart rate and palpitations, flushing, chilling, arrhythmia

General: Fever

different toxic compounds may be additive. If a person eats a pound of rhubarb with a half pound of potatoes, is the likelihood of toxicity greater because both foods have toxic compounds? The effects do not seem to be additive (IFT Expert Panel, 1975), due primarily to the vast chemical diversity in foods. In fact, the adage that there is safety in numbers is one that applies to food. Experience has shown that the larger the number of different foods a person eats, the less likely it is that a hazardous level of any toxin will be ingested. Those at highest risk are those in subsistence economies who are forced to rely on a few food staples and those who place themselves on a bizarre diet of only a few food items.

A LOOK AT THE TOXICANTS

Toxicants are not new. Early humans learned by a potentially fatal Russian roulette game of trial and error which plants were and were not food. This vital information was transmitted through the cultural food practices of the group. Early groups recognized the alkaloids in plants for their for religious, medical, and social (opiate) uses. These same alkaloids from plants are currently our drugs of abuse.

Some of the earliest nutritional studies pointed out that products used as food had toxic factors. Early studies on rats fed raw soybeans showed that the rats would fail to grow unless the soybeans were first cooked (Johns and Finks, 1920). Other legumes and egg white showed similar toxicities when fed raw. These toxic factors were subsequently shown to be enzyme inhibitors. Other types of molecules were found to have a variety of toxic effects. The next portion of this chapter discusses known types of natural food toxicants.

Enzyme Inhibitors

The first group of food toxicants to be studied extensively was the enzyme inhibitors. All plants, especially the legumes, contain protease inhibitors. These inhibit trypsin, chymotrypsin, and other protein-digesting enzymes (Leiner, 1980). If the diet of a rat contains high levels of these compounds, the protein in the diet is poorly utilized and tissue growth and repair are impaired. The pancreas also becomes enlarged and overactivated. This is due either to the limited availability of the amino acid methionine (Rackis and Gumbmann, 1981) or to the inactivation of protein digestive enzymes. This, in turn, causes the pancreas to overproduce these enzymes in an attempt to compensate for the lack of protein digestion (Bender, 1987).

Common foods other than the legumes also have protease inhibitors. These include barley, beets, buckwheat, corn, lettuce, oats, peas, peanuts, potato, rice, rye, sweet potato, turnip, and wheat. The potato has a very high concentration of enzyme inhibitors. In fact, as much as 15% of the potato protein may be inhibitors.

It is ironic that low-cost, low-fat, widely available products such as legumes and potatoes contain factors that impair protein utilization. In most cases, but not all, cooking destroys the inhibitory effects. However, not all cooking methods are equally effective at this. In the case of the potato, microwave cooking and boiling are more effective than baking. One of the potato enzyme inhibitors, carboxypeptidase inhibitor, is stable to all three methods of cookery (Ryan and Hass, 1981). Since many vegetables are eaten raw or after very short cooking times, their enzyme-inhibiting factors may not be inactivated.

The actual adverse effects of protein inhibitors, if any, on humans are not well established. Humans are less susceptible to the effects of protease inhibitors than are some experimental animals, especially the rat (Rackis and Gumbmann, 1981; Xavier-Filho and Campos, 1989). Furthermore, there is some evidence that protease inhibitors offer a benefit as well was a risk in that they may inhibit tumor formation (Merz, 1983).

Certain enzyme inhibitors interfere with enzymes other than protein digestive enzymes. For instance, components in various beans inhibit plasmin, a factor necessary for the formation of fibrin (a blood clotting factor); components in potatoes inhibit kallikrein, a factor that aids in the formation of antibodies; and components in beans, wheat, rye, and sorghum inhibit amylases, enzymes necessary for starch digestion (Silverstone, 1985).

Another type of enzyme inhibitor prevents the breakdown of acetylcholine, an important neurotransmitter, in the cell. Acetylcholine is held in the synaptic vesicles until the arrival of an impulse, whereupon it is released. It then acts upon the membrane receptors of the postsynaptic cell, which in turn sets off a nerve impulse in that cell. After excitation, the acetylcholine must be quickly destroyed by the enzyme acetylcholinesterase. If too much acetylcholine remains in the synapse, the nerve will remain excited, leaving it unready to accept another impulse.

A wide array of foods contain inhibitors of cholinesterase. These include the edible parts of raw asparagus, broccoli, carrot, cabbage, celery, radish, pumpkin, raspberry, vegetable marrow, orange, pepper, strawberry, tomato, turnip, apple, eggplant, and potato (Silverstone, 1985). The most active inhibitor is in the potato (Concon, 1988).

Other Toxic Factors Primarily Present in Legumes

Hemagglutinins and lectins. These are naturally occurring constituents that are mainly in seeds but may also be found in other parts of plants. They are present at high levels in all legumes and grain products. Fifty-three plants have lectin activity (Pusztai, 1989). Castor beans contain so much of one of the most toxic lectins (ricin) that these beans are unsuitable for use as food. Lectins can destroy the epithelia of the gastrointestinal tract; interfere with cell mitosis; cause local hemorrhages; damage kidney, liver, and heart; and agglutinate red blood cells, the last of these being the effect after which they are named. A diet of raw black beans can kill rats in four to five days.

Since the toxicity of these compounds is dramatically reduced by cooking with moist heat, their use in human diets is little cause for concern except at high altitudes, where the boiling point is reduced, or in situations where heat transfer is terribly uneven. Toxicity is also reduced because many toxic compounds are destroyed or neutralized in a normal digestive tract and most are poorly absorbed.

Poor absorption means that lectins reach the colon in a biologically intact form and thus can have a beneficial effect (Leiner, 1981). They appear to protect the human body against colon cancer, either by causing hypersecretion of intestinal mucus or by exerting a direct toxic effect on tumor cells.

Saponins. Saponins have a soaplike quality from which they get their name. They are found in soybeans, as well as in alfalfa, spinach, asparagus, broccoli, potatoes, apples, eggplant, and ginseng root. Saponins are capable of disrupting red blood cells and producing diarrhea and vomiting (Oakenfull and Sidhu, 1989). On the plus side, they may have the pharmacological function of complexing with cholesterol and reducing serum cholesterol (Birk and Peri, 1980). Small concentrations are generally harmless to warm-blooded animals because intestinal microflora destroy them and blood plasma inhibits their action. Ginseng (a common component of herb teas), with its high saponin content, has been shown to cause nervousness, mood alterations, anorexia, hypertension, edema, sleeplessness, amenorrhea, and diarrhea (Siegal, 1979; Wahlquist, 1982). Whether these effects are due to saponins or to other factors in the ginseng is unknown.

Vicine and covicine in fava beans. Fava or broad beans are widely eaten in England and countries around the Mediterranean and are associated with the disease favism. Favism falls into the first scenario category; that is, it results from a normal food eaten in normal amounts by an individual with a genetic sensitivity. It is an inherited, sex-linked metabolic disturbance occurring in some people of Mediterranean and Asiatic origin. It is also seen in parts of Africa and among African-Americans. Eating of fava beans by genetically sensitive individuals results in a hemolytic anemia and hematuria (blood in the urine) caused by the rupturing of the older red blood cells by the fava bean nucleosides, vicine and covicine. Children are especially susceptible and favism is occasionally fatal (Marquardt, 1989).

Other factors in fava beans. Due to their high levels of monoamine oxidase inhibitors, which can cause headaches, palpitations, and sharp increases in blood pressure, fava beans can also affect individuals who do not have favism. They have also been implicated as having a carcinogenic compound possibly related to a high incidence of certain types of cancer in Latin America (Ames, 1983; Nagao et al, 1986).

Lathyrogens. Lathyrism is a neurological disease that occurs in India, China, parts of Africa, and the areas around the Mediterranean, especially Spain. Its incidence increases during periods of drought when people are forced to eat sweet pea (*Lathyrus sativus*) and vetch, which contain lathyrogens, unusual amino acids. In lathyrism, altered metabolism of the connective tissue results in skeletal and aortic abnormalities. Symptoms occur after several months on a diet of these foods. The nervous system is also affected because the metabolism of the neurotransmitter glutamic acid is impaired. This results in weakness, irritability, tremors, spasticity, jerky motions, and even convulsions (Wilson, 1979). Neurological effects are seen only in humans, and males are much more susceptible than females (Roy and Spencer, 1989).

Other factors in beans. An as-yet-unidentified factor of common beans has been reported to cause pulmonary damage when fed to rats. The damage decreased lung capacity and caused lung tissue damage as seen under a microscope (Palecek, 1969).

Goitrogens. Soybeans, pine nuts, peanuts, and millet, as well as fruits and vegetables, contain goitrogens, substances that cause goiters. The genus *Brassica*, which is the primary dietary source of goitrogens (glucosinolates), includes broccoli, Brussels sprouts, cabbage, cauliflower, horseradish, kale, kohlrabi, mustard, and mustard seed. Two members of this family, turnips and rutabaga, are particularly high in goitrogens (Silverstone, 1985). Carrot, peach, pear, radish, and strawberry may also contain goitrogens. Milk may also be a source if one of the *Brassica* species such as turnips was used as fodder. Cooking and freezing reduce the amount in milk or vegetables. A high intake of these vegetables with an inadequate intake of iodine could precipitate goiter. The use of soy milk by infants also could be a potential problem if iodine is not part of the diet or formulation.

Goitrogenic foods are thought to be responsible for 4% of the world's goiter cases. In Zaire and the Sudan, the use of the staples cassava and millet, together with lack of adequate iodine, results in over half the population having goiter (Gaitan, 1990). Protein-poor diets appear to increase the antithyroid effects of goitrogen. The risk of large amounts of cooked vegetables containing these compounds in otherwise adequate diets is unknown (Fenwick et al, 1989).

Cyanogenic glycosides. These compounds, as the name implies, yield hydrogen cyanide when acted upon by stomach acid or certain plant enzymes. Hydrogen cyanide is notorious as a potent respiratory inhibitor. As little as 30–250 milligrams is lethal to the adult male. The amounts of cyanide released from some common plant tissues are given in Table 5.1.

Several types of these glycosides are present in plants. Linamarin is the cyanogen in cassava, flax, and lima beans (10–300 milligrams per gram). Cassava is a potential hazard in countries where it is a major source of calories. To use cassava safely, it must be adequately processed commercially or in the home. Soaking, grating, and fermenting allows the hydrogen cyanide to be removed before the cassava is used. In some countries such as Nigeria, chronic cyanide poisoning occurs because some methods of processing cassava root may leave as much as 1,375 milligrams of cyanide per kilogram of dry weight of the cassava meal (Farrer, 1983). Chronic cyanide poisoning results in a degenerative neuropathy that causes problems with the optic nerve (amblyopia) and with moving (ataxia). Cell dysfunction and cell death can also occur due to poisoning of the cell respiration mechanism. Cyanide has also been implicated in a type of diabetes in which cyanide levels lead to loss of β-cell function of the pancreas (Tewe and Iyayi, 1989). In the United States, very little cassava is eaten, and then only in the processed form, as tapioca.

Other foods besides cassava cause a problem with cyanogenic glycosides. A steady diet of lima beans could create problems from cyanide ingestion. Dry beans cooked in crock pots have made people sick because the temperature was too low to destroy some of the toxic materials and the pot was tightly covered so that the cyanide could

Table 5.1. Cyanide in Various Foodstuffs[a]

Food	Cyanide Yield (mg/100 g)
Sorghum	
Mature seed	0
Leaves and young shoots	60–240
Almond	290
Apricot seed	60
Peach seed	160
Black beans	400
Pinto beans	17
Cassava	7–104

[a] Adapted from Conn (1981).

not escape harmlessly into the atmosphere (Slovut, 1982). Although cooking destroys the cyanide-releasing enzyme in the plant (emulsin), acids in the stomach still can release the cyanide. Other cyanogenic glycosides include the compounds dhurrin and amygdalin (laetrile). These compounds are found in almonds, bamboo shoots, sorghum, chokecherries, pin cherries, and wild black cherries and in the pits of apple, apricot, cherry, plum, quince, and peach (so-called bitter almonds) (Conn, 1981). Just 12 bitter almonds can kill a child! In some countries, marzipan and almond paste have been a source of cyanide. Australia reduced the allowable cyanide content of marzipan and almond paste from 50 milligrams per kilogram to 5 milligrams (Farrer, 1983). In some areas, livestock have become intoxicated and died from eating sorghum or the foliage of wild black cherry and other species (Wilson, 1979).

Another cyanogen, benzyl cyanide, is found in cress. A diet of cress is lethal to mice. Brassicas contain cyanogens as well as goitrogens. Allyl isothiocyanate, a principal flavor compound in these vegetables, causes chromosome aberrations at low concentrations in hamster cells and is a carcinogen to rats (Ames, 1983).

Figure 5.1. Young cassava plants. When mature, cassava is a woody shrub 1–5 meters high with large, elongated starchy roots. It is a significant source of calories for millions of people in developing countries, especially in Africa and South America. After the cyanogenic glucosides are removed, the roots can be boiled, baked, or ground to meal. In some areas of Africa, the leaves are also eaten. (Reprinted, with permission, from Thurston, 1984)

Plant Phenolics

Phenolic compounds are found in every plant, and in some they are in sizeable amounts. Diets that are primarily vegetarian have a higher intake of phenols. Some phenols are quite toxic and have caused grave effects in animals. Those, such as safrole and coumarin, that are toxic at any level or are carcinogenic have been banned from use as food additives. Some have been found to be mutagenic. Others appear to have desirable effects as drugs or as processing aids in that they are natural antioxidants and protect against light-induced cancer (Singleton, 1981). Some may exhibit both a beneficial and a toxic effect depending on the amount ingested. Phenolic nutrients such as menadione (vitamin K-3) or tyrosine cause adverse effects at levels greater than 3% of the diet (Singleton, 1981).

In most instances, humans and animals tend to select food with low phenol contents when they select on the basis of texture (lignin) and astringency (tannin), but they get high phenol when they select for color (anthocyanin) and freshness. For instance, a high tannin content of certain parts of some plants was thought to have evolved so that it would be less desirable as food for herbivores. Some phenols protect against microbial or fungal attack (Singleton, 1981).

Tannins. Tannins are found in almost every plant in the produce section. Tannins react to cause enzymatic browning on the cut or bruised surfaces of fresh fruits and vegetables. The tannin content of customary fruits and vegetables varies dramatically. Unripe persimmons, an unusually high source, contain 13 grams per kilogram or 6 grams per pound of fruit. Bananas, grapes and raisins, sorghum, spinach, red wine, and beer contain high levels. Other fruits contain 100–300 milligrams per kilogram. Most of us eat several grams per week from fruits and vegetables; coffee, chocolate, and tea may provide one gram of tannin per day. The amount of tannin in tea varies with the type of tea and its stage of growth. Green tea has 4% tannin and black tea may have as much as 33% (Kapadia et al, 1983).

Tannins are toxic. In the body, they bind protein, precipitate the protein of the epithelium, penetrate through the superficial cells, and cause liver damage (Mehansho et al, 1987). A 3–5% solution of tannin is toxic orally and retards growth. Tannins inhibit virtually every digestive enzyme (Butler, 1989) and reduce the bioavailability of iron and vitamin B-12 (Leiner, 1980). Even greater toxicity is seen if these compounds enter the blood stream. Tannins have been responsible for the fatal poisoning of domestic animals fed acorns and other high-tannin feeds. Toxic effects have been attributed to the consumption of excessive amounts of other high-tannin foods including carob, certain sorghums, rapeseed meal, and grapeseed meal. Human fatalities have occurred from high concentrations of tannic acid in enemas or burn remedies.

The amount of tannin absorbed is dependent on the types and amount of amino acids in the diet. For instance, drinking tea with milk in the British manner is more healthful than drinking tea by itself, as the milk binds the tannins and makes them less absorbable.

Tannins are both mutagenic and carcinogenic (Stoltz, 1982). From a regulatory perspective, they are Category I; that is, these substances are confirmed carcinogens to which we have wide exposure not only in food but in wood, soil, and other plant products.

The tannin quercitin has been identified as a mutagen and has been shown to be carcinogenic to two strains of rats, but not to hamsters or mice. Extracts of lettuce, string beans, paprika, and rhubarb have all been mutagenic in tests, and quercitin in these extracts was thought to be the cause. Onions have a very high level of quercitin. Consumption of large amounts of onions and members of the allium family can cause anemia and even death in dogs, horses, and cattle (Shamberger, 1984). It is estimated (Bryan and Pamukcu, 1982) that over a lifetime an adult consumes 25 kilograms (over 50 pounds) of quercitin! Some studies indicate quercitin may prevent the promotion phase of cancer (Birt, 1989).

Other tannins, rutin, kaempferol, and catechin, have also raised concern (Anonymous, 1984; MacGregor, 1986a, 1986b). Rutin from red wine becomes a mutagen through activation by liver and colonic enzymes (Yu et al, 1986). Some studies have shown rutin to protect certain types of tumors (Carr, 1985). It is difficult to predict when rutin or other tannins will become active carcinogens or anticarcinogens. Diet or nutritional status may influence the outcome (Weisburger and Barnes, 1986). Tannins and chlorogenic acid were not carcinogenic when present with adequate iron, copper, or manganese (Rosin and Stich, 1983a, 1983b).

Betel nuts, which are chewed after dinner in the Far East, contain 11–26% tannin and are believed to be responsible for high levels of cheek and esophageal cancer in that region, as is the tannin-containing beer and porridge eaten by the Bantus. In South America and the Caribbean, the use of sorghum as a staple and the heavy use of tannin-rich herb teas such as mate are thought to contribute to high levels of esophageal cancer (Butler, 1989; Morton, 1968; Oterdoom, 1985; Stich and Powrie, 1982).

The use of tannin-rich herb teas in the United States is of concern (Kapadia et al, 1978, 1983). Some herb tea constituents currently being sold are very high in tannins, including bayberry, comfrey, blackberry leaves, hawthorne berries, and mate. Herbs and herb teas are also being marketed in "natural" weight loss schemes. The presence of the tannins may actually speed weight loss by interfering with the protein utilization of the dieter. Considerations of other aspects of metabolism that may also be speeded would seem to be warranted.

Other phenolics. Phenolic compounds are ubiquitous in foods, spices, and drugs (especially some nasal sprays). They give plant foods many of their natural colors (e.g., flavonoids) and characteristic flavors. In the plant these are antimicrobial and antifungal.

Some common phenolics in foods and beverages are listed in Table 5.2. Two common ones include vanillin, from both natural and artificial vanilla, and catechol, from coffee, tea, and vegetables such as apples and potatoes that brown on cut surfaces.

Except for coumarin and safrole (which is carcinogenic), phenolics were originally thought to be nontoxic, but recent data have raised questions about these natural constituents. Phenols have been reported in some studies to be cancer promoters. They appear to speed the formation of the carcinogen nitrosamine by accelerating the reaction of amines with a common air pollutant, nitrogen dioxide. Nitrogen dioxide in the environment comes from cooking with gas, cigarette smoke, generation of power, and automobile exhaust (Cooney and Ross, 1986). In surprising contrast, some studies have shown that other phenolics such as chlorogenic and gallic acid inhibited nitrosation reactions (Ohshima and Bartsch, 1983).

Likewise, certain phenolics such as gallic, caffeic, and chlorogenic acid reduce mutagen formation under specified situations. They were at least as effective as vitamin C in this regard (Stich and Rosin, 1984). For example, coffee and tea, which contain both caffeic and chlorogenic acid, reduced mutagens caused by salt-cured fish (Fahey and Jung, 1989; Nagao and Sugimura, 1983; Rosin and Stich, 1983a, 1983b; Stich

Table 5.2. Some Common Phenolics in Foods and Beverages[a]

Compound	Common Sources	Amount (mg/100 g)
Cinnamic acid	Plums, cherries, apples, pears, grapes	90–182
Caffeic acid	Brussels sprouts, radishes, cabbage	1–30
	Eggplant	36–44
	Carrot, celery, lettuce	2–90
Chlorogenic acid	Apples	89
	Crab apples	46–205
	Cup of coffee	250–260
	Tea shoots	559–674
Gallic acid	Cup of black tea (200 ml)	5–10
	Instant tea	64
	Red wine	3,500
	White wine	200
Coumarin	Cabbage, radish, spinach	2–15

[a] Source: Stich and Powrie (1982), used by permission. Copyright CRC Press, Inc., Boca Raton, FL.

and Rosin, 1984). Much more information is needed for a full understanding of when phenolics increase mutagen formation and when they reduce it.

The phenolic coumarin is found in herb teas made from tonka beans, melilot, and woodruff (Hogan, 1983), in the flavoring oil of bergamot, and in the spice cassia (sometimes sold as cinnamon). Coumarin impairs blood clotting, causes menstrual irregularities, damages the liver and other organs, and causes adverse symptoms in children identified as hyperkinetic (Desphande et al, 1985). Coumarin fed to experimental animals induced bile duct carcinomas, primarily because the DNA repair process was inhibited (Shamberger, 1984).

Safrole is a phenolic that makes up 80% of the essential oil extracted from sassafras tree root and bark. Prized since its discovery in the New World, sassafras was used in teas, tonics, and cure-alls. The bark of the sassafras tree is made into filé powder, a gumbo ingredient in the New Orleans area. Safrole is also a minor component of the spices nutmeg, star anise, mace, and cinnamon. It was established as being carcinogenic to rats and was banned as a food additive in 1960. Until that time, natural and synthetic safrole were used to flavor root beers and other foods. Besides being carcinogenic, it has been subsequently shown to injure the liver (Alvares, 1984; Kapadia et al, 1978; Miller et al, 1982). Despite the banning of safrole, sassafras is still a popular ingredient in herb teas and preparations.

Estragole, the aromatic oil from tarragon, basil, and fennel, has a structure similar to that of safrole and is a weak carcinogen. Tarragon frequently flavors salad dressings, vinegars, and other dishes (Siegal, 1979). Basil also flavors vinegars and is used widely in many cuisines. Methyl eugenol, also similar to estragole in structure, is found in bay leaves, cloves, and lemon grass. It has weak to moderate carcinogenic activity in rodents. All these compounds, although carcinogenic, are found in human diets at very low levels. The potential carcinogenicity of these low doses to humans is unknown (Miller and Miller, 1986).

Black pepper, a seasoning widely used in nearly every cuisine in the world, also has some compounds that are suspect. The average intake of pepper is approximately 280 milligrams per person per day. Pepper contains low levels of safrole and tannins as well as a large amount of a related compound, piperidine, which can become a strong carcinogen (Ames, 1983). In addition, black pepper contains at least seven other substances that have been implicated as either inducing or promoting cancer.

Experiments with pepper fed in large amounts show it to be a tumor promoter for mice (Concon et al, 1981). However when fed at five to 20 times the normal human intake, it had no effect on the growth, feed efficiency, organ weight, or blood constituents of laboratory animals (Bhat and Chandrasekhara, 1986). Epidemiological data from countries such as Curacao, where consumption of red and black pepper is high, do not necessarily show the same effects of peppers as animal studies. In fact, components other than peppers accounted for the cancer incidence (Morton, 1968).

In addition to factors in black pepper that have been implicated as cancer-promoting (Concon, 1988), both red and black peppers significantly increase stomach acid, pepsin secretion, and potassium loss. Peppers can also cause tiny amounts of bleeding in the lining of the stomach, and in some individuals the bleeding can be significant (Graham, 1986).

On the plus side, capsicums, the active principle in red pepper, are more effective antioxidants than vitamin E. They are antifungal, anti-bacterial, and decrease mutagenicity in food (Nakatani et al, 1986). They also are useful in clearing the sinuses and in desensitizing the respiratory tracts of animals and have therefore been proposed for treatment of asthmatics (Brody, 1984). Hot peppers may also increase the basal metabolic rate by as much as 25%. That's good news for dieters but may be bad news for people in countries where highly spiced food is eaten but where an inadequate supply of calories is available.

Now, before we all swear off cinnamon, nutmeg, tarragon, and an occasional bowl of gumbo or pesto, it is useful to consider Miller's perspective. Miller et al (1982) state that despite exposure from food in the parts per million range throughout a large portion of the human life span, the risks from these weak carcinogens are small. Furthermore, we have data showing these compounds to be both mutagenic and antimutagenic, both carcinogenic and antioxidant (usually an anti-carcinogenic effect), and also antifungal, preventing growth of molds containing carcinogenic mycotoxins. So even if the food contains a carcinogenic compound, we do not know whether it would ever exert a carcinogenic effect, both because the level is small and because the other compounds in the food may not allow it to be carcinogenic.

Photosensitizers. These phenolic compounds behave just as their name implies—they increase sensitivity to light. This ability is not destroyed by cooking. Parsnips, carrots, celery, parsley, oranges, lemons, limes, figs, fennel, dill, mustard, and coriander all contain photosensitizers. These compounds have tested as being highly mutagenic and carcinogenic when activated by ultraviolet light (Poulton and Ashwood-Smith, 1983). When activated, these compounds can cause direct damage to DNA or can produce oxygen radicals (Ames, 1983).

Some of these compounds, known as psoralens, are toxic even when not activated by light. Less than 0.25 pounds of parsnips gives a 4-5-

milligram dose of psoralens, a dose that causes a physiological response (Ivey et al, 1981). If the plant has been subjected to some type of stress during growth or storage, the amount of these compounds increases. Ivey (1981) believes that psoralen-containing vegetables present some toxicological risk to humans.

Gossypol. Gossypol is a toxic phenolic compound found in cottonseed. Since each bale of cotton produces 840 pounds of seed, cottonseed is clearly a potential source of human and animal food. The protein is of high nutritional quality and is used on a limited basis for human food, as in the plant protein blend available in Guatemala as *Incaparina*. One of the limits on the usefulness of this untapped source of protein is the toxin gossypol. Its toxicity is dependent on the composition of the diet and on the species eating it. Gossypol inhibits the conversion of pepsinogen to pepsin and limits the bioavailability of iron. It may cause loss of appetite and weight loss, diarrhea, anemia, diminished fertility, pulmonary edema, circulatory failure, and hemorrhages of the small intestine, liver, and stomach. Different animals vary in their sensitivity to the effects of gossypol (Silverstone, 1985). Selective breeding programs have developed a cottonseed plant that produces much less gossypol. Unfortunately, low-gossypol cultivars are more susceptible to mold growth and may therefore have aflatoxin problems (Ames, 1983).

Box 5.3

Incaparina—A Food for Humans Utilizing Cottonseed

Incaparina is a cereal food comprised of various grains blended to give a protein nutritional value similar to that of higher-priced animal proteins. It was introduced by the Institute of Nutrition of Central America and Panama (INCAP) in 1961. The first mixture was 29% corn flour, 29% sorghum, 38% cottonseed flour, 3% torula yeast, and 1% $CaCO_3$. Later, alternate mixtures were developed with other grains and legumes, including soy, sesame, and peanut. Thus, the mixtures could be adapted to take into account the availability of various grains and the regional taste preferences. All mixtures were formulated to give proteins of the same biological value for children as milk proteins give (Scrimshaw, 1980).

Although cottonseed and other grain sources are successful in reducing and preventing protein calorie malnutrition, the real problem of getting the food to those who need it most has still not been fully solved, as the most needy in a population do not have even the small amount of money necessary to purchase this nutritious product (Wise, 1980).

Oxalates, Phytates and Other Alterers
of Mineral Balance

Oxalates. Oxalic acid (oxalate) is a strong, organic acid. Oxalates bind calcium and other needed trace minerals and make them unavailable for absorption. Increased oxalate intake can be responsible for decreased bone growth and for kidney stones. High levels can cause vomiting, diarrhea, blood-clotting problems, convulsions, and coma. No food regulatory authority would allow oxalates to be added to food in any level, yet they are present in low levels in peas, cocoa, spinach, beet greens, berries, rhubarb, carrots, lettuce, turnips, and beets.

Some foods have much higher levels of oxalate than others. In spinach and rhubarb the level is up to 1% of the fresh weight and in tea leaves up to 2%. The oxalate content of tea is high and accounts for 50% of the dietary oxalate in areas where tea drinking is high (Crampton and Charlesworth, 1975). If tea is made with hard water or served with milk, it has less oxalate because the calcium from the milk or the hard water precipitates oxalate. During World War II, the high incidence of renal stones in British troops in India was attributed to high intake of tea without milk and attendant dehydration from the heat (Crampton and Charlesworth, 1975).

Rhubarb leaves (and the house plant dieffenbachia) contain toxic levels of oxalate. The anthraquinone as well as the oxalate content of rhubarb leaves may also be responsible for their toxicity (NAS/NRC, 1989).

Phytates. Phytic acid is a simple sugar with many phosphoric acids attached. This structure makes it a very effective chelator, which means that it will effectively tie up divalent metals like zinc, iron, and calcium and make them unavailable for absorption. Phytates are primarily found in nuts, legumes, and the germ and bran of cereal grains but are also present in green beans, carrots, broccoli, potatoes, sweet potatoes, artichokes, blackberries, strawberries, and figs. Whole wheat and rye flours have 458–1,780 milligrams per kilogram and white flour 161–381 milligrams (Silverstone, 1985). The increased zinc content of whole wheat flour over white flour does not necessarily result in higher levels of zinc in the body because the much higher phytate content of the whole wheat flour makes the zinc less available (Hambraeus, 1982).

Breads like pocket bread that have short fermentation and baking times create a particular phytate problem. In this case, the phytase enzyme in wheat does not have adequate time to split the metal-phytate complex apart. In parts of the Middle East where the consumption of whole wheat pocket bread makes up 85% of the calories and where other sources of dietary zinc are expensive, human zinc deficiency occurs. In like manner, iron availability may be limited in populations that subsist on unrefined grains, and deficiencies may result (Reinhold et al, 1976).

Mineralocorticoid Stimulators

Licorice. Licorice contains glycyrrhizin, and as few as four twists per day for just one week (100–200 grams per day) can cause considerable and sustained mineralocorticoid action. This results in retention of sodium and water. Licorice really should carry a warning for people with hypertension and other circulatory disorders (American Academy of Allergy and Immunology, 1984; Epstein et al, 1977).

Other Alkaloids

Alkaloids are bitter components in plants that frequently have pharmacological properties. Two alkaloids from food, solanine and tomatine, have already been discussed. Caffeine and theobromine are alkaloids found in coffee, tea, and chocolate. Caffeine inhibits DNA

Thanksgiving Dinner
MENU

COURSE	CHEMICAL COMPOSITION INCLUDES
Appetizer	
Cream of Mushroom Soup	hydrazines
Fresh Vegetable Tray	
Carrots	carotatoxin, myristicin, isoflavones, nitrate
Radishes	glucosinolates, nitrate
Cherry Tomatoes	hydrogen peroxide, nitrate, quercetin glycoside, tomatine
Celery	nitrate, psoralens

Menu analysis prepared by ACSH staff and its Directors and Scientific Advisors, with technical assistance from Leonard T. Flynn, Ph.D., M.B.A., a scientific consultant.

Figure 5.2. Thanksgiving menu—appetizers. In this meal, as in all our food, there are many chemicals that would be toxic if eaten in excess amounts. (Reprinted, with permission, from ACSH, 1987)

repair under some conditions, and heavy coffee drinking has been implicated (but not proven) in epidemiological studies of cancer of the ovary, bladder, pancreas, and large bowel (Ames, 1986). Theobromine at high concentrations retards its own breakdown, and when given as chocolate has no effect, probably because of limited bioavailability from chocolate (Vesell, 1985). Caffeine is discussed in more detail in Chapter 11.

Another common alkaloid is quinine. Quinine is used as a drug and is found in the beverage tonic water. Tonic waters contain 15 milligrams of quinine per 250 milliliters (8 ounces) of water. In hypersensitive individuals, two 8-ounce glasses can produce a reaction. Some of the side effects include skin rash, tinnitis (ringing in the ears), slight deafness, vertigo, and slight mental depression. An acute oral dose is about 18 grams for an adult and 1 gram for a child (Fletcher, 1976).

Fats, Sugars, Proteins, and Vitamin Antagonists

Toxic fats. Rancid or oxidized fat is indeed quite toxic, especially when vitamin E is deficient or marginal (Gurr, 1988; Pryor, 1985). Carotatoxin is a triple-bonded lipid in celery. Erucic acid is a 22-carbon monounsaturated fatty acid found in high concentration in mustard and rapeseed. It causes growth depression in most species tested as well as liver degeneration and kidney nephrosis. As little as 15% erucic acid can cause a severe heart problem (myocarditis), which results in a pale-colored heart that is unable to adequately use oxygen. Plant breeding has developed a low-erucic-acid species of rapeseed called canola that can be used for cooking oils and margarines since the rapeseed oils with high erucic acid are not allowed for use in the United States.

Nondigestible sugars. Legumes may contain up to 4% of their dry weight as the trisaccharides raffinose and stachyose. The human digestive system does not have the enzymes necessary to split these sugars, so they move unabsorbed into the large bowel and are fermented by the bacteria, resulting in flatus (gas) and abdominal rumblings. In some persons, they can also cause headache, dizziness, and cramps (Olson et al, 1981).

Amino acids. Common amino acids have toxic effects at high levels (Hegarty, 1986) and are promoters of bladder cancer in tests with rats. Three essential amino acids—leucine, isoleucine, and valine—are tumor promoters of the same potency as saccharin. The degree of tumor promotion is dose dependent (Nishio et al, 1986). Trytophan is also a tumor promoter (NAS/NRC, 1989). Thus, diets high in protein, as in Western countries, may increase cancer risk.

Vasoactive amines. As previously mentioned, vasoactive amines are present in many foods, including cheeses such as Camembert and Stilton (200 milligrams of tyramine per 100 grams), wine, beer, yeast extracts such as marmite, broad beans, potatoes, plantain, bananas, avocados, pickled and fermented meats or fish, and sauerkraut. Vasoactive amines—histamine, tryptamine, tyramine, and serotonin—can cause facial flushing, rapid heart rate, increased blood pressure, urticaria (rash and hives), and headaches, especially in susceptible subjects. These substances act directly by constricting the blood vessels and indirectly by liberating adrenaline and noradrenaline from the nerve endings. Eating foods with tyramine while taking monoamine oxidase inhibitors can lead to severe complications, even fatalities (McCabe, 1986).

Thanksgiving Dinner
MENU

COURSE	CHEMICAL COMPOSITION INCLUDES
Entree	
Roast Turkey	*heterocyclic amines, malonaldehyde*
Bread Stuffing with onions, celery, black pepper, mushrooms	*benzo(a)pyrene, di- and tri-sulfides, ethyl carbamate, furan derivatives, hydrazines, psoralens, safrole*
Cranberry Sauce	*eugenol, furan derivatives*
Choice of Vegetable	
Lima Beans	*cyanogenetic glycosides*
Broccoli Spears	*allyl isothiocyanate, glucosinolates, goitrin, nitrate*
Baked Potato	*amylase inhibitors, arsenic, chaconine, isoflavones, nitrate, oxalic acid, solanine*
Sweet Potato	*cyanogenetic glycosides, furan derivatives, nitrate*
Rolls	*amylase inhibitors, benzo(a)pyrene, ethyl carbamate, furan derivatives*
with Butter	*diacetyl*

Menu analysis prepared by ACSH staff and its Directors and Scientific Advisors, with technical assistance from Leonard T. Flynn, Ph.D., M.B.A., a scientific consultant.

Figure 5.3. Thanksgiving menu—entree. Humans consistently eat naturally occurring toxicants. The fact that they are natural does not make them safe; the fact that they are eaten in very small amounts does. (Reprinted, with permission, from ACSH, 1987)

Areas where plantain is a staple show increased incidence of a heart disease that results from large amounts of an amine found in plantain, 5-indoleacetic acid (Silverstone, 1985). Phenylethylamine from chocolate and two compounds from citrus also can have a vasoactive effect (Joint Committee of the Royal College of Physicians and British Nutrition Foundation, 1984). This compound from chocolate also produces mild euphoria.

Vitamin antagonists. An enzyme, thiaminase, which destroys the vitamin thiamin, is found in 31 vegetables, 18 fruits, and several species of raw and cooked fish. Its highest concentration is in blackberries, black currants, blueberries, beets, brussels sprouts, red cabbage, bracken fern, tuna, and other raw fish.

Tea also destroys thiamin. It is interesting to note that human thiamin deficiency, beriberi, is most common in the Far East, where white rice (low in thiamin) is the staple food and where tea, bracken fern, and raw fish are common dietary components.

A vitamin B-6 antagonist, similar to the amino acid proline, is found in linseed meal and flaxseed. This amino acid forms a stable complex with vitamin B-6, which makes it unavailable to the body (Bender, 1987).

Citral from orange peel and lemon grass is thought to inhibit vitamin A and to produce changes in the eye similar to vitamin A deficiency. Lipoxidase, an enzyme in soybeans and other plant products, destroys vitamin A (Silverstone, 1985). Avidin from raw egg white binds the B vitamin biotin, but a deficiency is seen only if significant amounts of raw egg are ingested, as cooking renders the avidin inactive.

Natural Antibiotics

Tea, honey, grape juice, and wine are weak antibiotics. Tangeretin and nobiletin from the peels and oils of citrus fruits are both antiviral and antimicrobial and are somewhat toxic to mammals. They have little effect on adults but have been implicated in some stillbirths (Silverstone, 1985).

Spices, Herbs, and Teas

Currently, more than 400 herbs are commercially available in the United States (Tyler, 1987). Herbal teas and preparations represent the second largest segment of the health food market. Some of the herbs and spices can cause significant problems when consumed at high levels, as in the overconsumption of herb teas (Anonymous, 1977; Siegal, 1976).

Specific herbs and flavorings have a wide range of effects. For instance, the juniper berries used in certain herb teas and as flavoring for meats and gin can cause gastrointestinal irritation. Buckthorn, senna, dock, and aloe, frequently found in tea formulations, are cathartics. Indian tobacco can cause paralysis and hypothermia. Chamomile causes extreme allergic reactions in some individuals. Mate tea has been associated with liver damage (Kapadia et al, 1983).

A resurgence in the use of herb teas has increased poisonings by plants from the foxglove family, including foxglove, dogbane, wallflower, lily of the valley, and oleander. Teas from these plants can cause digitalis toxicity from as little as one cup of tea (Dickstein and Kunkel, 1980).

Many teas, spices, and herbs are also hallucinogens. Tea constituents burdock root, catnip, juniper, hydrangea, lobelia, jimson, persimmon,

Thanksgiving Dinner
MENU

COURSE	CHEMICAL COMPOSTION INCLUDES
Dessert	
Pumpkin Pie with cinnamon & nutmeg	myristicin, nitrate, safrole
Apple Pie with cinnamon	acetaldehyde, isoflavones, phlorizin, quercetin glycosides, safrole
Beverages	
Coffee	benzo(a)pyrene, caffeine, chlorogenic acid, hydrogen peroxide, methylglyoxal, tannins
Tea	benzo(a)pyrene, caffeine, quercetin glycosides, tannins
Red Wine	alcohol, ethyl carbamate, methylglyoxal, tannins, tyramine
Water available upon request	nitrate
Assorted Nuts Mixed Nuts	aflatoxins

Menu analysis prepared by ACSH staff and its Directors and Scientific Advisors, with technical assistance from Leonard T. Flynn, Ph.D.,M.B.A., a scientific consultant.

Figure 5.4. Thanksgiving menu—dessert and beverages. Potentially, the danger of carcinogenic and toxic effects from naturally occurring chemicals is greater than from pesticide chemicals because we eat more of the natural ones. (Reprinted, with permission, from ACSH, 1987)

and passionflower all contain hallucinogens (Siegal, 1976; Tyler, 1987; Wahlquist, 1982). Parsley contains the hallucinogen apiol; nutmeg, carrots, parsnips, and bananas contain myristicin. High doses (10 grams) of these hallucinogens can also cause liver damage.

Some herbs contain carcinogens. Pyrrolizidine is a carcinogenic alkaloid in the herb comfrey and in a vegetable used in Japan (Silverstone, 1985). The seeds of crotalarias, used to make the liqueur curacao and some herb teas, contain carcinogens (Concon, 1988).

Naturally Occurring Carcinogens

Cycasin. Nuts of cycad trees, which grow in tropical and subtropical climates, contain cycasin, which causes acute toxicity, increases the incidence of Lou Gerhig's disease (amyotrophic lateral sclerosis) 100-fold, and is carcinogenic. A detailed review is given in Concon (1988).

Nitrates and nitrites. Nitrates and nitrites themselves are not carcinogenic, but nitrates can be reduced to nitrites, which can be converted in the body to nitrosamines, potent liver carcinogens. Eighty percent of the nitrates and nitrites in the diet are from naturally occurring sources, with only 20% coming from food additives (primarily those used in smoked meats). Beets, celery, cucumber, lettuce, spinach, radish, and rhubarb have as much as 200 milligrams per 100 grams (Ames, 1986). Under certain storage conditions, nitrate in the vegetables can be converted to nitrite.

Human saliva can have as much as 8–12 milligrams of nitrite per liter. The concentration of nitrite in the saliva is affected by diet. For instance, ingestion of lettuce or cucumber can increase the salivary nitrite for several hours (Wasserman and Wolff, 1981).

Nitrites can react with amines present both in food and in the body to form nitrosamines. Despite their strong carcinogenicity in experimental animals, no direct evidence for their carcinogenicity in humans exists (Gross and Newberne, 1979). Furthermore, in studies with experimental animals, a no-effect dose of this substance has been determined (Wasserman and Wolff, 1981).

Epidemiological evidence shows that areas where nitrate levels are high in food, soil, and well water, as in parts of China or Iran, have a higher incidence of esophageal and stomach cancer. The dietary habits in these areas include not only high intakes of nitrate, but also low intakes of protein, fat, and fresh fruits. Specific vitamins noted to be inadequate include vitamins A and C and riboflavin. In some areas, a low intake of molybdenum has also been documented. The marginal diet may be important, as nitrosamine formation is inhibited in the presence of vitamin C and phenolics (Ames, 1986). Such diets are not only marginal in nutrients thought to inhibit cancer, but also include many preserved items implicated in the increase in rates of certain types of cancers. In addition, evidence of improper preservation of foods was found, which would increase likelihood of fungal mycotoxins, many of which are also carcinogenic (Cai, 1982; Gross and Newberne, 1979; Williams et al, 1982).

An interaction between nitrate and soy sauce in the diet may occur. Soy sauce mixed with nitrate forms a variety of potent mutagens. In the Orient, the average annual consumption of soy sauce is over 10 liters, which potentially contains 2–18 milligrams of mutagens (Nagao et al, 1986).

Estrogens. Many plant materials contain phyto- (plant-) estrogens; these include wheat, corn, barley, oats, rice, soybeans, potatoes, carrots, parsley, olives, peanuts, apples, plums, cherries, vegetable oils, and yams. The animal products liver and egg yolk also contain estrogen (Silverstone, 1985). The activity of food estrogen is one tenth to one thousandth that of the most common form of circulating human estrogen, 17-B-estradiol. The amount found in food is far below that required for a physiological response, but effects of long-term subphysiological doses remain unknown. So far, no problems with food sources of exogenous estrogens in humans are known; however, sheep in parts of the United States and Australia have had breeding problems attributed to grazing on crops high in estrogen.

Estrogens are known to promote tumor growth. The tumor-promoting activity of phytoestrogens is similar to that of 17-B-estradiol and diethylstilbestrol (DES) (Murphy, 1982). Phytoestrogens were not found to be mutagenic, substantiating the belief that they are tumor promoters, not tumor initiators. Murphy (1982) compared the the natural estrogenic activity level of soy phytoestrogens in normal soybean products with the activity of the synthetic estrogen feed additive DES found in the livers of improperly handled beef and determined that the natural activity level in soy was approximately 34 times greater.

Other Plant Carcinogens

Bracken fern. Bracken fern, often called fiddlehead, is produced commercially for human diets as greens and is eaten in Japan, Canada, and New England (Bryan and Pamukcu, 1982). Young plants have the most toxin. Toxicity is reduced but not eliminated by cooking and pickling (Hirono, 1983, 1989; Shamberger, 1984). The toxin can damage bone marrow.

Animal evidence indicates that this plant is carcinogenic as well as toxic. Cattle that forage on bracken fern as a major component of their diet develop cancer of the urinary bladder in two to five years. Dry bracken fern, fed as one third of the diet of rats for four to six months, caused the development of bladder and intestinal tumors (Hirono, 1989). Interestingly, the carcinogenicity of bracken is affected by other components in the rat diet. Addition of soy protein, the food additive BHA, calcium chloride, or the vitamin niacin reduced the number of tumors, whereas addition of thiamin or nitrate increased them.

Human intake of bracken fern is related in epidemiological studies to increased risk of esophageal cancer. This relationship was found to be true in studies in Japan but not in Canada. Dietary factors may have a role in protecting or increasing the carcinogenic factors in bracken fern in humans as well as in animals. For instance, a commonly used Japanese hot gruel drink has been implicated as a factor increasing the incidence of esophageal cancer (Shamberger, 1984). Perhaps this or the mutagens in soy sauce interacting with the components of bracken fern increase the cancer incidence. In addition, other dietary factors may prove to be of importance.

The compound in bracken fern that is the culprit is under study. It has been suggested to be quercitin (Bryan and Pamukcu, 1982) or maybe a glucoside, ptaquiloside (Hirono, 1989).

Parasorbic acid. Cranberries and berries of the mountain ash contain parasorbic acid, which is carcinogenic in experimental animals.

Halogenated compounds. Seaweeds eaten in Hawaii and Japan contain high concentrations of halogenated compounds that are suspected of being carcinogenic. Epidemiological data support this possibility. For example, Hawaiians have the highest cancer incidence of any ethnic minority in the 50 states. There is a particularly high incidence of liver cancer (Mower, 1983).

Hydrazines. The common cultivated mushroom of Western commerce (of which we in the United States alone eat 600 million pounds per year [Breene, 1990]), as well as the shiitake mushroom, contain several carcinogens of the hydrazine family. Humans eating a 100-gram serving (approximately one-fourth pound) of mushrooms daily would be eating the same dosage (50 ppm) as caused cancer in mice and hamsters.

Hydrazine levels vary both with processing and storage as well as with variety of mushroom. Fresh, raw mushrooms contain the most hydrazine. Both cooking and refrigerator storage for one week cause the loss of nearly one third of the hydrazine. All the hydrazine is lost in canning (Toth, 1983). Other dietary factors modulate carcinogenicity (Visek and Clinton, 1985), so the contribution that exposure to these carcinogens makes to cancers in humans is difficult to determine (NAS/NRC, 1989).

Ethyl carbamate (urethane). Urethane is a natural by-product of fermentation. Traces occur in foods that have been fermented such as yogurt, bread, beer, olives, soy sauce, and some cheeses. This compound has been shown to be carcinogenic in experimental animals (Shamberger, 1984). The role of ethyl carbamate from foods in the development of human cancers is unknown (NAS/NRC, 1989).

Recent attention has been focused on the urethane found in wines, sherries, port, bourbons, and liqueurs. Canada has set legal limits on the amount of urethane (40 ppb) allowed in alcoholic beverages. In the United States, as this was being written, the FDA was still studying the issue, as some data indicated that alcohol blocks the metabolism of urethane, possibly preventing its carcinogenic action (Monmaney et al, 1987).

Polycyclic Aromatic Hydrocarbons

This chapter has dealt so far with toxicants and carcinogens that occur naturally in food. Some of these, particularly from members of the legume family, are rendered less toxic by cooking or processing. However, cooking or processing can also create toxins, mutagens, and carcinogens. Polycyclic aromatic hydrocarbons (PAHs) are carcinogens produced from the heat-induced breakdown of organic matter. This family of compounds has long been recognized as potent carcinogens. They are widely distributed not only in food but also in the environment and thus are cause for concern. In the environment they result from combustion of organic matter. They are some of the major carcinogenic constituents isolated from cigarette tar and products of air pollution. In fact, nearly 90% of the total national emission of PAHs was attributed to combustion of coal, residential furnaces, and coke ovens (Harvey, 1982).

PAHs have been found in a wide variety of foodstuffs. They may be in some natural foods and are produced or increased by cooking and processing. Over 20 different PAHs have been isolated; the most common are benzo[a]pyrene and quinoline compounds.

Any heat treatment may produce some benzopyrene, but broiling and frying produce the most (Anonymous, 1991). Some form during the canning of protein foods and some, albeit to a lesser degree, in browning reactions such as occur during the baking process, the caramelizing of sugar, or the roasting of coffee (Krone et al, 1986; Powrie et al, 1986). Pickling and fermentation also can produce PAHs, and thus they are found in soy sauce and a pickle made from Chinese cabbage (*kimchee*) (Sugimura, 1986). Peanut, sesame, and safflower (cold-pressed) all carry 1–2 micrograms of benzopyrenes per kilogram (Achaya, 1981).

Paraffin wax, a petroleum derivative used on food packages or vegetables, can be a source of PAHs. In the United States, PAH contamination of food is controlled by mandating the purity of paraffin materials allowed for use in packaging materials and vegetable waxing. Shellfish living in petroleum-contaminated waters can also contain PAHs (Dunn, 1982).

Estimates from West Germany indicate that the annual per capita consumption of benzopyrene varies from 250 to 1,000 micrograms. Cereals, fruits, and vegetables supply 45–90% of the ingested amount, with leafy vegetables and the surface of fruits and vegetables having the most. Smoking and roasting only represent a small contribution to the total (Adrian et al, 1984).

Significant numbers of PAH mutagens were measured in pan-fried and broiled well-done beef, ham, bacon, beef sausage, pork chops, fried eggs (cooked at high temperature), chicken, and fish (Hatch et al, 1982). One study (McCann, 1983) showed that the mutagenic activity of meat cooked well-done was equal to that of five cigarettes! Low to intermediate levels of mutagens were produced by roasting and baking, with the levels dependent upon the temperatures and the cooking time. Meat loaves with higher fat levels surprisingly had fewer mutagens than those with lower fat because of shorter cooking times (Aeschbacher, 1986). Microwave cooking, stewing, boiling, and poaching showed little mutagen formation, presumably due either to lack of browning or to short cooking times (Bjeldanes, 1985; Bjeldanes et al, 1983; Powrie et al, 1982).

Broiling seems to be a particular concern for the production of mutagenic PAHs. Epidemiological evidence from Japan showed that the stomach cancer mortality rate for those consuming broiled fish twice a week or more was 1.67 versus 1.00 for those who consumed broiled fish less often. For Japanese Seventh Day Adventists (who are vegetarian), it was only 0.32 (Kuratsune, 1986). Broiling was indicated as a stomach cancer risk factor in this Japanese study. However, stomach cancer rates in the United States have declined steadily for the past 50 years (NCI, 1990), while U.S. consumption of broiled food has increased (Shamberger, 1984).

The amount of mutagen that is produced during broiling depends on a wide variety of factors. Foods that are cooked over charcoal, mesquite, electric element, or gas flame have lower mutagen levels 1) if the products are low in fat and trimmed of excess fat so that a minimum of drippings is redeposited on the food via smoke; 2) if flare-ups are extinguished; 3) if foods are kept on the grill a minimum time by precooking and then cooking only to rare, where appropriate; 4) if foods are cooked at the proper temperature (low) and the proper distance from the broiler to avoid excessive browning on the surface;

5) if the heat source is either above the food item or the food is separated from the heat source by a pan or aluminum foil; 6) if charcoal and gas are used as fuels rather than wood and crumpled paper; and 7) if hard rather than soft woods are used (AICR, 1986; CBNS, 1978; Larsson, 1986; NAS/NRC, 1989; Powrie et al, 1982).

The threat to human health posed by PAH mutagens remains unknown. However, their wide distribution, coupled with their demonstrated carcinogenicity in rodents, certainly raises some legitimate concerns. We have no data as yet to show how rodents and humans differ in their repair of DNA or in their susceptibility to PAHs. We have problems with dose extrapolation, as the average human intake is only 0.0002 of that fed to animals to produce cancer in half the animals (Sugimura, 1986) and may therefore not pose a serious risk (NAS/NRC, 1989). Disconcerting information came from a study on rats that showed that feeding a mixture of these compounds together was more carcinogenic than feeding an equivalent dose alone. Clearly, humans do get mixtures of these chemicals from the diet and the environment.

Further confounding the issue are test results that frequently conflict (Ames et al, 1987). For instance, browning reactions of sugars are reported as both antimutagenic and mutagenic (Mauron, 1989; NAS/NRC, 1989), and coffee is antimutagenic to aflatoxin but mutagenic under other circumstances (Powrie et al, 1986).

Factors in the body modulate the effect of these compounds. Vitamin A, saliva, nitrite, porphyrins, unsaturated fatty acids, thiols, soy protein, beef extracts, and vegetable extracts are just some of the compounds in food or in the body that inhibit the mutagenic activity of PAHs and other mutagens (Friedman, 1984a, 1984b; Miller 1985).

CONCLUSION

The material in this chapter forces each of us to realize that many carcinogens and toxicants are present in every meal, regardless of whether the food was "natural," raw, processed, or cooked. In addition, long-term food selections designed to eliminate small amounts of carcinogens or toxins are bound to create both an uninteresting and an inadequate diet. We may end up at the grocery store check-out line saying, as the Wizard of Id did, "I'll have a quart of gin and a bag of prunes." The effect of any one compound in a food is certainly dependent on the interaction of various food and dietary components. The moral for this story is a simple one suggested by Ben Franklin in an earlier century: "moderation in all things." By applying moderation to eating patterns, the amounts of any one toxic compound are kept to a minimum.

We also must remember that, despite many pages of information in this chapter on various toxic compounds in food, toxicological studies have been completed on only a small percentage of natural toxicants (Gormley et al, 1987). In fact, much less information exists on natural toxicants than on regulated food additives (Shank et al, 1989).

Finally, we should consider the wisdom of the statement from the American Council on Science and Health that

> the increasing body of evidence documenting the carcinogenicity of common, everyday substances in nature (at least under laboratory conditions) points up the contradiction we have created in our past regulatory approach to carcinogens. The contradiction consists of the huge discrepancy in the weight we have hitherto placed on manmade carcinogens—trying to purge our land of them—and natural carcinogens—which we have simply ignored. Our emphasis should be on the potency of the chemical and the level of human exposure rather than on the chemical's natural or artificial origin [ACSH, 1985].

Although this statement was written about carcinogens, it applies equally well to all toxicants that may occur in food.

REFERENCES

AAAI. 1984. Adverse Reactions to Foods. NIH Publication 84-2442. American Academy of Allergy and Immunology, Bethesda, MD.

Achaya, K. T. 1981. Regulatory aspects of food additives. Indian Food Packer 35(3):11-14.

ACSH. 1985. Does nature know best? Natural carcinogens in American food. American Council on Science and Health, New York.

ACSH. 1987. Thanksgiving Dinner Menu. American Council on Science and Health, New York.

Adrian, J., Billaud, C., and Rabache, M. 1984. Part of technological processes in the occurrence of benzo[a]pyrene in foods. World Rev. Nutr. Diet. 44:155-184.

Aeschbacher, H. V. 1986. Genetic toxicology of browning and caramelizing products. Pages 133-144 in: Genetic Toxicology of the Diet. A. R. Liss, ed. Alan R. Liss, New York.

AICR. 1986. Facts You Should Know about Outdoor Cooking. American Institute for Cancer Research, Washington, DC.

Alvares, A. P. 1984. Environmental influences on drug bio-transformations in humans. World Rev. Nutr. Diet. 43:45-49.

Ames, B. N. 1983. Dietary carcinogens and anticarcinogens. Science 221:1256.

Ames, B. N. 1986. Food constituents as a source of mutagens, carcinogens, and anticarcinogens. Pages 3-32 in: Genetic Toxicology of the Diet. A. R. Liss, ed. Alan R. Liss, New York.

Ames, B. N., Magaw, R., and Gold, L. S. 1987. Ranking possible carcinogenic hazards. Science 236:271-280.

Anonymous. 1977. Herb teas: How safe? Consumers' Res. Mag. 3:35-36.

Anonymous 1984. Rutin causes oral cancer—Hard to swallow. Food Chem. Toxicol. 22(6):484-485.

Anonymous. 1991. Carcinogen threat may mean cutting down on barbecues. Star Tribune (Minneapolis), April 24, p. 3T.

Bender, A. E. 1987. Effects on nutritional balance: Antinutrients. Pages 110-124 in: Natural Toxicants in Food: Progress and Prospects. D. H. Watson, ed. VCH Publishers, New York.

Bhat, G. B., and Chandrasekhara, N. 1986. Lack of influence of black pepper in the weanling rat. J. Food Saf. 7:215-223.

Birk, Y., and Peri, I. 1980. Saponins. Page 161 in: Toxic Constituents of Plant Foodstuffs, 2nd ed. I. E. Leiner, ed. Academic Press, New York.

Birt, D. F. 1989. Dietary inhibition of cancer. Pages 107-121 in: Carcinogenicity and Pesticides: Principles, Issues, and Relationships. N. N. Ragsdale and R. E. Menzer, eds. American Chemical Society, Washington, DC.

Bjeldanes, L. F. 1985. Hazards in the food supply. Nutr. Update 1:105-119.

Bjeldanes, L. F., Fenton, J. S., and Hatch, F. T. 1983. Mutagens in cooked food. Pages 149-168 in: Xenobiotics in Foods and Feeds. ACS Symposium 234. J. W. Finley and D. E. Schwass, eds. American Chemical Society, Washington, DC.

Breene, W. M. 1990. Nutritional and medicinal value of specialty mushrooms. J. Food Prot. 10:883-894.

Brody, J. 1984. Effect of spicy foods a hot topic. Minneapolis Star and Tribune, March 21.

Bryan, G. T., and Pamukcu, A. M. 1982. Sources of carcinogens and mutagens in edible plants. Pages 75-82 in: Carcinogens and Mutagens in the Environment, Vol. 3. H. F. Stich, ed. CRC Press, Boca Raton, FL.

Butler, L. 1989. Polyphenols. Pages 95-122 in: Toxicants of Plant Origin. Vol. 4. Phenolics. P. R. Cheeke, ed. CRC Press, Boca Raton, FL.

Cai, H.-Y. 1982. Etiology and prevention of esophageal cancer in China. Carcinogens and Mutagens in the Environment, Vol. 3. H. F. Stich, ed. CRC Press, Boca Raton, FL.

CBNS. 1978. Broiling and benzo[a]pyrene. CIP Bulletin 2. Center for the Biology of Natural Systems, St. Louis, MO.

Carr, C. J. 1985. Food and drug interactions. Pages 221-228 in: Xenobiotics in Foods and Feeds. ACS Symposium 234. J. W. Finley and D. E. Schwass, eds. American Chemical Society, Washington, DC.

Concon, J. M. , Swerczek, T. W., and Newburg, D. S. 1981. Potential carcinogenicity of black pepper (*Piper nigrum*). Pages 359-374 in: Antinutrients and Natural Toxicants in Foods. R. L. Ory, ed. F&N Press, Westport, CT.

Concon, J. M. 1988. Food Toxicology. Parts A and B. Marcel Dekker, New York.

Conn, E. E. 1981. Unwanted biological substances in foods. Pages 105-121 in: Impact of Toxicology on Food Processing. J. Ayres and J. Kirschman, eds. AVI, Westport, CT.

Cooney, R. V., and Ross, P. D. 1986. Phenols help form nitrosamines from NO_2. Chem. Eng. News. 65(Sept. 29):32.

Crampton, R. F., and Charlesworth, F. A. 1975. Occurrence of natural toxins in food. Br. Med. Bull. 31(3):209-213.

Davies, A. M. C., and Blincow, P. J. 1984. Glycoalkaloid content of potatoes and potato products sold in the UK. J. Sci. Food Agric. 35:553-557.

Deinzer, M. L., Thomson, P. A., Burgett, D. M., and Isaacson, D. L. 1977. Pyrrolizidine alkaloids: Their occurrence in honey from tansy ragwort (*Senecio jacobaea* L.). Science 195:497-499.

Desphande, S. S., Sathe, S. K., and Salunke, D. K. 1985. Chemistry and safety of plant polyphenols. Adv. Exp. Med. Biol. 177:457-495.

Dickstein, E. S., and Kunkel, F. W. 1980. Foxglove tea poisoning. Am. J. Med. 69:167-

169.

Dunn, B. P. 1982. Polycyclic aromatic hydrocarbons. Pages 175-178 in: Carcinogens and Mutagens in the Environment, Vol. 1. H. F. Stich, ed. CRC Press, Boca Raton, FL.

Epstein, M. T., Espiner, E. A., Donald, R. A., and Hughes, H. 1977. Effect of eating liquorice on the renin-angiotensin-aldosterone axis. Br. Med. J. 1:488-490.

Fahey, G. C., and Jung, H.-J. G. 1989. Phenolic compounds in forages and fibrous feedstuffs. Pages 123-191 in: Toxicants of Plant Origin. Vol. 4. Phenolics. P. R. Cheeke, ed. CRC Press, Boca Raton, FL.

Farrer, K. T. H. 1983. Fancy Eating That! Melbourne University Press, Melbourne, Australia.

Fenwick, G. R., Heaney, R. K., and Mawson, R. 1989. Glucosinolates. Pages 2-42 in: Toxicants of Plant Origin. Vol. 2. Glycosides. P. R. Cheeke, ed. CRC Press: Boca Raton, FL.

Fletcher, D. C. 1976. Can the quinine in tonic water be hazardous? J. Am. Med. Assoc. 236(3):305.

Friedman, M., ed. 1984a. Nutritional and Toxicological Aspects of Food Safety. Plenum Press, New York.

Friedman, M. 1984b. Sulfhydryl groups and food safety. Adv. Exp. Biol. Med. 177:31-63.

Gaitan, E. 1990. Goitrogens in food and water. Annu. Rev. Nutr. 10:21-39.

Gormley, T. R., Downey, G., and O'Beirne, D. 1987. Food, Health and the Consumer. Elsevier, New York.

Graham, D. 1986. Red and black pepper can irritate the stomach. Speech to American Gastroenterological Association, San Francisco, May 18, 1986.

Grivetti, L. E. 1982. Coturnism: Poisoning by European migratory quail. Pages 51-58 in: Adverse Effects of Foods. E. F. P. Jelliffe and D. B. Jelliffe, eds. Plenum Press, New York.

Gross, R. L., and Newberne, P. M. 1979. Naturally occurring toxic substances in foods. Clin. Pharmacol. Thera. 22(5, pt. 2):680-698.

Gurr, M. I. 1988. Lipids: Products of industrial hydrogenation, oxidation and heating. Pages 139-158 in: Nutritional and Toxicological Aspects of Food Processing. R. Walker and E. Quattrucci, eds. Taylor & Francis, New York.

Hambraeus, L. 1982. Naturally occurring toxicants in food. Pages 13-36 in: Adverse Reactions to Foods. E. F. P. Jeliffe and D. B. Jeliffe, eds. Plenum Press, New York.

Harvey, R. G. 1982. Polycyclic hydrocarbons and cancer. Am. Sci. 70(4):386-392.

Hatch, F. T., Felton, J. S., and Bjeldanes, L. F. 1982. Mutagens from the cooking of food. Pages 147-164 in: Carcinogens and Mutagens in the Environment, Vol. 1. H. F. Stich, ed. CRC Press, Boca Raton, FL.

Havender, W. R., and Meister, K. A. 1986. Does nature know best? Natural carcinogens in American foods. American Council on Science and Health, Summit, NJ.

Hegarty, M. D. 1986. Toxic amino acids in foods of animals and man. Proc. Nutr. Soc. Asut. 11:73-81.

Hirono, I. 1983. Carcinogens of plant origin. Pages 75-80 in: Carcinogens and Mutagens in the Environment, Vol. 3. H. F. Stich, ed. CRC Press, Boca Raton, FL.

Hirono, I. 1989. Carcinogenic bracken glycosides. Pages 230-252 in: Toxicants of Plant Origin. Vol. 2. Glycosides. P. R. Cheeke, ed. CRC Press: Boca Raton, FL.

Hogan, R. P. 1983. Hemorrhagic diathesis caused by drinking an herbal tea. J. Am. Med. Assoc. 249(19):2679-2680.

Hopkins, H. 1984. Mexican Coumarin No Bargain. DHHS Publ. (FDA)84-1105.

IFT Expert Panel on Food Safety and Nutrition. 1975. Naturally occurring toxicants in foods. Food Technol. (Chicago) 29(3):67-72.

Ivey, G. W., Holt, D. L., and Ivey, M. C. 1981. Natural toxicants in human foods: Psoralens in raw and cooked parsnip root. Science 213:909-910.

Johns, A. N. 1980. Beware of the bay leaf. Br. Med. J. 281:1682.

Johns, C. O., and Finks, A. J. 1920. Studies in nutrition. II. The role of cystine in nutrition as exemplified by nutrition experiments with the proteins of navy beans, *Phaseolus vulgaris.* J. Biol. Chem. 41:379-389.

Joint Committee of the Royal College of Physicians and the British Nutrition Foundation. 1984. Food intolerance and food aversion. J. R. Coll. Physicians Lond. 18(2):83-123.

Kapadia, G. J., Chung, E. B., Ghosh, B., Shukla, Y. N., Basak, S. P., Morton, J. F., and Pradhan, S. N. 1978. Carcinogenicity of some folk medicinal herbs in rats. J. Natl. Can. Inst. 60(3):683-686.

Kapadia, G. J., Rao, G. S., and Morton, J. F. 1983. Herbal tea consumption and esophageal cancer. Pages 3-12 in: Carcinogens and Mutagens in the Environment, Vol. 3. H. F. Stich, ed. CRC Press, Boca Raton, FL.

Keeler, R. F. 1980. Toxins in plants. Pages 591-624 in: Safety of Foods. H. D. Graham, ed. AVI, Westport, CT.

Krone, C. A., Yeh, S. M. J., and Iwaoka, W. T. 1986. Mutagen formation during commercial processing of foods. Environ. Health Perspect. 67:75-88.

Kuratsune, M., Ikeda, M., and Hayashi, T. 1986. Epidemiological studies of possible health effects of intake of pyrrolyzates of foods. Environ. Health Perspect. 67:143-146.

Larsson, B. K. 1986. Formation of polycyclic aromatic hydrocarbons during the smoking and grilling of food. Pages 169-180 in: Genetic Toxicology of the Diet. A. R. Liss, ed. Alan R. Liss: New York.

Leiner, I. E. 1980. Toxic Constituents of Plant Foodstuffs. Academic Press, New York.

Leiner, I. E. 1981. The nutritional significance of the plant lectins. Pages 143-158 in: Antinutrients and Natural Toxicants in Foods. R. L. Ory, ed. F&N Press, Westport, CT.

Leopold, A. C., and Ardrey, R. 1972. Toxic substances in plants and the food habits of early man. Science 176:512-514.

MacGregor, J. T. 1986a. Genetic toxicology of dietary flavonoids. Pages 33-43 in: Genetic Toxicology of the Diet. A. R. Liss, ed. Alan R. Liss, New York.

MacGregor, J. T. 1986b. Mutagenic and carcinogenic effects of flavonoids. Pages 411-424 in: Plant Flavonoids in Biology and Medicine. A. R. Liss, ed. Alan R. Liss, New York.

Marquardt, R. R. 1989. vicine, covicine, and their arglycones—Divicine and isouramil. Pages 161-200 in: Toxicants of Plant Origin. Vol. 2. Glycosides. P. R. Cheeke, ed. CRC Press: Boca Raton, FL.

Mauron, J. 1989. Browning reaction products and food quality. In: Trends in Food Science and Technology. M. R. R. Rao, N. Chandrasekhara, and K. A. Ranganath, eds. Sharada Press, Mangalore, India.

McCabe, B. J. 1986. Dietary tyramine and other pressor amines in MAOI regimens: A review. J. Am. Diet. Assoc. 86:1059-1064.

McCann, J. 1983. Mutagens and cancer. Pages 137-140 in: Diet, Nutrition and Cancer. D. Roe, ed. Alan R. Liss, New York.

Mehansho, H., Butler, L. G., and Carlson, D. M. 1987. Dietary tannins and salivary proline-rich proteins. Annu. Rev. Nutr. 7:423-440.

Merz, B. 1983. Adding seeds to the diet may keep cancer at bay. J. Am. Med. Assoc. 249:2746.

Miller, A. J. 1985. Processing-induced mutagens in muscle foods. Food Technol. (Chicago) 39(2):75-78 ff.

Miller, E. C., and Miller, J. A. 1986. Carcinogens and mutagens that may occur in foods. Cancer 58:1795-1803.

Miller, J. A., Miller, E. C., and Phillips, D. H. 1982. The metabolic activation and carcinogenicity of alkenylbenzenes that occur naturally in many spices. Pages 83-

96 in: Carcinogens and Mutagens in the Environment, Vol. 3. H. F. Stich, ed. CRC Press, Boca Raton, FL.

Monmaney, T., Hager, M., and Katz, S. 1987. The dangers in your drink. Newsweek on Health (Spring):12.

Morgan, M. R. A., and Coxon, D. T. 1987. Tolerances: Glycoalkaloids in potatoes. Pages 221-230 in: Natural Toxicants in Food: Progress and Prospects. D. H. Watson, ed. DCH Publishers, New York.

Morton, J. 1968. Plants associated with esophageal cancer cases in curacao. Cancer Res. 28:2268-2271.

Mower, H. F. 1983. Mutagenic compounds contained in seaweed. Pages 81-86 in: Carcinogens and Mutagens in the Environment, Vol. 3. H. F. Stich, ed. CRC Press, Boca Raton, FL.

Murphy, P. A. 1982. Phytoestrogen content of processed soybean products. Food Technol. (Chicago) 36(1):60-64.

Nagao, M., and Sugimura, T. 1983. Suppression and enhancement of the mutagenicity of coffee. Pages 101-106 in: Carcinogens and Mutagens in the Environment. Vol. 2. H. F. Stich, ed. CRC Press, Boca Raton, FL.

Nagao, M., Wakabayaski, F., Fujita, Y., Tahura, A., Ochrai, T., and Sugimura, T. 1986. Mutagenic compounds in soy sauce, Chinese cabbage, coffee and herbal teas. Pages 55-62 in: Genetic Toxicology of the Diet. A. R. Liss, ed. Alan R. Liss, New York.

Nakatani, N., Inatani, R., Ohta, H., and Nishiota, A. 1986. Chemical constituents of peppers and application to food preservation. Environ. Health Perspect. 67:135-142.

NAS/NRC. 1989. Diet and Health: Implications for Reducing Chronic Disease Risk. National Academy Press, Washington, DC.

National Cancer Institute (NCI). 1990. Cancer Statistics Review 1973–1987. The Institute, Bethesda, MD.

Nishio, Y., Kakizoe, T., Ohtani, M., Sato, S., Sugimura, T., and Fukushima, S. 1986. Leucine and isoleucine promote bladder cancer. Science 231:843-845.

Ohshima, H., and Bartsch, H. 1983. A new approach to quantitate endogenous nitrosation in humans. Pages 3-15 in: Carcinogens and Mutagens in the Environment, Vol 2. H. F. Stich, ed. CRC Press, Boca Raton, FL.

Okenfull, D., and Sidhu, G. S. 1989. Saponins. Pages 97-142 in: Toxicants of Plant Origin. Vol. 2. Phenolics. P. R. Cheeke, ed. CRC Press, Boca Raton, FL.

Olson, A. C., Gray, G. M., Gumbmann, M. R., Sell, C. R., and Wagner, J. R. 1981. Flatus causing factors in legumes. Pages 275-294 in: Antinutrients and Natural Toxicants in Foods. R. L. Ory, ed. F&N Press, Westport, CT.

Oterdoom, H. J. 1985. Tannin, sorghum, and oesophageal cancer. Lancet 2:330.

Palacek, F. 1969. Pulmonary damage in rats fed by beans (*Phaseolus vulgaris*). Experentia 25(3):285.

Poulton, G. A., and Ashwood-Smith, M. J. 1983. Photosensitizing plant products. Pages 77-98 in: Carcinogens and Mutagens in the Environment, Vol. 3. H. F. Stich, ed. CRC Press, Boca Raton, FL.

Powrie, W. D., Wu, C. H., and Stich, H. F. 1982. Browning reactions as sources of mutagens and modulators. Pages 121-133 in: Carcinogens and Mutagens in the Environment, Vol. 1. H. F. Stich, ed. CRC Press, Boca Raton, FL.

Powrie, W. D., Wu, C. H., and Molund, V. P. 1986. Browning reaction systems as sources of mutagens and antimutagens. Environ. Health Perspect. 67:47-54.

Pryor, W. A. 1985. Free radical involvement in chronic diseases and aging. Pages 77-96 in: Xenobiotics in Foods and Feeds. ACS Symposium 234. J. W. Finley and D. E. Schwass, eds. American Chemical Society, Washington, DC.

Pusztai, A. 1989. Lectins. Pages 29-72 in: Toxicants of Plant Origin. Vol. 3. Proteins and Amino Acids. P. R. Cheeke, ed. CRC Press, Boca Raton, FL.

Rackis, J. J., and Gumbmann, M. R. 1981. Protease inhibitors: Physiological properties and nutritional significance. Pages 203-237 in: Antinutrients and Natural Toxicants in Foods. R. L. Ory, ed. F&N Press, Westport, CT.

Reinhold, J. G., Faradji, B., Abadi, P., and Ismail-Beigi, F. 1976. Decreased absorption of calcium, zinc, and phosphorus by humans due to increased fiber and phosphorus consumption as wheat bread. J. Nutr. 106:493-503.

Rivera, M., and Salcedo, J. R. 1982. Worcestershire sauce-induced hematuria. J. Urol. 127(3):554.

Rosin, M. P., and Stich, H. F. 1983a. The combined inhibitory effect of chlorogenic acid, gallic acid, and ascorbic acid on the mutagenicity resulting from a model nitrosation reaction. Pages 107-114 in: Carcinogens and Mutagens in the Environment, Vol. 2. H. F. Stich, ed. CRC Press, Boca Raton, FL.

Rosin, M. P., and Stich, H. F. 1983b. The identification of antigenotoxic/anticarcinogenic agents in food. Pages 141-154 in: Diet, Nutrition and Cancer. D. Roe, ed. Alan R. Liss, New York.

Roy, D. N., and Spencer, P. S. 1989. Lathrogens. Pages 169-202 in: Toxicants of Plant Origin. Vol. 3. Proteins and Amino Acids. P. R. Cheeke, ed. CRC Press, Boca Raton, FL.

Ryan, C. A., and Hass, G. M. 1981. Structural, evolutionary, and nutritional properties of proteinase inhibitors from potatoes. Pges 169-185 in: Antinutrients and Natural Toxicants in Foods. R. L. Ory, ed. F&N Press, Westport, CT.

Scrimshaw, N. 1980. A look at the Incaprina experience in Guatemala: The background and history of Incaprina. Food Nutr. Bull. 2(2):1-2.

Shamberger, R. J. 1984. Nutrition and Cancer. Plenum Press, New York.

Shank, F. R., Carson, K. L., and Willis, C. A. 1989. Evolving food safety. Pages 296-307 in: Carcinogenicity and Pesticides: Principles, Issues, and Relationships. N. N. Ragsdale and R. E. Menzer, eds. American Chemical Society, Washington, DC.

Sharma, R. P., and Salunke, D. K. 1989. Solanum glycoalkaloids. Pages 179-236 in: Toxicants of Plant Origin. Vol. 1. Alkaloids. P. R. Cheeke, ed. CRC Press, Boca Raton, FL.

Siegal, R. K. 1976. Herbal intoxication. J. Am. Med. Assoc. 235(5):473-476.

Siegal, R. K. 1979. Ginseng abuse syndrome. J. Am. Med. Assoc. 241:1614-1615.

Silverstone, G. A. 1985. Possible sources of food toxicants. Pages 44-60 in: Diet-Related Diseases: The Modern Epidemic. S. Seely, D. L. Freed, G. A. Silverstone, and V. Rippere, eds. AVI, Westport.

Singleton, V. L. 1981. Naturally occurring food toxicants: Phenolic substances of plant origin common in foods. Adv. Food Res. 27:149-242.

Slovut, G. 1982. Cyanide in lima bean soup? Minneapolis Star Tribune, Nov. 11.

Stich, H. F., and Powrie, W. D. 1982. Plant phenolics as genotoxic agents and as modulators for the mutagenicity of other food components. Pages 135-146 in: Carcinogens and Mutagens in the Environment, Vol. 3. H. F. Stich, ed. CRC Press, Boca Raton, FL.

Stich, H. F., and Rosin, M. P. 1984. Naturally occurring phenolics as antimutagenic and anticarcinogenic agents. Adv. Exp. Med. Biol. 177:1-29.

Stoltz, D. R. 1982. The health significance of mutagens in foods. Pages 75-82 in: Carcinogens and Mutagens in the Environment, Vol. 3. H. F. Stich, ed. CRC Press, Boca Raton, FL.

Sugimura, T. 1986. Past, present, and future of mutagens in cooked food. Environ. Health Perspect. 67:5-10.

Tewe, O. O., and Iyayi, E. A. 1989. Cyanogenic glycosides. Pages 43-60 in: Toxicants of Plant Origin. Vol. 2. Glycosides. P. R. Cheeke, ed. CRC Press, Boca Raton, FL.

Thurston, H. D. 1984. Tropical Plant Diseases. The American Phytopathological Society, St. Paul, MN.

Toth, B. 1983. Carcinogens in edible mushrooms. Pages 100-108 in: Carcinogens and Mutagens in the Environment, Vol. 3. H. F. Stich, ed. CRC Press, Boca Raton, FL.

Tyler, V. E. 1987. The New Honest Herbal. George F. Stickley Co., Philadelphia, PA.

Vesell, E. S. 1985. Effect of dietary factors on drug disposition in normal human subjects. Pages 61-76 in: Xenobiotics in Foods and Feeds. ACS Symposium 234. J. W. Finley and D. E. Schwass, eds. American Chemical Society, Washington, DC.

Visek, W. J., and Clinton, S. K. 1985. Dietary protein and the carcinogenesis, metabolism and toxicity of 1,2-dimethylhydrazine. Pages 293-307 in: Xenobiotic Metabolism. ACS Symposium 277. American Chemical Society, Washington, DC.

Wahlquist, M. L. 1982. Social toxicants and nutritional status. Pages 227-238 in: Adverse Reactions to Foods. E. F. P. Jeliffe and D. B. Jeliffe, eds. Plenum Press, New York.

Wasserman, A. E., and Wolff, I. A. 1981. Nitrates and nitrosamines in our environment. Pages 35-52 in: Antinutrients and Natural Toxicants in Foods. R. L. Ory, ed. F&N Press, Westport.

Weisburger, J. H., and Barnes, W. S. 1986. Influence of composition of diet on the formation of mutagens. Pages 227-235 in: Genetic Toxicology of the Diet. A. R. Liss, ed. Alan R. Liss, New York.

Williams, G., Weisburger, J. H., and Wynder, E. L. 1982. Lifestyle and cancer etiology. Pages 53-64 in: Carcinogens and Mutagens in the Environment, Vol. 3. H. F. Stich, ed. CRC Press, Boca Raton, FL.

Wilson, B. J. 1979. Naturally occurring toxicants of foods. Nutr. Rev. 37(10):305-312.

Wise, R. P. 1980. The case of Incaprina in Guatemala. Food Nutr. Bull. 2(2):3-8.

Xavier-Filho, J., and Campos, F. A. P. 1989. Proteinase inhibitors. Pages 1-28 in: Toxicants of Plant Origin. Vol. 3. Proteins and Amino Acids. P. R. Cheeke, ed. CRC Press, Boca Raton, FL.

Yu, C., Swaminathan, B., and Butler, L. 1986. Isolation and identification of rutin as the major mutagen in red wine. Mutat. Res. 170:103-113.

Bacteriological Problems Occurring in Food

CHAPTER 6

Microbiological contamination of food should cause more concern than any other food safety hazard. Of the microbes, bacteria are by far the most common and cause many different foodborne diseases. The FDA has estimated the incidence of foodborne disease in the United States to be between 25 and 81 million cases annually. Estimates are used because in only a small minority of the cases is medical attention sought and the case actually reported (Bean et al, 1990). Often the person who has a foodborne disease is unaware of it because the symptoms of most foodborne pathogens are difficult to distinguish from those of flu. In most cases, symptoms are of short duration and considered neither severe nor unusual enough to seek medical attention.

Incidence Increasing

Both the actual and the reported incidences of foodborne disease in the United States and in other developed countries are thought to be increasing. Statistics from the federal Centers for Disease Control (CDC) show that the reported incidence of salmonellosis has doubled in the United States in the last 16 years from 20,000 to 40,000 cases annually, while the FDA estimates the actual incidence to be 2-4 million cases annually (Lecos, 1986). Other estimates put the numbers even higher, 6-30 million per year (Roberts, 1990; Todd, 1989). The annual cost of foodborne disease may be as high as $8.4 billion (Todd, 1989).

Incidence of microbial diseases caused by other organisms also appears to be increasing. One half to one third of all diarrhea cases in the United States are of foodborne origin (Nightingale, 1987). In 1986, the largest outbreak of shigellosis ever reported in the United States occurred in Texas. Table 6.1 gives some recent data on the number of cases of foodborne illness as estimated by the CDC.

Reasons for Increased Incidence

Many reasons are given for the increasing incidence of foodborne disease, among which are the following.

1. Better methods of detection and identification together with better epidemiological capacity have contributed to a rise in reported cases.

2. Lifestyle factors may be responsible for a rise in actual incidence. People are eating out more, traveling more, and choosing exotic foods (such as sushi) more frequently than in the past. Vegetables and fruits available during the winter increasingly come from other countries with different food sanitation standards and perhaps different strains of organisms (Osterholm, 1991).

3. As people get further away from learning about correct food handling, they have greater likelihood of handling food incorrectly.

4. Current food trends—with increased food being purchased from restaurants and delis and in prepared and refrigerated forms—result in food being handled by more people, treated and distributed in stages, and held before being sold, all factors that increase the chance of food becoming contaminated or being held at improper temperatures.

5. Many new products are designed to meet the consumer's demand for fresh products with no additives. These may be pasteurized and stored under controlled atmospheres and sold in the refrigerator case. If they are mishandled at any point, heat-resistant spores that survived pasteurization may proliferate, whereas organisms that would normally indicate spoilage are killed during pasteurization. Thus the product is believed to be safe, is consumed, and causes foodborne disease.

6. Consolidation of many small food processing operations into larger ones could contribute to an increase in foodborne disease. Although larger operations are likely to be even more conscious of sanitation than smaller ones, one food handling mistake reaches large numbers

Table 6.1. Estimated Annual Cases of Foodborne Illness in the United States[a]

Causative Organism	Cases	Deaths
Salmonella	3,000,000	3,000
Campylobacter jejuni	2,100,000	2,100
Toxoplasma gondii[b]	1,500,000	300
Staphylococcus aureus	1,155,000	<10
Clostridium perfringens	650,000	<10
Shigella	300,000	600
Enteric *Escherichia coli*	200,000	400
Norwalk virus[b]	180,000	0
Trichinella spiralis[b]	100,000	1,000
Bacillus cereus	84,000	0
Vibrio species	<40,000	<900
Listeria monocytogenes	25,000	1,000
Yersinia enterocolitica	<20,000	0
Other	<10,000	<500

[a] Adapted from Todd (1989).
[b] These organisms are discussed in later chapters.

of people and results in a massive outbreak. A 1985 salmonella milk incident in Chicago affected an estimated 18,500 people (Kvenberg and Archer, 1987), the largest single salmonella outbreak ever recorded.

7) The concentration of animals into larger production units and of animal slaughter into fewer and larger plants increases the possibility of cross-contamination among meat carcasses.

8) Microorganisms are developing resistance to antibiotics, which may contribute to increased incidence and decreased treatability of foodborne disease. For example, the organism involved in the Chicago salmonella incident was a resistant strain.

Placing the Blame

Faulting the purveyor or producer of food for mishandling is common. However, the blame may be misplaced and may rest with the consumer who purchased the food and did not handle it correctly. Data on common food handling practices in the home show that, in many cases, our own food handling practices put us at greater risk for foodborne disease than those introduced by industry. Improperly cleaned kitchen equipment (such as blender and can opener blades) and kitchen work surfaces (such as wooden cutting boards) provide fertile breeding grounds for all sorts of bacteria. Careless hand washing and inadequate general sanitation both introduce microorganisms into the foods, and allowing food to stay in the temperature danger zone—between 40° and 160°F— encourages their proliferation. Some people still believe the old-fashioned idea that foods should be cooled to room temperature before being refrigerated. This myth originated when food was cooled in iceboxes. Hot food melted the ice and caused all the food in the icebox to spoil. Thus it was important to first cool the food if one had an icebox, but few people have iceboxes in the 1990s.

DEFINITIONS OF FOODBORNE DISEASE

Foodborne diseases come from eating foods that contain a sufficient quantity of either pathogenic microorganisms or their toxins to cause symptoms. Pathogenic microorganisms cause foodborne infection, and toxins cause foodborne intoxication.

Foodborne infection can cause illness in three ways. In one, ingested pathogenic microorganisms penetrate the intestinal mucosa and colonize the gastrointestinal (GI) tract, producing an adverse effect on the body. Examples of this type include *Salmonella* species, *Shigella* species, and some strains of enteropathogenic *Escherichia coli*. In another method of infection, the microorganism travels from the GI tract to other tissues, where it lodges; for instance, the Hepatitis A

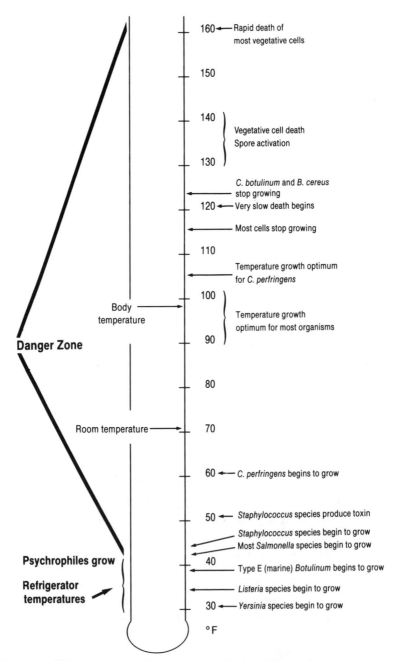

Figure 6.1. The danger zone and temperatures of microbial growth.

virus goes to the liver and *Trichinella spiralis* to muscle. A third type of infection occurs when toxins are released as the infecting organism multiplies or lyses in the intestinal tract. This is the case with *Vibrio cholerae, Clostridium perfringens*, and some strains of enterotoxigenic *E. coli*.

Foodborne intoxication occurs when enterotoxins produced by microbial growth (such as those from *Staphylococcus aureus*) are ingested and exert adverse effects on the body. Some of these toxins are heat labile, as is the case with *Clostridium botulinum* toxin; others, such as *S. aureus* toxin, are heat stable.

Symptoms of foodborne disease can range from mild to severe. Even death can result in the very young, the very old, or those already compromised by other disease. In other instances, the organism may actually colonize but no symptoms occur. Symptomless people can unknowingly transmit the disease.

The mere presence of pathogens is not usually sufficient to cause foodborne disease. In fact, many foods test positive for a variety of pathogens. Thus, as with other toxicants, "only the dose makes the poison." The dose required to cause disease can vary markedly for different organisms. Many require populations of at least 10,000 and sometimes as many as a million before any symptoms appear. On the other hand, some require frighteningly few. For example, it took nearly 100,000 salmonellae to produce disease in volunteers, but as few as 10 shigellae. For staphylococcal enterotoxin to be produced, nearly 500,000 organisms were needed (Snyder, 1986; Bryan, 1979). Furthermore, individuals vary in their susceptibility to these microorganisms just as they vary in susceptibility to a variety of other disease organisms.

Records of Foodborne Disease

Reporting of foodborne disease in the United States began in the 1920s when the Public Health Service first prepared annual summaries of milkborne disease. In 1938, reporting was broadened to include diseases transmitted through all food and water. Data from these early surveys led to widespread adoption of public health measures such as pasteurization of milk and chlorination of water.

Outbreaks of foodborne disease are now reported in annual summaries published by the CDC in Atlanta. An outbreak is defined as two or more persons having a similar illness after ingestion of food at a specific time and place. Only botulism is an exception, where a single case is all that is necessary to constitute an outbreak (Bryan, 1979).

In the period between 1983 and 1987, 2,397 outbreaks involving 91,678 cases of foodborne disease were reported to the CDC (Bean et al, 1990).

Less than half the reported outbreaks had an identified cause, indicating the need for improved methodologies to identify pathogens.

Of those instances in which the cause was determined, 66% of the outbreaks and 92% of the cases were of bacterial origin, 26% of the outbreaks and 2% of the cases were from chemical agents, 4% of the outbreaks and under 1% of the cases were due to parasites, and 5% of both outbreaks and cases were viral in origin. Salmonella accounted for 57% of the bacterial disease. Meat and poultry foods were involved in 22% of the outbreaks and fish and shellfish in 26%. In most outbreaks of bacterial origin, food was stored at improper holding temperatures, or the food handlers had improper hygiene. Of the outbreaks of chemical origin, fish accounted for 73% because of toxins that either are ingested by fish under certain conditions (ciguatoxin) or are formed in fish not properly refrigerated (scrombotoxin). Food service and commercial establishments accounted for three fourths of the *reported* outbreaks (Bean and Griffin, 1990). Certain foodborne diseases, such as botulism, were more likely to occur in the home.

FOODBORNE MICROORGANISMS

This chapter considers the major microorganisms contributing to human foodborne disease. The older, more well-known ones are discussed first, followed by lesser-known ones and those described as problems more recently.

Salmonella

The salmonellosis outbreak in the Chicago area, coupled with concerns over the safety of poultry, brought this organism back into the headlines in the late 1980s. However, this was certainly not because salmonellosis was or is a new problem. Data from many industrialized nations suggest that it is one of the major causes of foodborne disease in industrialized nations and that its incidence is increasing (Figure 6.2).

The main reservoir for salmonellae is the intestinal tract of animals. Meat and poultry products are thus prime offenders. As many as half of healthy poultry and one quarter of healthy cattle have been shown to harbor this organism. Poor food handling practices can also introduce the organism. An estimated 0.2–5.0% of people in the United States are chronic carriers.

Salmonellae can thrive in many foods because of their simple nutritional requirements and ability to grow under both aerobic and anaerobic conditions. Furthermore, they can exist over a diverse range of pH and temperature. Most strains are heat sensitive, although some strains isolated from eggs and meat exhibit heat resistance. Drying or freezing does not kill all of them. Some strains can grow slowly at moderate refrigeration temperatures, e.g., over 45°F (Box 6.1).

Sources. Some of the most common sources of salmonellae are egg and poultry products or products containing dried or processed eggs. *S. enteriditis* has become a problem because of infected flocks (St. Louis et al, 1988). The egg is contaminated with the organism when it is laid and causes a problem unless the organism is later killed by heat. Television news reports have made consumers aware that poultry is a source of contamination. One company produced a home-use product called Chik Chek (Diversified Diagnostic Industries) with which, they claimed, consumers could test their poultry for salmonellae. However, a report from the USDA showed that the product performed very unreliably, reacting positively to sterile water and failing to react to known *Salmonella* cultures (Caldwell, 1988).

Other sources of salmonella foodborne infections include commercial ice cream prepared from processed eggs and homemade ice cream made with raw eggs. Deli salads, sandwiches, cold roast meats, and reheated sauces and gravies are also prime culprits. Infected food handlers have appeared to be responsible for several outbreaks of salmonellosis in fast food restaurants.

The organism is found in many foods because of cross-contamination. Improperly cleaned countertops, cutting boards, and cutting utensils used for uncooked meat and poultry may serve to inoculate other foods.

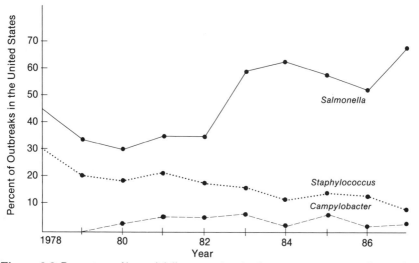

Figure 6.2. Percentage of bacterial disease outbreaks due to some common and emerging pathogens, 1978-1987.

Recent studies of sprouting kits showed that salmonellae quickly reached hazardous levels during the growing period (Harmon et al, 1987). Most outbreaks occur between May and December.

Milk and *Salmonella* species have always been linked. *Salmonella* species were found in 60% of raw milk samples tested (McManus and Lanier, 1987). Because of the widespread use of pasteurization, most of the recent outbreaks were associated with raw milk, while certified

---

Box 6.1

Salmonella **Growth at Various Temperatures**

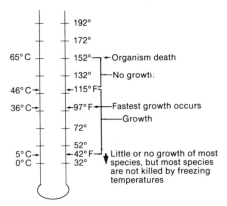

Note that the fastest growth time is at body temperature (37°C or 98.6°F). At this temperature, salmonellae double in only 23 minutes. Thus, if there were 10 organisms in food at breakfast, 8 to 10 generations could form by lunch. This could be a pathogenic level for susceptible individuals.

Time	Minutes at 97°F	Theoretical Number of Organisms
8:00	0	10
8:23	23	20
8:46	46	40
9:09	69	80
9:32	92	160
9:55	115	320
10:18	138	640
10:41	161	1,280
11:04	184	2,560
11:27	207	5,120
11:50	230	10,240

raw milk and pasteurized milk were thought to be safe. However, certification of milk has been shown not to ensure that the milk is pathogen-free. Certified milk was determined to be the source for a 1986 outbreak of the dangerous *S. dublin* in California and Washington states. This outbreak consisted of 114 cases with two deaths. Both deaths were in individuals with existing chronic disease. Subsequent outbreaks from *S. dublin* caused the FDA to declare it a life-threatening hazard. The term *hazard* was chosen for this organism because it constitutes a significant threat for persons with over 40 different underlying diseases and syndromes. Death occurred in 25% of the cases and hospitalization in 80% (Calif. Dept. of Health, 1986; Hooper, 1989).

Although salmonellae have been associated with raw milk, the aging of cheese made from raw milk for over 60 days was thought to inhibit the growth of this pathogen—hence the FDA requirement that cheeses made from raw milk be aged a specific length of time. However, an outbreak of salmonellosis affecting nearly 2,000 people in Canada was traced to cheddar cheese made from raw milk. Subsequent studies show that salmonella from raw milk can survive in cheddar and other cheeses aged at 13°C for 112–210 days (Johnson, 1987).

The fact that pasteurized and not raw milk was involved in the Chicago 1985 salmonellosis outbreak puzzled the special task force and investigators from the FDA working to establish the cause of the outbreak. The antibiotic-resistant strain of *S. typhimurium* infected over 16,000 people and killed 10. The source of contamination was determined to be a small connecting piece in the milk tank that allowed milk to collect and the microorganisms to multiply and contaminate the pasteurized milk (Lecos, 1986). The outbreak was tragic and shocking. Shocking because we simply were not used to hearing about such massive problems in the U.S. food supply. The fact that we were shocked underscores our underlying expectations of safe food and emphasizes the incredible safety record of the dairy industry (Russo, 1985).

Symptoms. Symptoms usually occur 12–24 hours after eating and subside in 24–48 hours. They include diarrhea, cramps, nausea, and vomiting and can also include chills and fever. Stools may contain mucus or blood.

In severe cases, bacteremia (bacteria in the blood) may result and cause severe localized infections such as meningitis or pneumonia, which can result in permanent impairment or death. The most severe salmonella infection is caused by *S. typhimurium*. Its symptoms mimic those of influenza and other viral illnesses, and they worsen with a progressive rise in fever. In contrast to other *Salmonella*-induced gastroenteritis, there is usually no diarrhea. In fact, constipation is a common feature. Treatment with antibiotics should be used only in severe cases because the organism quickly develops resistance (Patten, 1981).

Age and individual susceptibility are important determinants of infection. Infants seem to possess little resistance to this infection and, if infected, become seriously ill. An adult is much more likely to have intestinal colonization without signs of disease. Susceptibility to *Salmonella* varies with both the strain of the organism and the individual. Studies with healthy male prisoner volunteers showed that the infecting dose varied from 125,000 to 16,000,000 microorganisms (Bowner, 1965).

Staphylococcus aureus

Staphylococcal intoxication is a major cause of foodborne disease in many parts of the world (Picard et al, 1987). In some years it is the second most commonly reported foodborne disease in the United States. The estimated annual incidence in the United States is 1,555,000 cases, resulting in 5,400 deaths (Todd, 1989). Like many foodborne diseases, it is very unpleasant but usually not fatal. Humans appear to be the species most sensitive to the toxin.

The organism is a common inhabitant of the body and can be introduced by food handlers from their skin. Two thirds of the outbreaks occur in foods served in restaurants, delis, schools, church suppers, and the like, and one third occur in the home. Peak time for this disease is from mid-August through December (Figure 6.3).

Sources. Previously cooked protein foods are the most frequently named sources, with baked ham being the most common source of intoxication in the United States. Cooking kills many of the harmless competitive organisms that would customarily keep this organism in check (Newsome, 1988). Poultry, egg products, prepared salads and sandwich fillings, salads, and cream-filled bakery goods are other vehicles. Foods that require a great deal of handling during preparation pose a greater risk. Slicing machines in restaurants and delis sometimes harbor the organism (Holmberg and Blake, 1984).

Milk can also be a source of *S. aureus* intoxication. If staphylococci contaminate milk after pasteurization, the organisms proliferate more readily than in raw milk, as the competitive organisms that check their growth have been killed by the pasteurization (Hayes and Campbell, 1986).

Although the organism is killed by heat, the toxin is very heat resistant. Heating the food after the toxin is present does not ensure safety. Some amount of toxin will even survive the boiling and canning process (Bennett and Berry, 1987; Bergdoll, 1989). Unfortunately, it is impossible to detect the toxin in the food by appearance, taste, or smell. With this organism, it is crucial to prevent toxin formation by minimizing food contamination and by keeping food out of the danger zone.

Symptoms. The rapid onset of symptoms distinguishes this type of food poisoning from salmonellosis. Symptoms usually occur within 2-3 hours after eating, with a range of 1-6 hours. They include nausea, vomiting, retching, abdominal cramping, diarrhea, headache, weakness, chills, and fever. The illness lasts until the toxin is expelled from the system, usually in 24 hours. Occasionally symptoms last for several days. Hospitalization results in 10% of the reported cases, and the

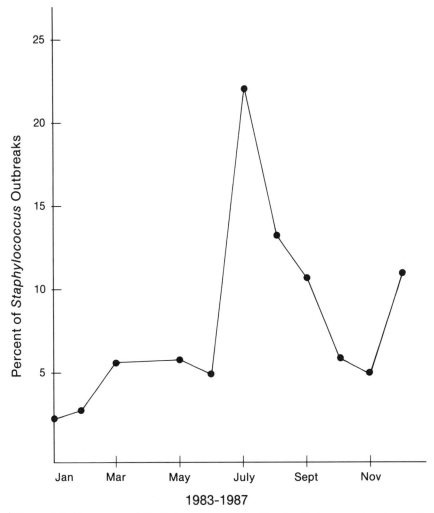

Figure 6.3. Percentage of *Staphylococcus aureus* outbreaks that occurred each month in 1983-1987.

mortality rate is 4%. The toxins show an immunosuppressive effect in vivo. It has been postulated that immune problems such as systemic lupus erythematosus, rheumatoid arthritis, or allergic reactions may be caused even by subemetic amounts of toxin (Archer, 1978).

Sensitivity of individuals can vary significantly. In studies with volunteers, one became ill after ingesting just 0.5 milliliter of infected filtrate, while another did not become ill after ingesting 13 milliliters of the same filtrate. Persons eating the same amount of contaminated food may show variations not only in their sensitivity but also in the apparent amount of toxin ingested, since toxin distribution within the food can vary substantially (Kubista, 1985).

Clostridium perfringens

Food poisoning from *C. perfringens* occurs primarily from food served in institutions. The chance for contamination with this organism is great, as it is found widely distributed in nature and has been recovered in all types of soil except those from the desert. Between 1973 and 1987, this organism was responsible for 190 outbreaks and 12,234 cases in the United States (Bean and Griffin, 1990). The spore form of this organism is highly heat resistant and may survive some cooking procedures. Spores germinate to form a large number of vegetative cells. When contaminated food is ingested, the organism grows in the small intestine and produces a toxin (Tranter et al, 1987).

Sources. The organism's requirement for many amino acids makes meat, meat products, gravies, and casseroles good candidates for this type of food poisoning. Conditions of little oxygen (such as occur inside a roast) favor its growth. Illness is most frequently associated with those items held for a long time before cooking or with precooked meat items either eaten cold or reheated. Holding food in the danger zone allows this microorganism to proliferate (Foster, 1981). Preventive measures include the rapid refrigeration of roasts and other meat items so that the organism cannot grow. Cured meats are rarely the vehicle for this type of infection, as several factors in the cured product prevent the survival of spores (Labbe, 1988).

Food containing the enterotoxin may be different in odor, color, and texture (Al-Obaidy et al, 1985), but differences that would be detected by taste panelists may be overlooked by consumers.

Symptoms. Symptoms usually occur 8–24 hours following the ingestion of the food and usually last 1 day. Abdominal pain and diarrhea are common, but vomiting is rare. The illness is caused by ingestion of a large number of vegetative cells. When the vegetative cells reach the intestine, they form spores and release an enterotoxin that causes the symptoms (Labbe, 1989).

Clostridium botulinum

Botulism is caused by ingesting a toxin produced by *C. botulinum*. It has been recognized for over 1,000 years as a foodborne disease. The word *botulism* actually means disease that comes from sausage (Latin *botulus*, sausage). In fact, nitrate was first added to sausage and meats to reduce the risk of botulinum poisoning.

In the United States between 1899 and 1973, 688 outbreaks of foodborne botulism occurred. Of these, 71% were from home-canned food, 8.6% from commercial products, and 20% of unknown origin. These outbreaks resulted in 1,784 cases and 978 deaths—a death rate of 50%. The mortality later decreased substantially (Troller, 1986) with the use of antitoxin and better recognition of the symptoms.

Sources. The organism is found everywhere and commonly inhabits the soil. Types A and B are the variants in the soils in the United States. Like *C. perfringens*, this organism is a spore-former. While in the spore form, it is resistant to heat and other common treatments that customarily destroy microorganisms in their vegetative phase. Neither the organism nor its spores are harmful, but the toxin produced under anaerobic conditions is so lethal that the food industry insists that food must be handled correctly every time.

Historically, this was not the case. In the United States in 1913, botulism affected 12 people who ate canned string beans. Between 1913 and 1920, many outbreaks of botulism from canned vegetables occurred. About 50% of the cases were due to commercially canned food. The USDA began recommending heating foods for 1 hour in a boiling water bath. This process still resulted in some deaths, as the temperatures were not high enough to kill all the spores. By 1940, things had changed. Foods with pH over 4 were canned under pressure so that the temperature would be high enough (240°F) to kill the botulism spores. Outbreaks from underprocessed home-canned food still occurred, but those from commercially canned food became rare. There were only five botulism outbreaks, with three deaths, from commercially canned foods in the United States between 1956 and 1982 (Denny, 1982). This figure becomes particularly impressive when one realizes that 1 trillion cans of food were sold in this country during the same period.

Luckily, the toxin is a protein that is not heat stable. It is denatured and rendered completely harmless by a few minutes of boiling. Accordingly, it is recommended that all home-canned products be boiled before being tasted. Most reported outbreaks are associated with canned foods that are often eaten without heating, such as tuna, mushrooms, and vichyssoise.

Botulism has also been implicated in outbreaks from food items that were not canned. Investigators were initially puzzled by this. One outbreak in New Mexico in 1978 involved leftover foil-wrapped baked potatoes held at room temperature for several days and then used to make potato salad. In Illinois in 1983, onions sauteed, kept on the stove in a covered pan below 140°F for an extended period of time, and then used as a hamburger topping were implicated. An outbreak in Vancouver in 1985 involved garlic butter made from a chopped garlic product that was not refrigerated according to label directions (IFT's Office of Scientific and Public Affairs, 1986).

One botulism incident occurred from the ingestion of turkey loaf that was inadvertently left for 24 hours in an oven with a pilot light after being cooked. The taste was apparently not impaired, and so the loaf was heated (approximately 20 minutes) to serving temperature and served. Another outbreak was associated with the consumption of stew left unheated on the stove top for 16 hours after cooking. One taste landed the victim in the hospital with a need for mechanical respiratory assistance (Calif. Dept. of Health, 1985).

Concern has been expressed over refrigerated, processed foods with extended shelf life that are hermetically sealed (Notermans et al, 1990). Since these products have been heated to temperatures high enough to kill the vegetative organisms but inadequate to kill the spores, if food is not kept out of the danger zone spores might germinate and toxin might be produced.

A type of botulism not associated with canned food was first reported in the 1960s when seven deaths occurred from smoked fish from the Great Lakes region of the United States. This was caused by the Type E strain of the *C. botulinum* organism, which is associated with marine life. In the Pacific Northwest, an annual average of eight cases and one death are caused by fermented fish products shown to contain the Type E toxin (Hauschild, 1986; Hauschild and Gavreau, 1985; Todd, 1988). Some cases caused by smoked fish of several ethnic types have been reported in New York City (Anonymous, 1989a, 1989b). Unfortunately, this strain can grow and produce toxin at refrigerator temperature. Further, it is disturbing to learn that more toxin is produced if the fish is first irradiated (Palumbo, 1986), possibly because of the elimination of competitive organisms.

Symptoms. Botulism symptoms develop 12–36 hours after ingestion of the food. Shorter incubation periods follow the ingestion of larger amounts of toxin. Fatalities have resulted from a single bite! Symptoms, nonspecific at first, include weakness, dizziness, and somnolence; this is followed by blurred or double vision, impaired salivation, hoarseness, and extreme thirst. Finally, the diaphragm is paralyzed, resulting in death. Clinical diagnosis is difficult due to the rarity of the problem and nonspecificity of the symptoms. An antitoxin is available and effective if it is administered soon enough and supplemented by mechanical respirators.

Infant botulism. In 1976, infant botulism was recognized as a cause of unexplained deaths in infants under 14 months old. Since this recognition, more than 500 cases have been reported, but fortunately only two deaths resulted (IFT's Office of Scientific and Public Affairs, 1986). Affected infants show signs of neurological distress, including the inability to cry normally, to suckle properly, and to hold up their heads. Bottle-fed and breast-fed babies seem equally likely to contract the disease. Breast milk antibody was shown to agglutinate the *C. botulinum* and thus delay the onset of the disease, whereas bottle-fed babies appeared to be stricken at a younger age than breast-fed babies.

A record number (88) of infant botulism cases were reported in 1986 (Busta, 1988). Of these, 30% could be traced to use of honey and 33% to corn syrup. A random sampling of honey for *C. botulinum* spores detected their presence in two out of 100 samples of honey and eight out of 40 samples of corn syrup in the Washington, DC, area. Another study of corn syrup detected spores in five of the 961 bottles sampled. Other nonsterilized foods and nonfood items in the infant's environment may also be a source of spores (Pierson and Reddy, 1988). To minimize the risk of botulism poisoning, it is recommended that food not essential for nutrition, such as honey and corn syrup, not be given to infants (Arnon, 1986; Can. Disease Weekly Report, 1986; Hauschild et al, 1988; Kautler, 1982).

Interestingly, adults ingest honey containing spores with no illness occurring, either because the acidity of the stomach does not allow the spores to grow and produce toxin or because the competitive microorganisms in the adult gut prevent *C. botulinum* growth. Since the infant's stomach does not acquire the adult acidity until around 1 year of age, the organism evidently can thrive in the near-neutral pH of a young infant's stomach and proliferate toxin.

For infants 9–14 months of age, botulism may not be linked to food. It may be the result of an intestinal infection with in vivo production of toxin (Labbe, 1988).

Shigella

Shigella species cause a bacterial dysentery called shigellosis. In the United States it is far less common than salmonellosis and accounts for only 2% of the reported outbreaks, although it is estimated that only one in 20 outbreaks are reported. Shigellae are spread by person-to-person contact, flies, water, and food. When food is involved, the disease is associated with conditions of poor personal hygiene and general sanitation and/or with food that is not subsequently cooked or is held at temperatures that allow the organism to grow. Most outbreaks occur in food service operations. Prime months for this infection are the late spring to early fall.

Sources. Foods implicated in outbreaks include various salads (potato, shrimp, and tuna), beans, and milk products. In the United States, this is primarily a foodborne and not a waterborne disease (Smith, 1987). Food served on a major U.S. airline caused an international outbreak of *Shigella sonnei* that was blamed on handling by infected workers (Osterholm, 1989).

Symptoms. The disease is more severe than salmonellosis and can be due to a toxin (Table 6.2). It is characterized by an incubation period of 1 to 7 days (with fewer than 4 days being the most common). Symptoms include diarrhea, abdominal pain, fever, vomiting, and tenesmus. Shigellae penetrate the intestinal lumen, invade the colonic epithelial cells, multiply, and produce a toxin; then they pass into adjacent cells. The result is an acute inflammatory reaction. Death of damaged cells produces ulceration and bloody diarrhea containing mucus and pus. Children aged 1–4 are more susceptible than older children and adults (Kubista, 1985). Repeated bouts of bacterially induced diarrhea from *Shigella* species may be the cause of chronic rheumatoid arthritis and Reiter's syndrome (Smith, 1987).

Bacillus cereus

Before the 1970s, *Bacillus cereus* was not thought to be a cause of outbreaks of food poisoning in the United States, although it had been recognized as causing foodborne disease in Europe about 20 years earlier. Public health officials in this country simply regarded it as an essentially nonpathogenic aerobic spore-former (Terranova and Blake, 1978). Even its characteristics of being fairly heat resistant and widely distributed in a variety of foods, soil, air, and water did not raise any alarm. However, two types of food poisoning result from the different strains of this organism. One type produces a heat-labile toxin, the other a heat-stable one. Both the symptoms and the incriminated foods vary for the different strains.

Table 6.2. Adverse Effects Associated with Toxin from *Shigella*[a]

System Affected	Effect
Nervous	Paralysis Destruction of neurons
Gastrointestinal	Fluid accumulation
RNA	Inhibited protein synthesis
General	Cell death

[a] Modified from Smith (1987).

Sources. Food products implicated in this type of illness have included cereal dishes containing corn and cornstarch, potatoes, cold soups, sauces, puddings, meats, cooked vegetables, ice cream, and fresh and powdered milk (Doyle, 1988a; Snyder and Poland, 1991). In 1973, a food poisoning incident was associated with vegetable sprouts. Tests of seed sprouting kits have since shown that 56 of 98 units tested positive for *B. cereus* contamination. Although washing with warm water three times for 10 minutes decreased the population of the organism in mung bean sprouts, it did not reduce the numbers of organisms in wheat sprouts to safe levels. Substantial numbers in wheat sprouts even survived three washes and 20 minutes of cooking (Harmon et al, 1987). The FDA issued seizure orders on such sprouting kits.

Symptoms. The symptoms produced by the heat-labile toxin or the organism producing it occur after 8–16 hours and are similar to those caused by *Clostridium perfringens* (Table 6.3). Symptoms are primarily diarrhea with abdominal pain and fever occurring occasionally. Symptoms usually subside in 24 hours.

The heat-stable toxin is found in fried rice. In Oriental restaurants, large batches of rice are frequently left to cool at room temperature. Refrigeration is said to make the rice sticky, yielding a less desirable fried rice product, so rice is cooled at room temperature. This cooling period creates ideal conditions for the organism to grow and proliferate toxin (Harmon et al, 1987; Troller, 1986). Pasta products have also been implicated as sources (Snyder and Poland, 1991).

The heat-stable toxin produces a severe emetic (vomiting) reaction accompanied by gastric pain, with onset 1–6 hours after ingestion of the food. Illness usually lasts 6–24 hours.

Campylobacter jejuni

C. jejuni has exploded in recent years from relative obscurity to being

Table 6.3. Comparison of Aspects of Two *Bacillus cereus* Food Poisoning Types with Aspects of Other Common Food Poisoning Organisms[a]

Feature	*B. cereus* Diarrheal	*Claustridium perfringens*	*B. cereus* Emetic	*Staphylococcus aureus*
Onset, hr	8–16	8–22	1–5	2–6
Duration, hr	12–14	12–24	6–24	6–24
Diarrhea or cramps	Predominant	Predominant	Common	Common
Nausea or vomiting	Occasional	Rare	Predominant	Predominant
Principal foods	Meat products, soup, vegetable, sauces, puddings	Cooked meat and poultry	Cooked rice and pasta	Cooked meat and poultry, milk products

[a] Adapted from Kramer and Gilbert (1990).

a recognized major cause of human diarrhea in the United States and throughout the world. Campylobacteriosis is estimated at over 4 million cases annually, which surpasses the incidence of salmonellosis (Kvenberg and Archer, 1987). A 15-month study in eight hospitals in different parts of the United States isolated this microorganism from stool samples more frequently than *Salmonella* species and *Shigella* species combined (Doyle, 1985).

Sources. One of the major sources is intestinal flora. Since in small animals like poultry, edible parts of the body touch the gastrointestinal tract, as much as 50% of poultry meat may be contaminated. Undercooked chicken, processed turkey, cake icing, raw clams, drinking water, raw milk, raw hamburger, and contact with cats have all been implicated as vehicles of transmission (Blaser, 1982; Deming et al, 1987; Doyle, 1985; Foster, 1981). Outbreaks due to this microorganism occur primarily in the spring and the fall (Finch and Blake, 1985).

Survival of the organism in food is fairly poor. Most processing techniques such as heat, acid, salt, and drying destroy it. Gastric pH of 2.3 rapidly kills the organism. It grows best with limited (5–10%) oxygen.

Despite its high lability, it is a major cause of disease, as levels of contamination are high and very small numbers of organisms are needed to produce the infection (less than 500 cells). Such extreme virulence explains why this sensitive bacterium has become recognized as a leading cause of human enteric infections. Some strains produce a toxin, and some evidence points to the production of a cytotoxin (Walker et al, 1986).

Symptoms. Symptoms are not very distinctive and thus are not easily differentiated from those resulting from other enteric pathogens. They vary from brief insignificant enteritis to an enterocolitis with abdominal pain and profuse diarrhea. If symptoms are sufficient to cause the person to seek medical treatment, there is also usually malaise, fever, vomiting, and, in severe cases, grossly bloody stools. Other non-gastrointestinal symptoms include meningitis in infants, bacteremia, urinary tract infections, Reiter's syndrome, and Guillain-Barre syndrome. Males up to age 45 are more vulnerable than females (FSIS, 1989).

Onset of the disease is usually somewhat slower than that from some of the other enteric pathogens, with an incubation period of 2 to 5 days. Most of those affected recover in less than a week, but 20% have prolonged illness.

Not all persons exposed to the organism develop the disease; a water supply contaminated by *Campylobacter* species in a Vermont town resulted in illness of only 20% of the town's population (Walker, 1986).

Further, there is evidence that immunity can be acquired as a consequence of one or two infections.

Yersinia enterocolitica

Y. enterocolitica was first recognized as a human pathogen in 1939 but did not become a concern until the mid-1970s, as evidenced by the increase in the number of documented cases in various parts of the world. Worldwide, 23 cases were reported in 1966, 4,000 in 1974. By 1979, a report indicated that this was the most frequent type of enteric infection in Denmark, with 20,000 cases occurring in a single year (Zink et al, 1982). In parts of Germany and Canada, incidence of *Y. enterocolitica* is comparable to that of *Salmonella* species.

An outbreak in the United States was first reported in upstate New York in 1967. It involved 220 persons, primarily school children, and during its course, 16 unneeded appendectomies were performed. The organism was isolated from chocolate milk. Over 10 years later, an outbreak occurred in a coeducational summer camp in New York state and involved 239 campers and staff members. The outbreak was related to powdered milk and turkey chow mein. The organism was thought to have been introduced by a food handler. A Washington state outbreak occurred in late 1981 and January of 1982 among 87 persons and was associated with ingestion of tofu packed in untreated spring water. A very large outbreak that occurred in the summer of 1982 in Tennessee, Arkansas, and Mississippi was associated with pasteurized milk from a specific plant. The actual incidence rate was unknown but was thought to be quite high because of the extensive distribution of milk into the three states.

Sources. This facultative anaerobe is frequently found in the environment in streams, lakes, and well water. The largest reservoir is the alimentary tract of warm-blooded animals. Swine have been recurringly implicated as carriers of this pathogen. Many dairy products have been identified as vehicles, including raw goat's and cow's milk, pasteurized milk, reconstituted dry milk, cheese curd, and ice cream. In one study (Schiemann, 1987), nearly half the samples of raw milk tested positive for *Y. enterocolitica* or other *Yersinia* species, but many strains present were not virulent. A scientist at USDA has patented a dye test that distinguishes between the virulent and nonvirulent strains (van Pelt, 1987). However, the presence of *Y. enterocolitica* in raw milk in many samples underscores once more the necessity of pasteurization (McManus and Lanier, 1987).

Other confirmed food sources include vacuum-packed meat, chicken, pork, beef, lamb, wild game, oysters, shrimp, crab, fish, tofu, vegetables, and drinking water (Doyle, 1985; Foster, 1981; FSIS, 1989; Stern, 1982; Zink et al, 1982).

This microorganism is a psychrotroph, which means that it grows in foods at refrigerator temperatures (Table 6.4). It has even been found to grow at 1°C (34°F) in raw beef or pork (Palumbo, 1986). Refrigerator storage is thus ineffective in controlling its growth in foods. This scary fact means that care must be taken to prevent contamination or that foods must be heated adequately to kill the organisms. Moreover, *Y. enterocolitica* appears to be a psychrotroph with a fair degree of heat resistance, and several studies have found that some strains produce a heat-stable toxin, so heating the food may not afford protection either (Schiemann, 1987; Troller, 1986). Most documented outbreaks have been associated with postprocessing contamination (IFT's Office of Scientific and Public Affairs, 1986). Interestingly, outbreaks are most prevalent in fall and winter, not summer as with many other organisms.

Symptoms. The onset of the disease known as yersiniosis is usually 2 to 5 days after eating the infected food. It is commonly characterized by diarrhea, which usually lasts 1 to 2 days but may last for several weeks. The organism penetrates intestinal epithelial cells and produces pain. The abdominal pain may be so severe as to be frequently misdiagnosed as appendicitis. Serious cases may also cause fever, dermatitis, inflammation of the lymph, abscesses of the liver and spleen, septicemia, acute carditis, meningitis, and arthritis (FSIS, 1989). Children are most at risk in contracting yersiniosis, but the aged or persons with immune deficiencies are at risk for complications. The disease is rarely fatal.

Table 6.4. *Yersinia* Doubling Time at Various Temperatures[a]

Temperature[b]		Approximate Doubling Time
°C	°F	
−1	30	...[c]
0	32	48 hr
1	35	36 hr
7	45	12 hr
10	50	8 hr
21	70	2 hr
26	80	1 hr
32	90	40 min
38	100	50 min
44	110	...[d]

[a] Adapted from Snyder (1990).
[b] 90–94°F is the organism's temperature optimum, but note its growth at refrigerator temperatures.
[c] Some growth.
[d] No growth.

Listeria monocytogenes

This organism was first reported as causing human disease in 1929, although infection in animals was recognized earlier. Like *Y. enterocolitica,* this microorganism was not considered a major problem until recently, in this case the 1980s. Furthermore, listeriosis was not a reportable disease, so no precise data about its incidence in the past are known.

In 1981, a large outbreak occurred in the Maritime provinces of Canada, with contaminated coleslaw as the probable vehicle. The cabbage was grown in fields fertilized with manure from a flock of sheep subsequently found to carry listeriosis. Before use, the cabbage was held in cold storage, which allowed the microorganism to proliferate (Schlech et al, 1983).

The Canadian outbreak was followed by several in the United States. The first was in Massachusetts in 1983, in which 49 patients were hospitalized with septicemia or meningitis. Nearly 40% of those affected died (14 patients). Pasteurized milk was the vehicle of transmission; the organism was found in a bulk tank of milk from one of the supplying farms. Since the organism is fairly heat-resistant, if there was a very large initial population of organisms, a few organisms could have survived pasteurization and subsequently proliferated (Fleming et al, 1985). The second occurred in Los Angeles and surrounding Orange County in 1985. In this outbreak, 181 mother-infant pairs were involved. All but 4% of the cases were of Hispanic origin. The vehicle was determined to be Mexican-style cheese (Donnelly et al, 1987). The organism could have been from infected raw milk or from postprocessing contamination of the cheese, as the organism was found in the plant. In this outbreak, the mortality rate was again almost 40% (Janes, 1985). One million gallons of ice cream contaminated with *L. monocytogenes* were recalled in 1986, but in that case no consumer was affected. Also in 1986, Brie cheese imported from France resulted in several outbreaks of listeriosis in Canada and England (Farber, 1986). The incriminated cheese was seized before any cases occurred in the United States.

Sources. Water, soil, and sewage are reservoirs of this facultative anaerobic, gram-positive rod-shaped organism (Gellin and Broome, 1989). Many healthy humans are carriers, as are healthy wild and domestic animals. Sheep, goats, and cows are the most common sources of human infection (Donnelly et al, 1987), although *L. monocytogenes* has also been isolated from household pets, mammals, fowl, trout, ticks, and crustaceans. The organism passes from humans into urine, feces, and milk. Even human milk has been found to contain the organism (Brackett, 1988).

Food vehicles of *L. monocytogenes* include leafy vegetables fertilized with manure containing the organism, chocolate milk, raw milk, raw fish, raw meat and chicken, soft cheeses, and fermented sausage. Incidence of contamination is given in Table 6.5. The fact that the organism survives pasteurization heat treatments, repeated freezing and thawing, direct sunlight, and long-wave ultraviolet light means that it is not easily destroyed by common processing techniques (Doyle 1988b). It also survives in shredded vegetables, even though chlorine-treated and stored under modified atmosphere (Beuchat et al, 1990). It is killed by short-wave ultraviolet light (Yousef and Marth, 1988).

L. monocytogenes has been shown to survive the manufacturing process for dry and creamed cottage cheese. The organism was found in cheddar cheese made from contaminated raw milk even after the required 60 days of aging. Both the chocolate and the sugar in chocolate milk aid the growth of *L. monocytogenes* (Rosenow and Marth, 1987). The other bad news is that, like *Y. enterocolitica*, it is a psychrotroph and grows at refrigerator temperatures (Palumbo, 1986).

Symptoms. In humans, the primary symptoms are a mononucleosis-like infection or meningitis. At risk are immunocompromised individuals, pregnant women, and infants. During the first trimester of pregnancy, puerperal sepsis or nonspecific "flulike" illness usually results in spontaneous abortion. If the infection occurs later in the pregnancy, premature delivery of stillborn or acutely ill infants (with perinatal or neonatal septicemia) results. In immunocompromised individuals, infections with *L. monocytogenes* can result in arthritis; inflammation of the bone marrow, gall bladder, or lining of the abdominal cavity; or spinal or brain abscesses (Marth, 1988).

Much FDA and public and private research activity currently surrounds this distinctively different pathogen. The concern stems from the high rate of brain damage and mortality among high risk groups and the fact that the organism both grows at refrigerator temperature and survives many common processing techniques (Ryser and Marth, 1987). CDC data show that the organism has a high death-to-case ratio (Bean and Griffin, 1990), so the concern is certainly justified.

Table 6.5. Incidence of *Listeria* Contamination[a]

Food	Contaminated Samples (%)
Raw ground beef, pork, veal	50–100
Raw chicken	50–100
Potatoes, fresh	26
Radishes, fresh	33

[a] From data cited by Snyder and Poland (1990).

Pathogenic Streptococci in Foods

Pathogenic streptococci were identified as responsible for 19 outbreaks and over 2,000 cases of foodborne disease in the United States in the 15 years before 1990 (Bean and Griffin, 1990). Certain streptococci have been isolated in large numbers from incriminated foods following outbreaks of foodborne disease. Dairy products that undergo heat treatment, such as dried milk and infant formulas, may be vehicles (Batish et al, 1988). The toxins produced by these strains of *Streptococcus* cause symptoms similar to those caused by staphylococcal enterotoxins. Strains of hemolytic streptococci have been found in foods, and the symptoms include those of the upper respiratory as well as gastrointestinal tracts (Tranter et al, 1987).

Pathogenic *Escherichia coli*

E. coli is a common resident of the intestinal tract of warm-blooded animals. For many years, the organism was thought to be harmless but was used as a marker organism to provide evidence of the nonsanitary handling of food and equipment. Later it was recognized that some strains of *E. coli* cause enteric disease. In fact, pathogenic strains of *E. coli* are now thought to be prime offenders in travelers' diarrhea and gastrointestinal illness in developing nations and in other areas with poor sanitary conditions.

The first recognized U.S. outbreak was traced to the consumption of soft fermented cheeses. The next series of outbreaks in this country was in 1982. In the first half of that year, two outbreaks of gastroenteritis, characterized by the sudden onset of severe abdominal cramps and grossly bloody diarrhea, occurred in Oregon and Michigan. Both outbreaks were linked with ground beef sandwiches from chain restaurants. A third outbreak that year involved 31 residents in a home for the aged. Food was the only common factor. In 1983, gastrointestinal illness in several states and in Europe was attributed to pathogenic *E. coli* from imported Brie cheese. All the 1983 outbreaks were tied to soft cheese from the same factory in France (IFT's Office of Scientific and Public Affairs, 1986). In Canada, hamburgers were the source of hemorrhagic colitis in 125 persons in 1984 (Pai et al, 1984).

Sources. Food vehicles include raw ground beef, perhaps chicken, and some imported soft cheeses.

The incidence of contamination of ground meat with this organism is under 4%, but the concern is great because of the severity of the illness (Synder and Poland, 1990). Some strains can survive long periods of frozen storage but are very sensitive to thermal inactivation (Doyle, 1984). Frozen precooked beef patties have been identified as a source (Minn. Dept. of Health, 1988).

Sanitary food handling is essential to minimize infection by this organism, as the human gut is the only source for certain strains (Frank, 1988). The animal's gut may be the source of meat contamination. Even under the best slaughter conditions, animal carcasses may regularly be contaminated with *E. coli* from the animal's bowel. If the organism happens to be a pathogenic strain, it may cause an outbreak (Pinegar and Cooke, 1985). Person-to-person transmission in day care centers has also been documented (Anonymous, 1989c).

Codes of good manufacturing practice that have been set down for the U.S. food industry are useful in preventing contamination by this microorganism, since some strains can grow at refrigerator temperatures and some may have picked up antibiotic resistance (Palumbo, 1986). Adequate cooking is the other way to avoid this infection. As with all raw meat products, it is crucial to avoid cross-contamination from cutting boards and equipment.

Symptoms. The symptoms of this foodborne disease vary significantly depending on the strain. Four types of pathogenic *E. coli* have been identified.

1) Enteropathogenic strains cause the sudden onset of severe abdominal cramps, followed by the development of a watery, then bloody diarrhea.

2) Enterotoxigenic strains produce toxins that cause mild to severe diarrhea with profound dehydration and shock without fever; both heat-labile and heat-stable toxins have been found.

3) Enteroinvasive strains penetrate the epithelium and cause fever, chills, headache, muscle pain, abdominal cramps, and profuse watery diarrhea.

4) Colohemorrhageic strains cause changes in the colon similar to those of colitis (Palumbo, 1986). Verocytotoxin is thought to be the causative agent that results in frank bloody diarrhea and severe abdominal pain. Vomiting may occur, but fever is rarely seen (IFT's Office of Scientific and Public Affairs, 1986). The syndrome usually last 7 days in adults and somewhat longer in children. However, the toxin can create a life-threatening situation (Pai et al, 1984).

Vibrio cholerae

V. cholerae—O1 strain. This organism has long been recognized as a source of human disease and still causes thousands of deaths each year in Asian countries (Madden, 1988). Cholera-01 caused seven pandemics during the 1800s and early 1900s, in which several hundred thousand Americans died. By contrast, from 1911 until 1973, no cases were reported in North America. Then in 1973, a case was reported in Texas followed by 11 cases in 1978 in Lousiana. Cooked crabs from the Lousiana marshes were identified as the vehicle. The crabs were well-cooked by traditional criteria, but the cooking apparently was not long enough to kill the organism. In 1986, 12 cases of *V. cholerae* were reported in Louisiana and associated with eating of crabs or shrimp. Boiling of crabs for 8 minutes appeared to be inadequate to kill the microorganism. It appears that 10 minutes is necessary (CDC, 1986).

An outbreak on a Texas oil rig in the Gulf Coast caused 17 cases. Infection was associated with eating rice prepared on a day when an open valve permitted the rig's drinking water system to be contaminated by canal water. The rice was rinsed after cooking. (This practice not only contaminated the rice, but also is not a good cooking procedure as it washes away water-soluble nutrients.) Two cases were reported among residents of the Texas Gulf and one case in a person visiting the Mexican Caribbean. Between 1973 and 1987, six outbreaks of this strain occurred. Five of the six were due to fish or shellfish. Oysters were implicated as the vehicle in an infection in the summer of 1988 (CDC, 1989).

Box 6.2

Cholera Resurgence in the 1990s

Cholera, an epidemic disease of the 19th century, reemerged in January 1991 in Peru and quickly spread to other countries in Central and South America. The epidemic strikes hardest in areas plagued by poverty, where fuel and equipment to adequately heat food and water are difficult to afford. Water used to clean food in many of these areas is contaminated by human waste, and such water is the main vehicle for the spread of cholera. Local preference for a raw seafood dish called *ceviche* also helps to spread the disease.

Within seven months, 270,000 cases and 3,000 deaths were reported, with the largest number occurring in Peru. These figures are particularly tragic because, with adequate heat treatment, the bacteria can be killed.

The epidemic is unlikely to spread into the United States because bacteria cannot survive in contaminated produce for more than 10 days, and more time that that is needed to ship produce here. The only concern is seafood, in which the organism can survive for a longer time. As a precautionary measure, the FDA has increased its testing of imported seafood and produce from Peru and other Latin American countries.

(Adapted from Williams, 1991)

Humans are the main reservoir of the infection, which is usually spread by human excrement that contaminates food and water. The appearance of this organism over a period of years and at a variety of locations suggests that it is endemic along the Gulf and southeastern Atlantic coasts (Doyle, 1985; Morris and Black et al, 1985; Rippey and Verber, 1988). It also has been found in coastal waters around Maryland and California (Madden et al, 1982).

V. cholerae—**Non-01 strains.** Other strains classified as non-01 have been recognized as pathogenic and have been associated with several outbreaks since 1965. The first U.S. outbreak of the non-01 type occurred in Florida in 1979 and was associated with the eating of raw oysters. An FDA survey in that same year documented the presence of non-01 organisms in 14% of 790 samples of raw oysters.

Figure 6.4. Map showing the number of 1991 cholera cases in the continental United Sates and in Latin America reported by the Pan American Health Organization by July 26, 1991. Reported deaths were 2,343 in Peru, 505 in Ecuador, 74 in Colombia, and one in Chile. (Reprinted from Williams, 1991)

It has also been found in shrimp and crabs.

Non-01 *V. cholerae* is normally found in bays and estuaries and in brackish inland lakes. Surveys of coastal waters in Florida, California, Louisiana, and Maryland not only establish the presence of non-01 as well as 01 strains but also indicate that non-01 is more prevalent than 01. It should be noted that not all non-01 isolates appear to be capable of causing disease.

Non-01 gastroenteritis has been associated with the eating of raw oysters and with traveling to Mexico. Water, ice, eating utensils, soft drinks, and fruits and vegetables washed with polluted water have all been implicated as vehicles of transmission (Hoover, 1985). Outbreaks of both types of cholera may increase with the increased popularity of raw fish. Months when the incidence is the highest are August and September.

Symptoms. Both types of cholera have an onset time of 6 hours to 5 days after the ingestion of contaminated food or water. They produce a gastroenteritis characterized by diarrhea, nausea, and vomiting. Cases of non-01 have also included soft tissue infection and septicemia (Ryser and Marth, 1989). Type 01 is more severe. Those stricken can experience severe dehydration and electrolyte imbalance through stool losses of greater than a liter per hour. If dehydration is not controlled, death results. With type non-01, diarrhea occurs in all cases, and bloody diarrhea occurs in 25% of the cases (Hackney and Dicharry, 1988).

Vibrio parahemolyticus

V. parahemolyticus is a common marine isolate found in coastal waters throughout the world, especially those adjacent to warm countries. First recognized as a pathogen in the early 1950s, it is a significant source of foodborne illness in Japan, perhaps responsible for 50% of the reported cases of such illness in Japan.

It was not until 1969 that this organism was thought to be a problem in the United States. Since that time only a few outbreaks have been attributed to it, principally cases around the Gulf and southeastern Atlantic coasts and in the Pacific Northwest (Liston, 1990). However, most public health authorities around the world acknowledge it as a pathogen of concern even though accurate figures on the disease incidence are not available. At least two factors complicate accurate figures on disease incidence. One is that a special culture medium is necessary to identify it. The other is that not all environmental isolates are capable of causing disease—a situation similar to that for non-01 *V. cholerae*.

V. parahemolyticus outbreaks are most likely to occur in the summer, when water temperatures encourage the growth of this organism.

Ingestion of the organism itself results in a foodborne infection. The dose required to cause illness has been estimated at 1 million to 1 billion cells. The organism is killed by heating and drying. Although its growth is arrested at low temperatures, it can survive refrigeration and freezing.

Sources. Foods implicated include shellfish, raw fish, and salted cucumbers. Cross-contamination has been implicated in several outbreaks caused by this organism (Beuchat, 1982; Foster, 1981; IFT's Office of Scientific and Public Affairs, 1986). The incidence in Japan is high because of the large amount of raw fish eaten there. The incidence in the United States may increase with the increasing popularity of raw fish and sushi bars.

Symptoms. Clinical symptoms include diarrhea (98% of the cases), severe abdominal pain, cramps, nausea, vomiting, headache, fever, and chills. Incubation time ranges from 4 to 96 hours, with symptoms usually occurring 2–48 hours after consumption of the food. The disease seems to be self-limited, with a median duration of 3 days (Hackney and Dicharry, 1988; Morris and Black, 1985).

Vibrio vulnificus

V. vulnificus is the most dangerous of the vibrios and is truly a pathogen of emerging significance (Liston, 1990). It has been associated with gastrointestinal illness only since 1976. Surveys indicate that this organism is common in marine environments. It grows especially well in warm temperatures, which accounts for the higher incidence of this disease in summer months (IFT's Office of Scientific and Public Affairs, 1986; Morris and Black, 1985). Not all strains are pathogenic.

Sources. The primary vehicles are raw or undercooked seafood, particularly oysters and clams. The organism is heat sensitive, so adequate cooking minimizes any risk from this organism. However, as with other vibrios, steaming just to facilitate opening of the shell is insufficient to kill the organism.

Symptoms. The organism causes two forms of disease. One occurs 24–48 hours after ingestion of contaminated seafood. The organism penetrates the intestinal tract and may cause septicemia. The illness usually begins with malaise followed by chills, fever, and prostration. Vomiting and diarrhea occur but are not common. Blood pressures of less than 80 millimeters of mercury occur in about one third of the cases. Death occurs in 40–60% of those affected. However, infection usually only occurs in persons with underlying disease, such as liver disease, gastric disease, malignancy, iron storage disease, and chronic renal insufficiency. Health authorities have urged people with liver disease or immunodeficiencies not to eat raw seafood (Hackney and

Dicharry, 1988). Because of the high risk to certain population subgroups, the public health effort to inform consumers is of high priority (Roderick, 1990).

The other form is a rapidly progressing cellulitis resulting from seawater-associated wounds. It is frequently associated with a laceration on the hand sustained while cleaning shellfish.

Other Vibrios

V. mimicus is similar to *V. cholerae*. Food poisoning by this organism peaks in the summer months. Documented vehicles include raw oysters and boiled crayfish. Symptoms include diarrhea, which may be bloody and may last 1-6 days. Nausea, vomiting, and abdominal pain occur in two thirds of the cases (Hackney and Dicharry, 1988).

The species *V. hollisae* is different from other vibrios in that it grows on different media. It has been the documented cause of 36 cases of food poisoning associated with raw oysters, clams, and shrimp. Diarrhea is the major symptom. Vomiting and fever may occur.

V. furnissii has been documented as a source of food poisoning characterized by diarrhea and cramping, sometimes with nausea and vomiting (Hackney and Dicharry, 1988).

Plesiomonas shigelloides

P. shigelloides has been suspected of causing foodborne disease for over 40 years. The infection occurs more frequently when waters are warm. Raw and undercooked oysters are the most frequently implicated vehicles. Symptoms occur 1-2 days after ingestion of the food and include diarrhea, abdominal pain, and nausea. Chills, fever, vomiting, and headache have also been reported (Hackney and Dicharry, 1988).

Aeromonas Species

A. hydrophilia and *A. sobria* are associated with fecal matter and the spoilage of refrigerated animal products. They are well known pathogens of freshwater fish. These facultative anaerobes grow well at 1-5°C but would be killed by most food processing operations (Palumbo et al, 1987). They produce both a heat-stable and a heat-labile toxin (Buchanan and Palumbo, 1985; Palumbo, 1986). They have also been isolated from treated and untreated drinking water.

Symptoms. These organisms have been associated with traveler's and infantile diarrhea. Children under seven and adults over 60 are most often affected (Snyder and Poland, 1991). The symptoms can vary from extremely mild to life-threatening choleralike dehydration. The actual incidence of disease due to this organism is unknown at present, but several incidents associated with the eating of raw oysters have been reported (Anonymous 1986).

SUMMARY

I hope that this chapter has emphasized adequately that foodborne bacterial disease is a real hazard. In addition, emerging evidence indicates that continuous exposure to foodborne disease not only causes gastroenteritis but also may put persons at risk for much more serious diseases. There are even links with chronic diseases such as arthritis and atherosclerosis (Archer, 1987).

This chapter has further emphasized that the foodborne disease hazard can be minimized by maintaining high standards of sanitation during all the stages along the production, distribution, and consumption chain. Maintaining high standards of cleanliness for the food preparer, the equipment, and the food-preparation surfaces is crucial. Foods must also be kept out of the danger zone (40–160°F), even though we now are aware that certain microorganisms grow at refrigerator temperatures. The latter realization should underscore the importance of practices that minimize contamination of the food.

In the food processing industry, research is needed to identify sources of problems and ways to eliminate them. An example of a change in procedure that can eliminate a problem is the addition of acetic acid (vinegar) to the scald water of chickens—this substantially reduces cross-contamination in poultry slaughter plants (Brown, 1986). Increased use of a program of hazard analysis and critical control points (HACCP) is crucial (Snyder, 1990).

One way to minimize contamination is provided by well-designed packaging. The return to barrels of bulk foods at retail may offer an inexpensive and environmentally sound alternative to packaging, but it also increases the risk of microbial contamination. Once again we face a risk-benefit trade-off.

Another strategy for reducing risk is to purchase foods from a restaurant, deli, or grocery where principles of food sanitation are obvious from the surroundings. Operations that harbor danger include those that utilize well-meaning but uninformed and untrained volunteers to do tasks that could allow the inadvertent inoculation of the food with a variety of microorganisms or those with blatant breaches in sanitation practice.

Still another strategy is to minimize the use of raw or undercooked chicken and fish. Many diseases that are endemic in the Orient have been increasing in the United States as raw fish enjoys a wave of popularity.

Foodborne diseases from molds and mycotoxins are discussed in the next chapter.

REFERENCES

Al-Obaidy, H. M., Khan, M. A., Blaschek, H. P., and Klein, B. P. 1985. Early detection of *Clostridium perfringens* enterotoxin and its relationship to sensory quality of cooked chicken. J. Food Saf. 7:43-55.

Anonymous 1986. The truth about raw fish. Reprint from Emerg. Med., July 15.

Anonymous 1989a. Botulism. Dairy Food Environ. Sanit. 9:458.

Anonymous 1989b. Botulism associated with processed fish. Dairy Food Environ. Sanit. 9:459.

Anonymous 1989c. On the rise in Minnesota: *E. coli* 0157:H7. Food Prot. Rep. 5(2):4-5.

Archer, D. L. 1978. Immunotoxicology of foodborne substances: An overview. J. Food Prot. 41:983-988.

Archer, D. L. 1987. Foodborne Gram-negative bacteria and atherosclerosis: Is there a connection? Food Prot. 50:783-787.

Arnon, S. S. 1986. Infant botulism: Anticipating the second decade. J. Infect. Dis. 154:201-206.

Batish, V. K., Chandler, H., and Ranganathan, B. 1988. Heat resistance of some selected toxigenic enterococci in milk and other suspending media. J. Food Sci. 53:665-666.

Bean, H. G., Griffin, P. M., Goulding, J. S., and Ivey, C. B. 1990. Foodborne disease outbreaks, 5-Year Summary, 1973-1987. J. Food Prot. 53:711-728.

Bean, N. H., and Griffin, P. M. 1990. Foodborne disease outbreaks in the United States, 1973-1987: Pathogens, vehicles, and trends. J. Food Prot. 53:804-817. Also MMWR 39(SS-1):15-57.

Bennett, R. W., and Berry, M. R. 1987. Serological reactivity and in vivo *Staphylococcus aureus* enterotoxins A and D in selected canned foods. J. Food Sci. 52:416-418.

Bergdoll, M. S. 1989. *Staphylococcus aureus*. Pages 463-523 in: Foodborne Bacterial Pathogens. M. P. Doyle, ed. Marcel Dekker, New York.

Beuchat, L. R. 1982. *Vibrio parahaemolyticus*: Public health significance. Food Technol. (Chicago) 36(3):80-83 ff.

Beuchat, L. R., Berrang, M. E., and Brackett, R. E. 1990. Presence and public health implications of *L. monocytogenes* on vegetables. Pages 175-181 in: Foodborne Listeriosis. A. L. Miller, J. L. Smith, and G. A. Somkuti, eds. Elsevier Science Publ., Amsterdam.

Blaser, M. J. 1982. *Campylobacter jejuni* and food. Food Technol. (Chicago) 36(3):89-92.

Bowner, E. J. 1965. Salmonellae in food—A review. J. Milk Food Technol. 28:74.

Brackett, R. E. 1988. Presence and persistence of *Listeria monocytogenes* in food and water. Food Technol. (Chicago) 42(4):162-164 ff.

Brown, W. L. 1986. Current and future environmental issues as seen from the private sector. Cereal Foods World 31:800-801.

Bryan, F. L. 1979. Epidemiology of food-borne disease. Pages 3-69 in: Food-borne Infections and Intoxications, 2nd ed. H. Riemann and F. L. Bryan, eds. Academic Press, New York.

Buchanan, R. L., and Palumbo, S. A. 1985. *Aeromonas hydrophilia* and *Aeromonas sabria* as potential food poisoning species: A review. J. Food Saf. 7:15-29.

Busta, F. F. 1988. Risks and hazards in the food supply. Presented at the NW Section of the American Association of Cereal Chemistry, Feb 23, 1988, Minneapolis.

Caldwell, E. 1988. Salmonella screening tests rated. Cereal Foods World 33:243.

Calif. Dept. of Health. 1985. Unusual botulism cases—California. Calif. Morbid. 4(Feb. 1):1.

Calif. Dept. of Health. 1986. FDA declares *Salmonella dublin* "life threatening hazard." Calif. Morbid. 38(Sept. 26):1.

Can. Disease Weekly Report. 1985. 1984—A record year for infant botulism in California. Dairy Food Sanit. 6:253.

Centers for Disease Control (CDC). 1986. Cholera in Lousiana—An update. MMWR 35:687-688.

Centers for Disease Control (CDC). 1989. Toxigenic *Vibrio cholerae* 01 infection acquired in Colorado. MMWR 38(2):19-20.

Deming, M. S., Tauxe, R. V., Blake, P. A., Dixon, S. E., Fowler, B. S., Jones, T. S., Lockamy, E. A., Patton, C. M., and Sikes, R. O. 1987. *Campylobacter enteritis* at a university: Transmission from eating chicken and from cats. Am. J. Epidemiol. 126:526-534.

Denny, C. B. 1982. Industry's response to problem solving in botulism prevention. Food Technol. (Chicago) 36(12):116-117.

Donnelly, C. W., Briggs, E. H., and Donnelly, L. S. 1987. Comparison of heat resistance of *Listeria monocytogenes* in milk determined by two methods. J. Food Prot. 50:14-17.

Doyle, M. P. 1984. Hemorrhagic *E. coli.* J. Food Prot. 47:824-825.

Doyle, M. P. 1985. Food-borne pathogens of recent concern. Annu. Rev. Nutr. 5:25-41.

Doyle, M. P. 1988a. *Bacillus cereus.* Food Technol. (Chicago) 42(4):199-200.

Doyle, M. P. 1988b. Effect of environmental and processing conditions on *Listeria monocytogenes.* Food Technol. (Chicago) 42(4):169-171.

Farber, J. M. 1986. A review of foodborne listeriosis. Can. Dis. Weekly Rep. 12:98-100.

Finch, M. J., and Blake, P. A. 1985. Foodborne outbreaks of campylobacteriosis: The United States experience, 1980–1982. Am. J. Epidemiol. 122:262-268.

Fleming, D. W., Cochi, S. L., MacDonald, K. L., Brondum, J., Hayes, P. S., Plikaytis, B. D., Holmes, M. B., Audurier, A., Broome, C. V., and Reingold, A. 1985. Pasteurized milk as a vehicle of infection in an outbreak of listeriosis. New Engl. J. Med. 312:404-407.

Foster, E. M. 1981. Impact of microorganisms and their toxins on food processing. In: Impact of Toxicology on Food Processing. J. C. Ayers and J. C. Kirschmann, eds. AVI, Westport, CT.

Frank, J. F. 1988. Enteropathogenic *Escherichia coli.* Food Technol. (Chicago) 42(4):192-193.

FSIS. 1989. Preventable foodborne illness. USDA FSIS Facts (FSIS-34).

Gellin, B. G., and Broome, C. V. 1989. Listeriosis. JAMA 261.

Hackney C. R., and Dicharry, A. 1988. Seafood-borne bacterial pathogens of marine origin. Food Technol. (Chicago) 42(3):104-109.

Harmon, S. M., Kautler, D. A., and Solomon, H. M. 1987. *Bacillus cereus* contamination of seeds and vegetable sprouts grown in home sprouting kits. J. Food Prot. 50:62-65.

Hauschild, A. 1986. Botulism in Canada. Canada Dis. Weekly Rep. 12:53-54.

Hauschild, A., and Gauvreau, L. 1985. Food-borne botulism in Canada, 1971–1984. Can. Med. Assoc. J. 133:1141-1145.

Hauschild, A., Hilsheimer, R., Weiss, K. F., and Burke, R. B. 1988. *Clostridium botulinum* in honey, syrups, and dry cereal. J. Food Prot. 51(11):892-894.

Hayes, J. R., and Campbell, T. C., 1986. Food additives and contaminants. Pages 771-800 in: Casarett and Doull's Toxicology, 3rd ed. C. D. Klaassen, M. O. Amdur, and J. Doull, eds. Macmillan, New York.

Holmberg, S. D., and Blake, P. A. 1984. Staphylococcal food poisoning in the United States: New facts and old misconceptions. JAMA 251:487-489.

Hooper, A. J. 1989. Foodborne illnesses of tomorrow are here today. Dairy Food Environ. Sanit. 9:549-551.

Hoover, D. G. 1985. Review of isolation and enumeration methods for *Vibrio* species of food safety significance. J. Food Prot. 7:35-42.

IFT's Office of Scientific and Public Affairs. 1986. New bacteria in the news: A special symposium. Food Technol. (Chicago) 40(8):16-26.

Janes, S. M. 1985. Listeriosis outbreak associated with Mexican-style cheese. MMWR 34:357-359.

Johnson, R. W. 1987. Microbial food safety. Dairy Food Environ. Sanit. 7:174-176.

Kautler, D. A. 1982. *Clostridium botulinum* spores in infant foods: A survey. J. Food Prot. 45:11 ff.

Kramer, J. M., and Gilbert, R. J. 1989. *Bacillus cereus* and other *Bacillus* species. Pages 21-70 in: Foodborne Bacterial Pathogens. M. P. Doyle, ed. Marcel Dekker, New York.

Kubista, R. 1985. Bacterial food poisoning. Hazelton Food Sci. Newsl. 9(May/June):1-3.

Kvenberg, J. E., and Archer, D. L. 1987. Economic impact of colonization control on foodborne disease. Food Technol. (Chicago) 41(7):77-81 ff.

Labbe, R. G. 1988. *Clostridium perfringens.* Food Technol. (Chicago) 42(4):195-196.

Labbe, R. G. 1989. *Clostridium perfringens.* Pages 191-235 in: Foodborne Bacterial Pathogens. M. Doyle, ed. Marcel Dekker, New York.

Lecos, C. 1986. Of microbes and milk: Probing America's worst *Salmonella* outbreak. Dairy Food Environ. Sanit. 6:136-140.

Liston, J. 1990. Microbial hazards of seafood consumption. Food Technol. (Chicago) 44(12):56-62.

Madden, J. M. 1988. *Vibrio.* Food. Technol. (Chicago) 42(4):191-192.

Madden, J. M., McCardell, B. A., and Read, R. B. 1982. *Vibrio cholerae* in shellfish from U.S. coastal waters. Food Technol. (Chicago) 36(3):93-96.

Marth, E. H. 1988. Disease characteristics of *Listeria monocytogenes.* Food Technol. (Chicago) 42(4):165-168.

McManus, C., and Lanier, J. M. 1987. *Salmonella, Campylobacter jejuni,* and *Yersinia enterocolitica* in raw milk. J. Food Prot. 50:51-55.

Minn. Dept. of Health. 1988. Recent occurrence of infection with *Escherichia coli* 0157:H7 in Minnesota and request for case reports. Dis. Control Newsl. 175(8):56-61.

Morris J. G., and Black, R. E. 1985. Cholera and other vibrioses in the United States. New Engl. J. Med. 312:343-350.

Newsome, R. L. 1988. *Staphylococcus aureus.* Food Technol. (Chicago) 42(4):194-195.

Nightingale, S. L. 1987. Foodborne disease: An increasing problem. Am. Fam. Physician 35:353-354.

Notermans, S., Dufrenne, S. J., and Lund, B. M. 1990. Botulism risk of refrigerated, processed foods of extended durability. J. Food Prot. 53:1020-1024.

Osterholm, M. 1989. An international outbreak of shigellosis associated with consumption of food served by a Minnesota-based commercial airline. Dis. Control Newsl. (Minn. Dept. of Health) 17(1):1-4.

Osterholm, M. 1991. Food safety. Presentation by state epidemiologist to Minnesota Beef Council, Jan. 11, 1991.

Pai, C. H., Gordon, R., Sims, H. V., and Bryan, L. E. 1984. Sporadic cases of hemorrhagic colitis associated with *E. coli* 0157:H7. Ann. Intern. Med. 101:738-742.

Palumbo, S. A. 1986. Is refrigeration enough to restrain foodborne pathogens? J. Food Prot. 49:1003-1009.

Palumbo, S. A., Williams, A. C., Buchanan, R. L., and Phillips, J. G. 1987. Thermal resistance of *Aeromonas hydrophilia.* J. Food Prot. 50:764-763.

Patten, R. C. 1981. Salmonellosis. Am. Fam. Physician 23:112-117.

Picard, F. J., Jetté, L. P, Rochefort, J., and Brazeau, M. 1987. *Staphylococcus aureus* strains associated with food poisoning in Quebec. Can. J. Pub. Health 78(1):21-24.

Pierson, M. D., and Reddy, N. R. 1988. *Clostridium botulinum.* Food Technol. (Chicago) 42(4):196-198.

Pinegar, J. A., and Cooke, E. M. 1985. *Escherichia coli* in retail processed food. J. Hyg. 95:39-46.

Rippey, S. R., and Verber, J. L. 1988. Shellfish-borne disease outbreaks. Dept. of Health and Human Services, Public Health Service, Food and Drug Administration, Shellfish Sanitation Branch, North East Technical Service Unit, Davisville, RI.

Roberts, T. 1990. Bacterial foodborne illness in USA. Food Lab. News 19:53.

Roderick, G. E. 1990. *Vibrio vulnificus* (letter). Dairy Food Environ. Sanit. 10:93.

Rosenow, E. M., and Marth, E. H. 1987. Addition of cocoa powder, cane sugar, and carrageenan to milk enhances growth of *Listeria monocytogenes*. J. Food Prot. 50:726-729.

Russo, J. R. 1985. Food poisoning. Prep. Foods (May):9.

Ryser, E. T., and Marth, E. H. 1987. Behavior of *Listeria monocytogenes* during the manufacture and ripening of cheddar cheese. J. Food Prot. 50:7-13.

Ryser, E. T., and Marth, E. H. 1989. "New" food-borne pathogens of public health significance. J. Am. Diet. Assoc. 89:948-956.

Schiemann, D. A. 1987. *Yersinia enterocolitica* in milk and dairy products. J. Dairy Sci. 70:383-391.

Schlech, W. F., Lavigne, P. M., Bortolussi, R. A., Allen, A. C., Haldane, E. V., Wort, J. A., Hightower, A. W., Johnson, S. E., King, S. H., Nicholls, E. S., and Broome, C. V. 1983. Epidemic listeriosis. New Engl. J. Med. 308:203-206.

Smith, J. L. 1987. *Shigella* as a foodborne pathogen. J. Food Prot. 50:788-801.

Snyder, O. P. 1986. Microbial quality assurance in food service operations. Food Technol. (Chicago) 40(7):122-130.

Snyder, O. P. 1990. Food Safety Through Quality Assurance Management. Hospitality Institute of Technology and Management, Inc., St. Paul, MN.

Snyder, O. P., and Poland, D. M. 1990. America's "safe" food, part 1. Dairy Food Environ. Sanit. 10:719-724.

Snyder, O. P., and Poland, D. M. 1991. America's "safe" food, part 2. Dairy Food Environ. Sanit. 11:14-20.

St. Louis, M. E., Morse, D. L., and Potter, M. E. 1988. The emergence of grade A eggs as a major source of *Salmonella enteritis* infections. JAMA 259:2103-2107.

Stern, N. J. 1982. *Yersinia enterocolitica*: Recovery from foods and virulence characterization. Food Technol. (Chicago) 36(3):84-88.

Terranova, W., and Blake, P. A. 1978. Current concepts: *Bacillus cereus* food poisoning. New Engl. J. Med. 298:143-144.

Todd, E. C. D. 1988. Botulism in native peoples—An economic study. J. Food Prot. 52:581-588.

Todd, E. C. D. 1989. Preliminary estimates of costs of foodborne disease in the United States. J. Food Prot. 52:595-601.

Tranter, H. S., Modi, N. K., abd Hambleton, P. 1987. New quality control methods: Dectecting bacterial toxins in food. Pages 169-220 in: Natural Toxicants in Food: Progress and Prospects. D. H. Watson, ed. Ellis, Horwood, and Weinham, Chichester, U.K.

Troller, J. A. 1986. Water relations of foodborne pathogens. J. Food Prot. 49:656-670.

van Pelt, D. 1987. Dye reveals danger of bacteria in food. Insight 3(July 6):52.

Walker, R. I., Caldwell, M. B., Lee, E. C., Guerry, P., Trust, T. J., Ruiz-Palacias, G. M. 1986. Pathophysiology of *Campylobacter enteritis*. Microbiol. Rev. 50(1):81-94.

Yousef, A. E., and Marth, E. H. 1988. Inactivation of *Listeria monocytogenes* by ultraviolet energy. J. Food Sci 53(2):571-573.

Williams, R. D. 1991. FDA helps prevent spread of cholera to U.S. FDA Consumer 25(7):14-17.

Zink, D. L., Lachia, R. V., and Dubel, J. R. 1982. *Yersinia enterocolitica*: Their pathogenicity and significance in foods. J. Food Saf. 4:223-241.

Molds and Mycotoxins

CHAPTER 7

Toxins produced by molds are known as mycotoxins (*myco* = fungal). Problems from molds and mycotoxins have considerable worldwide significance in terms of public health, agriculture, and economics. The Council on Agricultural Sciences and Technology (CAST), a nonprofit group of scientific societies and agricultural scientists from across the United States, has estimated that in the United States alone $20 million is lost annually on just one crop, peanuts, contaminated with just one mycotoxin, aflatoxin (CAST, 1989). Obvious mold growth causes rejection of the food, whereas mold growth that is not obvious can leave noxious mycotoxins that render the food harmful when eaten.

Mold Growth and Mycotoxins

Molds grow at moisture levels lower than will support the bacterial growth discussed in Chapter 6, although heat and high relative humidity sharply increase growth rates. Any food on which molds have grown can potentially contain mycotoxins. Some examples include fruit juice, yogurt, sour cream, cheese, bread, grains, nuts, cured meats, and jams. However, not all molds produce mycotoxins, and the very same species of mold that produces mycotoxins under one set of conditions does not produce them under another. It is impossible to tell by appearance, taste, or smell which molds have produced mycotoxins.

Some foods may not carry any visible evidence of mold yet may still bear mycotoxins. Particular problems are foods that are ground, such as flour or peanut butter, or that come from animals fed mycotoxin-contaminated feed. Grinding a few moldy peanuts into peanut butter or moldy kernels of grain into feed or flour may contaminate the whole batch. Milk and dairy products may be contaminated if cattle are fed moldy rations, as aflatoxins can be secreted into the milk (Stoltz, 1983).

HISTORY OF MOLD AS A CAUSE OF DISEASE

Moldy food has long been known to be a cause of disease. Ergotism or "holy fire" is the earliest and best known mold-associated disease.

Bread prepared from grains infected by *Claviceps purpurea*—especially blackened rye—causes ergotism. A gangrenous form of ergot poisoning was common in central Europe from the ninth to the 14th century. Descriptions of those afflicted noted that the first symptoms were prickly sensations in the limbs. The limbs then became swollen, inflamed, and subject to sensations of intense heat and cold (the disease thus became known as St. Anthony's fire), followed by gangrene and eventual loss of the limb (Smith and Moss, 1985). The outbreaks were so frequent in France that the Order of St. Anthony was formed to provide hospitals for those suffering from St. Anthony's fire (Floss and Anderson, 1980).

A convulsive form of ergotism involving the nervous system was also described as occurring in Europe from the late 16th to the late 19th century. It was also reported in the United States and has been implicated in the Salem witchcraft trials of 1692 (Matossian, 1982). Ergot mold has since been shown to contain derivatives of the hallucinogen lysergic acid (better known as LSD).

In Japan as early as the 17th century, an acute form of cardiac beriberi was associated with yellow rice toxins (Ueno, 1983), which belong to the *Penicillium* genus of molds and have also been associated with liver cancer. Other rice molds, noted as causing diseases such as "red mold" or "scabby disease," have been found to belong to the *Fusarium* group of molds (Beuchat, 1978).

A particularly severe and often fatal mycotoxicosis, alimentary toxic aleukia (ATA), is caused by the ingestion of grains left outside during the winter. It was first recognized in Russia in 1913. Food shortages near Siberia forced overwintered wheat, millet, and barley to become a significant part of the Soviet diet. The melting snow raised the moisture content of the grains and favored mold growth (Smith and Moss, 1985) (Box 7.1).

ATA reached epidemic proportions in the Soviet Union during and just after World War II. The disease is characterized by three stages. The first stage, consisting of mouth and throat inflammation, gastroenteritis, and vomiting, is followed by an asymptomatic second stage during which immunodepression is observed (Ueno, 1987). The third and often fatal stage is characterized by pinpoint-sized hemorrhages in the skin (petechiae) and necrotic ulcers in various parts of the body. In the Soviet Union outbreak, mortality was as high as 60%. Highly toxic T-2 toxin along with other trichothecenes were later confirmed as the cause (Bullerman, 1979; Ueno, 1987). A less severe form of ATA that also occurred there had the unusual name of "drunken bread syndrome." Symptoms included headache, vertigo, trembling of extremities, and gastrointestinal symptoms (Joffe, 1983).

Box 7.1

Alimentary Toxic Aleukia (ATA) in Humans

Symptoms occur after the ingestion of overwintered grain, i.e., grain that is left in the field through the fall and the winter. These conditions allow mold containing mycotoxins of the *Fusarium* type to form on the grain. Winters that are characterized by frequent freezing and thawing foster abundant mold growth. Grains harvested during the thaw are highly contaminated.

At least 2 kilograms of food prepared with the moldy grain must be ingested. Blood abnormalities occur 2-3 weeks after the ingestion of the grains. Death occurs after 6-8 weeks.

Wheat and prosomillet are the most toxic.

The disease affects persons aged 8 to 50 to the greatest extent. Breast-fed infants are unaffected, as the toxic compounds don't pass into the milk.

(Adapted from Joffe, 1983)

RECENT UNDERSTANDING OF MYCOTOXINS

Despite their role in history, awareness of mold-related disease was virtually nil until the 1960s. Most people gave little thought to the health consequences of mold contamination of food and feed. The scenario quickly changed in the early 1960s with the unexplained liver necrosis and death of thousands of turkeys in England (Siller and Ostler, 1961). The problem was subsequently traced to imported peanut meal used as feed.

The peanut meal from a tropical country was analyzed and found to contain four toxins of the mold *Aspergillus flavus*. During analysis, the toxins in the meal were found to fluoresce blue and green, and so they were named to reflect their colors, aflatoxin blue (AFB) and aflatoxin green (AFG). After the analysis, the poultry were found to have suffered acute aflatoxicosis poisoning from the mold-produced metabolites (Blount, 1961).

Although poultry is highly susceptible to aflatoxicosis, the disease has been shown to occur in nearly all species tested, including humans (Wogan, 1965). Human outbreaks of acute aflatoxicosis appear to follow the same patterns as were seen historically in that they occur when food supplies are limited and people are forced to eat moldy grains. Incidents in India and Africa indicate that acute symptoms can occur

after just 3 weeks of ingesting either moldy rice or moldy cassava
(Shank, 1981). Other dietary factors may modify the toxicity of aflatoxin.
Symptoms of aflatoxicosis include flabby heart, edema, abdominal pain,
liver necrosis, and a palpable liver. Death can result. Because of their
small size and possible increased susceptibility, the lethal dose in
children is as little as 200 micrograms of aflatoxin per kilogram of
body weight for as little as 3 weeks (Bullerman, 1986; Concon, 1988;
Diener, 1981; Hayes and Campbell, 1986; Jelliffe and Jelliffe, 1982;
Rodricks et al, 1977).

The acute toxic effect of these metabolites is certainly cause for
concern, but the effects of chronic ingestion are equally alarming. Liver
tumors are produced in animal feeding studies when aflatoxins are
fed at the part per billion (ppb) level, making them among some of
the most potent liver carcinogens known. In addition to liver tumors,
they also induce kidney and colon tumors. Because of their high potency
and wide distribution in the environment, great concern about these
compounds exists (Campbell, 1983; Deger, 1976; Dichter, 1984; Miller
and Miller 1986; Williams and Weisburger, 1986; Stoltz, 1982).

Mycotoxins in Food and Feed

For some crops, especially those grown in warm, humid regions,
losses due to mold contamination can run as high as 45% of the annual
production, and contamination can be extensive (Table 7.1). Corn is
the United States grain crop with the greatest potential for contamination. In the Southeast, aflatoxin incidence for corn in various regions
has ranged from 42 to 93% in a growing season when the corn was
susceptible to mold growth because of drought and corn leaf blight.
In the Midwest, incidence has generally been under 10%. Eighty percent

Table 7.1. Occurrence of Aflatoxin in Selected Foods[a]

Food	Incidence[b]	Average Level (micrograms/kilogram)
Peanut butter		
U.S.	17/104	14
Phillipines	145/149	213
Corn		
U.S.	49/105	30
Phillipines	95/98	110
Wheat flour	20/100	0.25–150
Nuts imported into the United States		
Almonds	26/300	...
Brazil nuts	123/300	...
Peanuts	20/300	...

[a] Adapted from CAST (1989).
[b] Presence of detectable aflatoxin (number of samples with detectable aflatoxin/number
of samples).

of the corn grown in the United States is fed to animals, although in the Southeast much corn is consumed directly in human diets as cooked grits, cornbread, and other products made from corn grits and meal.

It is estimated that corn is responsible for most of the human aflatoxin exposure (Young and Fulcher, 1984; Smith and Moss, 1985), despite the facts that 1) in the last 20 years barley and wheat have also been affected to some extent and 2) a much smaller proportion of the corn than the wheat crop goes into edible food channels. Although not highly vulnerable to aflatoxin contamination, wheat may contain trichothecenes, e.g., deoxynivalenol, or ochratoxin.

Peanuts can also be an aflatoxin source in the U.S. diet (Figure 7.1) because the nuts and the ever-popular peanut butter are consumed in substantial quantities by certain segments of the population. Accordingly, this nut is carefully monitored by the regulatory agencies (National Peanut Council, 1988; Stoltz, 1982). Other nuts such as cottonseed, Brazil nuts, pistachio nuts, and pine nuts (Table 7.1) can also be heavily contaminated (CAST, 1989).

Since molds are abundant in the soil and air, some contamination of crops is unavoidable. However, the fungus is usually unable to penetrate the intact seed kernel. Other damage occurring to the plant enables the fungus to invade it. Drought or other stress to the plant

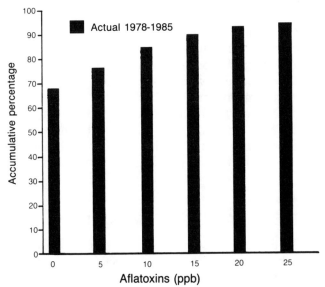

Figure 7.1. Percentage of aflatoxin found in U.S. shelled peanuts from 1978 to 1985. (Reprinted, with permission, from CAST, 1989)

also encourages insect and mold damage (Deiner, 1981). Thus, mycotoxin problems can be expected to be greater during years of extreme drought or infestation. As an example, the drought of the summers of 1988 and 1989 in the United States dramatically increased mycotoxin levels in grains (CAST, 1989). Biocides may be useful in reducing damage (Moss and Frank, 1987).

Reduction of Mycotoxins

Many of the mycotoxins are quite stable to heat and to other normal processing and cooking procedures, although moist heat, as in the roasting of peanuts, destroys some of the aflatoxin. However, heat treatment is not effective enough in itself, destroying only 20–80% of the mycotoxin. For corn products, about 20% is lost during baking or boiling, 50% during frying (Stoltz, 1982), and even more during alkali processing (Camou-Arriolo and Price, 1989).

Refining of peanut oil as practiced in the United States effectively removes any aflatoxin that may be present. In the Orient, where the flavor of crude peanut oil is preferred, the intake of aflatoxins through oils may be significant (Samarajeewa et al, 1990). Milling of grains also may reduce the content of deoxynivalenol in certain mill fractions and concentrate them in others (Moorman, 1990).

Since processing is not completely effective in reducing mycotoxins, it is crucial to prevent their formation whenever possible. Some food additives and natural constituents of food can inhibit mold growth and, in turn, reduce the possibility of mycotoxin contamination. Certain food additives such as sodium bisulfite, sorbate, propionate, and nitrate reduce the production of aflatoxin (Campbell, 1983; Tong and Draughon, 1985). Certain natural components of foods and spices such as peppers, mustard, cinnamon, and cloves also appear to be inhibitors of mycotoxin production (Madhyantha and Bhat, 1985).

Some attempts have been made to treat grain already contaminated by mycotoxin. Both ammonia and ozone treatments of grain appear to destroy several types of mycotoxins without producing any deleterious compounds or leaving any residues (Deiner, 1981; Meaba et al, 1988).

ADVERSE EFFECTS OF MYCOTOXINS

Nearly 100 different mycotoxins have been identified since the discovery of the aflatoxins in the early 1960s. Most are toxic to some degree, and some of them have been found to be carcinogenic. Those known to be carcinogenic or mutagenic include aflatoxin, cyclochlorotine, griseofulvin, luteoskyrin, ochratoxin A, patulin, penicillic acid, sterigmatocystin, T-2 toxin, and zearalenone (Hayes, 1985; Stoltz, 1982). A summary of some mycotoxins and their adverse effects are given in Table 7.2.

The Aflatoxins

Initially the words *mycotoxin* and *aflatoxin* were used interchangeably, but *mycotoxin* has now resumed its generic meaning of any toxin of fungal origin, whereas *aflatoxin* refers to one or more of the toxins from *Aspergillus flavus*. Of these, aflatoxin B1 (AFB1) is the most common, the most-studied, and the most toxic. Toxicity varies markedly among species (Concon, 1988). For example, the LD_{50} ranges from 0.5 milligrams per kilogram for the duckling to 60 milligrams per kilogram for the mouse. Rats, poultry (especially ducks), and trout are highly susceptible to effects of aflatoxin, whereas sheep, hamsters, mice, and pigs are fairly resistant (Christensen et al, 1977). Death usually results from liver damage.

AFB1 appears to be the most liver-carcinogenic compound known for the rat. For comparison, saccharin as a carcinogen for the rat is 50 million times less potent than AFB1 (Byard, 1984). The mycotoxin binds to nucleic acids and therefore could have great carcinogenic potential (Neal, 1987). Although this binding potential is seen in studies with the rat, it is apparently much lower in the mouse, which appears to be totally resistant to aflatoxin's carcinogenic effects (Schlatter, 1988). With such vastly different susceptibilities among experimental animals, it is difficult to make a meaningful assessment of human risk from this carcinogen.

Table 7.2. Mycotoxins, Their Producers, and Adverse Effects[a]

Mycotoxin	Possible Fungi	Adverse Effect
Aflatoxin	*Aspergillus flavus* *A. parasiticus*	Acute toxicity Cancer of the liver Reye's syndrome
Ergot alkaloids	*Claviceps purpurea*	Ergotism
Ochratoxin	*A. ochraceus* *Penecillium viridicatum*	Cancer Kidney disorders Liver damage
Trichothecenes	*Fusarium graminearum* *F. roseum*	ATA[b] Acute toxicity Cancer
Sterigmatocystin	*A. versicolor* *A. nidulans*	Cancer
Zearalenone	*F. graminearum*	Estrogenic effect

[a] Adapted from Goto (1990).
[b] Alimentary toxic aleukia.

This complexity becomes even more difficult, as studies with experimental animals indicate that diet also affects the toxicity of aflatoxin. Low-protein diets were shown to be protective against AFB1-induced carcinogenesis (Campbell, 1983; Concon, 1988). Rats deficient in B vitamins and fed aflatoxins showed signs of aflatoxicosis, whereas those replete with B vitamins did not. AFB1-induced colon cancer in rats was affected by vitamin A status. Diets deficient in vitamin A allowed the growth of tumors, whereas those replete with vitamin A did not (Hayes, 1985). Other factors such as indoles from cabbage family vegetables and antioxidants such as butylated hydroxytoluene inhibit aflatoxin-induced carcinogenesis (Wattenberg, 1985).

To further complicate matters, human response to aflatoxin appears to be modified by other factors such as sex and age. Males show greater susceptibility than females, children greater than adults.

In developing countries, there appears to be a direct correlation between dietary aflatoxin intake and incidence of liver cancer (Groopman et al, 1988). In fact, aflatoxin-related liver cancer often occurs in African males 20–30 years old (Williams and Weisburger, 1986). However, recent epidemiological data indicate that exposure to both AFB1 and to hepatitis B virus or comparable intestinal pathogens may be required before aflatoxin significantly increases cancer risk (Ueno, 1987).

Even with their great carcinogenic potential, aflatoxins may not pose a significant hazard in the United States. The major cancer site for this agent appears to be the liver, but liver cancer incidence is low in the United States and is not increasing. The incidence of hepatitis B and comparable intestinal infections is also fairly low.

One estimate of human cancer risk in the United States has been calculated based on the unavoidable contamination of peanuts with aflatoxin. The average concentration of aflatoxin in peanuts and peanut butter is thought to be below 2 ppb. The FDA defect action level (DAL) used to seize peanuts is 20 ppb, but in practice regulators reject any lots with levels greater than 15 ppb (National Peanut Council, 1988). The FDA estimates the average daily aflatoxin intake at 0.005 ppb from peanuts. At the current DAL, risk assessors estimate that under 2% of liver cancers in the United States are due to aflatoxin exposure (Dichter, 1984). Stoltz (1983) also felt that aflatoxin-contaminated corn did not pose a large risk. The CAST report stated that there is little evidence that contamination of food with aflatoxin represents a liver cancer hazard in the United States (CAST, 1979).

Cancer incidence at sites other than the liver has not been studied in a systematic way. The only report that indicated increased colon cancer risk was done in the 1970s. It did not deal with the ingestion of mold-contaminated food but rather with laboratory workers who studied aflatoxin-producing molds (Deger, 1976). However, while aflatoxin's actual role in human cancers remains to be determined, other disorders are thought to be made worse by the presence of aflatoxins. These include Reyes syndrome, cirrhosis, and kwashiorkor (Hayes and Campbell, 1986). Also, the immune system appears to be impaired by this and other mycotoxins.

Other Mycotoxins

Structurally related to aflatoxin, sterigmatocystin is also a liver carcinogen (Miller and Miller, 1986). It is found in heavily molded wheat and green coffee beans. Circumstantial reports link sterigmatocystin and gastric cancer in China, although low vitamin C and other dietary factors have also been implicated.

Other mycotoxins that are not structurally like aflatoxin vary widely in their toxic effects, carcinogenic potency, and mutagenic activity. Zearalenone and ochratoxin are found in moldy grain. Zearalenone, a toxin from *Fusarium* species, produces metabolites that act like the hormone estrogen and cause a hyperestrogenic syndrome. High levels of these metabolites cause infertility in both males and females (Gelderblom et al, 1988). Ochratoxin is toxic to the liver (hepatotoxic) and the kidney (nephrotoxic) (Campbell, 1983; Ueno, 1987).

Patulin from *Penicillium* and *Aspergillus* mold species may be found in high concentrations in apple and apple juice products. One product survey found this toxin in 50 of 136 consumer packs of apple juice, including organic apple cider. It is also found in moldy apples, plums, peaches, pears, apricots, cherries, and grapes. Studies have found it to be carcinogenic, mutagenic, and teratogenic. It has been shown to induce sarcomas in rat (Schlatter, 1988), although other tests indicate it may be a relatively weak carcinogen. However, since apple juice may be contaminated with up to several hundred parts per billion, human exposure may be higher than desired.

Epidemiological evidence has linked various cancers to mycotoxins. Ochratoxin A has been associated with kidney tumors in the Balkans (Ueno, 1987); zearalenone may be involved in cervical cancer in South Africa; and a toxin from *Fusarium moniliforme* is suspected in esophageal cancer in China and South Africa. High cancer rates have been related to high levels of consumption of moldy grain in Uganda and to peanuts in several African countries. In Kenya, higher liver cancer rates are found at low altitudes than at high altitudes (Ngindu et al, 1982). Warm, humid conditions at low altitudes are thought to favor mold growth in the grains and to account for the difference in liver cancer incidence (Okuda and MacKay, 1982; Concon, 1988).

Although one particular mycotoxin may be named as the cause of a problem, a given mycotoxin is rarely consumed alone, as several different molds and mold metabolites can occur simultaneously. These other products have the potential of modifying the effects of any one metabolite through either synergistic or inhibitory action (Concon, 1988).

COPING STRATEGIES

Constant testing by the agriculture and food processing industries, together with testing by the FDA and state authorities in the United States, carefully tracks the incidence of contamination of feed and food (CAST, 1989). In conditions such as the drought summers of 1988 and 1989, more grain is tested than in normal years because of the expected increase in affected crops. Seizure of contaminated products that exceed allowed levels and destruction of contaminated crop residues help to keep levels of mycotoxins in feed and food supplies well below problem levels. Agricultural techniques such as forced-air drying of crops and/or controlled storage conditions help to keep moisture below the critical level and thus reduce mold growth and the likelihood of mycotoxin contamination. Adequate moisture during crop growth (by rainfall or irrigation) and crop rotation both help to minimize mycotoxin contamination.

Mycotoxins create real hazards in some countries, but the likelihood of them causing disease in Western countries is minimal because the abundant food supply rarely forces the population to ingest moldy food (CAST 1989). Further, adequate vitamin nutriture appears to afford some protection against the adverse effects of mycotoxins. We once again are forced to make some risk-benefit decisions. If farmers decide not to dry grain to conserve natural gas or other fossil fuel, the moisture content of the grain may allow it to become moldy. If farmers decide to reduce pesticide use, the grain may suffer greater insect damage, which in turn has the potential to increase the growth of molds and proliferation of mycotoxins (Diener, 1981).

In the home, it is important to minimize exposure to moldy food. Most people readily throw away food with obvious signs of mold. Cheese is an exception in that the moldy surface can be trimmed. Since aflatoxin can penetrate as far as 1 1/2 inches (4 centimeters) from the surface, deep cuts must be made to ensure a safe product. If the cheese is moldy on all surfaces, it may be impossible to make the needed cuts and still have any cheese left. It would be safer to throw that particular piece away. Foods that are liquid such as juices or semiliquid like sour cream and jams or jellies should be discarded when molds are visible, as the toxins disperse throughout the food. Unfortunately, most toxins are heat stable, so subsequent heating of the food does not render the food safe. Nor does freezing change the potency of the toxins. Various detoxification strategies were discussed in a CAST (1989) report.

Bread sold without preservatives molds faster than bread containing preservatives, so once again we may have choices. We can buy bread with preservatives and face little or no risk from calcium propionate, or buy bread without it. The bread without preservatives molds faster, introducing the risk of the 100% natural carcinogen aflatoxin. On the other hand, we can refrigerate the bread to minimize mold and myco-toxin production, but then the bread texture is much less desirable because bread stales fastest at refrigerator temperatures. If there is

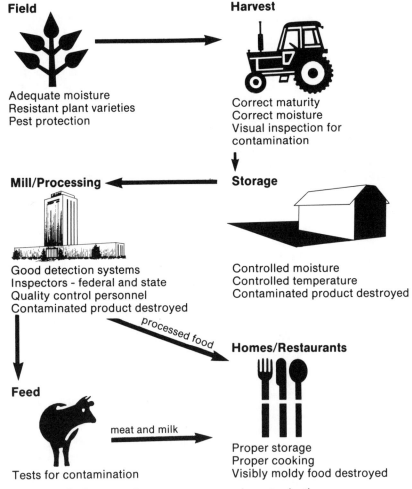

Field

Adequate moisture
Resistant plant varieties
Pest protection

Harvest

Correct maturity
Correct moisture
Visual inspection for
contamination

Mill/Processing

Good detection systems
Inspectors - federal and state
Quality control personnel
Contaminated product destroyed

Storage

Controlled moisture
Controlled temperature
Contaminated product destroyed

processed food

Feed

Tests for contamination

meat and milk

Homes/Restaurants

Proper storage
Proper cooking
Visibly moldy food destroyed

Figure 7.2. Coping strategies to prevent mycotoxin contamination.

adequate freezer space, frozen storage of nuts and grain products helps to keep molds from growing and to maintain their optimum quality.

So with molds, as with other food safety issues, we face some risk-benefit choices. We can choose which risk we want by making appropriate purchases in the marketplace and by exercising the responsibility of keeping the food safe in the home.

REFERENCES

Beuchat, L. R. 1978. Microbial alterations of grains, legumes and oilseeds. Food Technol. (Chicago) 32(5):193-198.

Blount, W. P. 1961. Turkey "X" disease. J. Br. Turkey Fed. 9(2):52 ff.

Bullerman, L. R. 1986. Mycotoxins and food safety. Food Technol. (Chicago) 40(5):59-66.

Byard, J. L. 1984. Metabolism of food toxicants: Saccharin and aflatoxin B1. Adv. Exp. Med. Biol. 177:147-151.

Campbell, C. T. 1983. Mycotoxins. Pages 187-197 in: Environmental Aspects of Cancer: The Role of Macro and Micro Components of Foods. E. L. Wynder, G. A. Leveille, J. H. Weisburger, and G. E. Livingston, eds. Food and Nutrition Press, Westport, CT.

Camou-Arriolo, J., and Price, R. L. 1989. Destruction of aflatoxin and reduction of mutagenicity of naturally contaminated corn during production of a corn snack. J. Food Prot. 52:814-817.

CAST. 1979. Aflatoxin and Other Mycotoxins. Report 80. Council for Agricultural Science and Technology, Ames, IA.

CAST. 1989. Mycotoxins: Economic and Health Risks. Report 116. Council for Agricultural Science and Technology, Ames, IA.

Christensen, C. M., Mirocha, C. J., and Meronuck, R. A. 1977. Molds, Mycotoxins, and Mycotoxicoses. Univ. Minn. Agric. Exp. Stn. Misc. Rep. 142.

Concon, J. M. 1988. Food Toxicology: Contaminants and Additives, Part B. Marcel Dekker, New York.

Deger, G. E. 1976. Aflatoxin—Human colon carcinogenesis? Ann. Intern. Med. 85:204.

Dichter, C. R. 1984. Risk estimates of liver cancer due to aflatoxin exposure from peanuts and peanut products. Food Chem. Toxicol. 22:431-437.

Diener, U. L. 1981. Unwanted biological substances in foods: Aflatoxins. Pages 122-150 in: Impact of Toxicology on Food Processing. J. C. Ayres and J. C. Kirschman, eds. AVI, Westport, CT.

Floss, H. G., and Anderson, J. A. 1980. Biosynthesis of ergot toxins. Pages 18-67 in: The Biosynthesis of Mycotoxins. P. S. Steyn, ed. Academic Press, New York.

Gelderblom, W. C. A., Jaskiewiez, K., Marasas, W. F. O., Thiel, P.G., Horak, R. M., Vleggaar, R., and Kriek, N. P. J. 1988. Fumonisins—Novel mycotoxins with cancer-promoting activity produced by *Fusarium moniliforme*. Appl. Environ. Microbiol. 54:1806-1811.

Goto, T. 1990. Mycotoxins: Current situation. Food Rev. Int. 6:265-290.

Groopman, J. D., Cain, L. G., and Kensler, T. W. 1988. Aflatoxin exposure in human populations: Measurements and relationship to cancer. CRC Crit. Rev. Toxicol. 19(2):113-145.

Hayes, J. R. 1985. Effect of nutrition on the metabolism and toxicity of mycotoxins. Pages 209-220 in: Xenobiotic Metabolism: Nutritional Effects. J. W. Finley, ed. American Chemical Society, Washington, DC.

Hayes, J. R., and Campbell, T. C. 1986. Food additives and contaminants. Pages 771-800 in: Casarett and Doull's Toxicology, 3rd ed. C. D. Klaassen, M. O. Amdur, and

J. Doull, eds. Macmillan, New York.

Jelliffe, E. F. P., and Jelliffe, D. B. 1982. Adverse Reactions to Food. Plenum Press, New York.

Joffe, A. Z. 1983. Foodborne diseases: Alimentary toxic aleukia. Pages 353-494 in: CRC Handbook of Foodborne Diseases of Bacterial Origin. CRC Press, Boca Raton, FL.

Madhyantha, M. S., and Bhat, R. V. 1985. Evaluation of substrate potentiality and inhibitory effects to identify high-risk spices for aflatoxin contamination. J. Food Sci. 50:376-378.

Maeba, H., Takamoto, Y., Kamimura, M., and Miura, T. 1988. Destruction and detoxification of aflatoxins with ozone. J. Food Sci. 53:667-668.

Matossian, M. K. 1982. Ergot and the Salem witchcraft affair. Am. Sci. 70:355-357.

Miller, E. C., and Miller, J. A. 1986. Carcinogens and mutagens that occur in foods. Cancer 58:1795-1803.

Moorman, M. 1990. Mycotoxins and food safety. Dairy Food Environ. Sanit. 10:207-210.

Moss, M. O., and Frank, M. 1987. Prevention: Effects of biocides and other agents on mycotoxin production. Pages 231-251 in: Natural Toxicants in Food: Progress and Prospects. D. H. Watson. VCH Publishers, New York.

National Peanut Council. 1988. U.S. peanut quality: Consensus Report of the peanut quality task force. The Council, Alexandria, VA. 48 pp.

Neal, G. E. 1987. Influences of metabolism: Aflatoxin metabolism and its possible relationships with disease. Pages 125-168 in: Natural Toxicants in Food: Progress and Prospects. D. H. Watson, ed. VCH Publishers, New York.

Ngindu, P., Kenya, P. R., Ocheng, D. M., Omondi, T. N., Ngare, W., Gatei, D., Johnson, B. K., Ngira, J. A., Nandwa, H., Jansen, A. J., Kaviti, J. N., and Siongok, T. A. 1982. Outbreak of acute hepatitis caused by aflatoxin poisoning in Kenya. Lancet I:1346-1348.

Okuda, K., and Mackay, J. 1982. Epidemiology of hepatocellular carcinoma. Pages 1-204 in: UICC Report 17. Union Internationale Contre le Cancer, Geneva, Switzerland.

Rodricks, J. V., Hesseltine, C. W., and Mehlman, M. A., eds. 1977. Mycotoxins in Human and Animal Health. Pathox Publishers, Inc., Park Forest South, IL.

Samarajeewa, U., Sen, A. C., Cohen, M. D., and Wei, C. I. 1990. Detoxification of aflatoxins in foods and feeds by physical and chemical methods. J. Food Prot. 53:489-501.

Schlatter, C. 1988. The importance of mycotoxins in foods. Bibl. Nutr. Dieta. 40:55-65.

Shank, R. C. 1981. Mycotoxins. Pages 107-140 in: Mycotoxins and N-nitroso compounds: Environmental risks. Vol. I. R. C. Shank, ed. CRC Press, Boca Raton, FL.

Siller, W. G., and Ostler, D. C. 1961. The histopathology of an enterohepatic syndrome of turkey poults. Vet. Rec. 73:134-138.

Smith, J. E., and Moss, M. O. 1985. Mycotoxins: Formation, analysis and significance. John Wiley, New York.

Stoltz, D. R. 1982. Carcinogenic and mutagenic mycotoxins. Pages 129-136 in: Carcinogens and Mutagens in the Environment, Vol. 3. H. F. Stich, ed. CRC Press, Boca Raton, FL.

Tong, C.-H., and Draughon, F. A. 1985. Inhibition by antimicrobial food additives of ochratoxin A production by *Aspergillus sulphreus* and Penicillium viridicatum. Appl. Environ. Micro. 49:1407-1411.

Ueno, Y. 1983. Trichothecenes: Chemical, Biological and Toxicological Aspects. Elsevier, New York.

Ueno, Y. 1987. Mycotoxins. Pages 139-204 in: Toxicological Aspects of Food. K. Miller, ed. Elsevier, New York.

Wattenberg, L. W. 1985. Chemoprevention of cancer. Cancer Res. 45:1-8.

Williams G. M., and Weisburger, J. H. 1986. Chemical carcinogens. Pages 99-173 in:

Casarett and Doull's Toxicology, 3rd ed. C. D. Klaassen, M. O. Amdur, and J. Doull, eds. Macmillan, New York.

Wogan, G. N. 1965. Chemical nature and biological effects of the aflatoxins. Bacteriol. Rev. 30:460.

Young, J. C., and Fulcher, R. G. 1984. Mycotoxins in grains: Causes, consequences, and cures. Cereal Foods World. 29:725-728.

Parasites, Viruses, and Toxins

CHAPTER 8

Although bacteria and molds are among the most frequent cause of foodborne disease, they certainly are not the only ones. Parasites, viruses, algae, and their toxins all create potential hazards. Around the world, 30 million people are infected with parasites, whereas in the United States 10 outbreaks and 60 cases are reported annually (Bean and Griffin, 1990). However, the increased popularity of raw fish presented in many ethnic ways and of undercooked meat and fowl may escalate the incidence of these diseases in the United States. In this chapter, common and emerging problems due to nonbacterial and nonfungal microorganisms and toxins are discussed.

PARASITES

Trichinella spiralis

T. spiralis is a parasite found worldwide in pigs, rats, and 40 other species of wild animals. Of game animals, bear causes the greatest problem in the continental United States, walrus in Alaska, and insufficiently cooked wild boar in Hawaii. The organism is especially prevalent where the animals feed on garbage.

Incidence. Although wild game is more likely to be contaminated, most cases of trichinosis in the United States result from the ingestion of raw or insufficiently cooked pork. In the past, the incidence of trichina-infested hogs was very much greater, as hogs were commonly fed garbage and slop. Currently, most hogs are on scientifically formulated, clean rations, a practice that has significantly reduced the incidence of trichina infestation. Data in Table 8.1 show that, of pigs testing positive for trichina, the greater percentage is for the garbage-fed hogs.

Even among garbage-fed hogs, the number infected is quite small. However, the possibility of human infection from a single hog multiplies because a hog can be cut into many different retail cuts. These various cuts may end up in 50 or more households.

The life cycle of the parasite is simple, as illustrated in the diagram in Figure 8.1. Encysted larvae contained in striated muscle are released in the host's stomach during digestion of infected meat. The larvae mature in the stomach, embed in the intestinal mucosa, and copulate. The female then deposits larvae in the lymphatics, enabling the parasite to reach all parts of the body.

The incidence of trichinosis has decreased markedly from 300–400 cases reported in the 1940s to an average of approximately 50 cases in the United States annually (Bean and Griffin, 1990) (Table 8.2). It is, however, a particular problem in some ethnic groups that eat raw pork (CDC, 1986a.) Although the numbers are small, trichinosis is still a public health concern, especially because proper handling will completely control it.

Control. The principal ways of killing trichinellae in muscle are

Table 8.1. Incidence of Trichina in U.S. Hogs

Hog Type	Number Tested	Number Positive	Percent Positive
Farm-raised breeder	1,858	0	0.00
Ration-fed	20,003	25	0.12
Garbage-fed	590	3	0.51

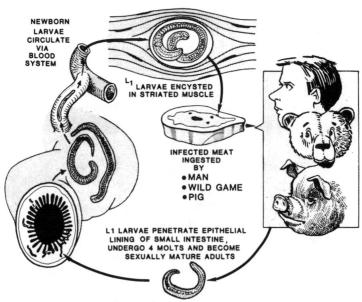

Figure 8.1. Life cycle of *Trichinella spiralis*. (Reprinted, with permission, from Murrel, 1985; modified by the author).

Table 8.2. Outbreaks and Cases of Illness Caused by Parasites and Fish Toxins in 1973–1987[a]

	Outbreaks		Cases	
	No.	Percent[b]	No.	Percent[b]
Giardia	5	<1	131	<1
Trichinella spiralis	128	5	834	1
Other parasites	7	<1	30	<1
Total parasites	140	5	1,004	1
Ciguatoxin	234	8	1,052	1
Paralytic shellfish poisoning	21	1	160	<1
Scombroid poisoning	202	7	1,216	1
Total fish toxins	457	16	3,428	2

[a] Data from Bean and Griffin (1990).
[b] In comparison with all outbreaks of known etiology during this period.

heating, freezing, and curing. During cooking, all parts of the pork must be raised to at least 137°F, although the recommended cooking temperature for pork is 170°F. Microwave cooking of pork can be a problem because of the uneven penetration of heat in this cooking system. Pork cooked in a microwave must be turned and rotated adequately so that all parts of the muscle reach the necessary temperature (Murrell, 1985).

Freezing also kills the organism. The time required to kill it varies with the temperature. For instance, pork frozen at −10°F for 20 days will be trichina-free. Colder temperatures require less time. Some countries in Europe guarantee trichina-free pork by dipping the pork in liquid nitrogen for a few seconds to kill the organism.

Symptoms. In the first week after the infection, transient gastrointestinal symptoms occur. In the second week, the larvae begin to penetrate the muscle. Fever, edema around the eyes, and muscle pain are cardinal features of the disease. Rash and respiratory symptoms may also be present. In severe cases, secondary pneumonia, congestive failure, toxemia, and encephalitis may occur. Death can result from trichina infections.

Toxoplasma gondii

T. gondii can be acquired by ingesting infected raw or very rare meat or from contact with infected cat feces. Most infections are subclinical and resemble the flu. The host's immune system slows growth of the organism and eventually forces it to change to the cyst form. The cysts may remain in the muscle of the individual, presumably for a lifetime.

Symptoms. For most people, this flu of short duration is no more than a minor annoyance. However, certain segments of the population are at great risk when they contract this parasitic infection. These include pregnant women, young children, the elderly, and immuno-suppressed persons. For example, in the United States alone, it is estimated that more than 3,000 babies are born each year infected with *T. gondii* because their mothers acquired an initial infection during pregnancy. Most infected infants are asymptomatic and remain normal for many years, although it has been suggested that many will develop problems at some time. Hospitalization, special education, and institutionalization may be required for some congenitally infected children. In the elderly and persons treated with immunosuppressive drugs, the immune system can no longer maintain the organism in the benign encysted stage in the muscle and the person develops acute toxoplasmosis, which can lead to blindness and even death.

Control. Cooking kills the organism, so it is the best method of

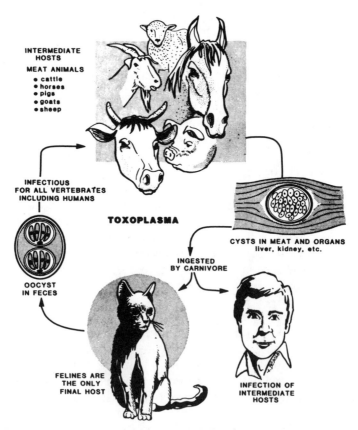

Figure 8.2. Life cycle of *Toxoplasma gondii*. (Reprinted, with permission, from Fayer and Dubey, 1985).

protection. Unlike trichina, toxoplasma cannot be killed by freezing in all instances (Fayer and Dubey, 1985).

Other Parasites

Tapeworms (cestodaria) and flukes (*Opisthorchis viverrini*) are also foodborne parasites. They come from raw or improperly cooked beef, pork, lamb, or fish. For most parasites, the course of the disease depends both on the parasite and on the host's resistance. Species of parasites vary in where they go and how they get there. They may invade the liver, lung, intestine, and other internal tissues (Ahmed, 1991). Beef tapeworm (*Taenia saginata*) is common in the Middle East, South America, Russia, Kenya, and Ethiopia, countries where the eating of raw or undercooked beef is popular. Well-nourished children usually have no symptoms except for evidence of infestation in the stool. These worms tend to have more dramatic effects on adults. They cause increased appetite and hunger pains, and malnutrition may be a side effect. Pig tapeworm (*T. solium*) may not cause any symptoms, or the symptoms may vary dramatically depending on the site where the parasite is feeding. For instance, it may cause muscle pain, convulsions, headache, and vomiting (Jelliffe and Jelliffe, 1982).

Roundworm infection (*Ascaris lumbricoides*) is found in children in tropical and subtropical areas. The worm is transmitted under conditions of poor sanitation. Vegetables from contaminated soil may contain the worms, and sewage-contaminated soil and water may also be vehicles. Children with these infections may show respiratory signs or an enlarged liver; malnutrition may be the presenting symptom. Sometimes the worm becomes so large as to cause an obstruction (Jelliffe and Jelliffe, 1982).

Ingestion of raw fish causes a large number of parasitic nematode infections, particularly in countries where raw fish is a delicacy. Fish tapeworm (*Diphyllobothrium latum* and *D. pacificum*) is common in areas such as Scandinavia and around the Baltic Sea, Alaska, and Canada. It can be prevented either by cooking the fish or by freezing it for 48 hours at −10°C. Another nematode disease common from undercooked fish is anisakiasis. This is a problem in Japan and Denmark, and a few cases have been reported in the United States.

Cod, Pacific salmon, and Pacific rockfish, some of the most popular fish in the United States, are among the most heavily infested with parasites (*Pseudoterranova dicipiens* and *Contraceacum*) (Ahmed, 1991; Roderick & Cheng, 1989). In some countries, salmon and other fish must be frozen before being used in sushi or other raw preparations. In this country, the fishing industry employs a monitoring technique called candling, in which fish are inspected visually for larval contamination. This technique works fairly well in fish with light flesh but not in those with dark flesh. Another strategy effective in avoiding parasites in sushi and other raw fish preparations is to use fish raised under controlled conditions (aquacultured) so that they are never contaminated by parasites (Burros, 1987).

VIRAL INFECTIONS FROM FOOD

Certain viruses have long been known to be transmitted by food. Media coverage of viruses such as those associated with acquired immunodeficiency syndrome (AIDS) has increased fears about viruses transmitted by food. According to the Centers for Disease Control, the AIDS virus is not transmitted through food (CDC, 1985). Despite this fact, some of the population is still fearful of such transmission.

An increase in raw fish consumption has increased concern about viruses in food. Unfortunately, viral survival in marine environments is enhanced by low temperatures, so virus outbreaks are more frequently associated with fish from colder northern waters (Liston, 1990). Furthermore, viruses seem to be more resistant to processes used to purify water such as relaying or depuration. The New York State Department of Health (1986) stated that viruses were the most-reported foodborne disease in that state. Foodborne outbreaks from viruses appear to be on the rise. The percentage of incidence for 1973–1987 is given in Table 8.3.

What Is a Virus?

Viruses are intracellular parasites and are not complete cells. They contain either RNA or DNA, but not both, and are surrounded by a lipid-containing envelope. Viruses can enter the food supply in several ways, such as through infected food handlers or through sewage. Over 100 known enteric viruses are excreted and ultimately find their way into domestic sewage. Sewage contamination of water and food can cause a wide variety of illnesses. Several main types of viruses are

Table 8.3. Outbreaks and Cases of Viral Illness During 1973–1987[a]

	Outbreaks		Cases	
	No.	Percent[b]	No.	Percent[b]
Hepatitis A	110	4	3,133	3
Norwalk virus	15	1	6,474	5
Other	10	<1	1,023	1
Total	135	5	10,630	9

[a] Data from Bean and Griffin (1990).
[b] In comparison with all outbreaks of known etiology during this period.

known to be foodborne—hepatitis A and non-A, non-B hepatitis, Norwalk-like agents including Snow Mountain type, and rotavirus. Certain other food-transmitted viruses such as astoviruses and caliciviruses are thought to be responsible for traveler's diarrhea and other gastroenteritis in children and adults (Blacklow and Cuker, 1981; Gerba, 1988). In the period from 1973 through 1987, there were 135 outbreaks and over 10,000 cases attributed to viral agents (Bean and Griffin, 1990).

Viral Stability in Food and Food Processing

Viruses are generally afforded greater stability in food than in water. For example, in water at near neutral pH, viruses are destroyed by temperatures above 65°C, but they can survive pasteurization temperatures in food. As a result, shellfish steamed "just to open" may be heated inadequately to kill many viruses, especially those that are heat resistant. The rotavirus, for instance, tolerates greater heat treatment than even heat-resistant forms of *Salmonella*.

While food appears to protect certain viruses, it is also true that some innate properties of food and certain processing techniques affect them. Acid foods and certain food additives such as sulfite and ascorbate accelerate virus inactivation, while fat slows it. Freeze-drying and ionizing or ultraviolet radiation can fully inactivate viruses. In most foods, partial inactivation occurs in spray-drying and freezing. In oysters inoculated with a virus, freezing has been shown to completely kill the virus. However, freezing doesn't kill all viruses and some viruses are actually transmitted by ice.

As with bacteria, it is very important to avoid contamination of food through good hand-washing practices and by taking precautions with ill workers. Since thermal processing is effective against viruses, it is important to cook food that may be contaminated to an internal temperature of 100°C. The process of depuration, in which shellfish are placed in sterile water for 48–72 hours to allow them to cleanse themselves, is effective against bacteria but not viruses (Anonymous, 1986). Thus, bivalves that have had their bacterial load reduced may still carry viruses such as hepatitis.

It is important to be aware that some viral outbreaks have involved items such as salads and frostings that are not heated before use. Furthermore, some typical water treatments such as chlorination may be inadequate to kill viruses in water heavily contaminated with sewage sludge. The actual concentration of virus necessary to cause infection is not known.

Specific foodborne viruses, including Hepatitis A, Norwalk virus, and rotavirus, are discussed in the following sections.

Hepatitis

Incidence. The hepatitis virus is primarily transmitted through fecal or oral contamination of food. It can also be introduced through sewage-polluted water. Shellfish is the most commonly implicated food. In just a 2-year period, 6,135 cases of hepatitis occurred in the United States from raw shellfish. Oysters and other mollusks are a potential source of the contamination as they not only collect, but actually concentrate, viruses that may be in the water. Since the incubation time is very long (15–50 days), some researchers think that there may be a bias toward incriminating oysters and other shellfish because the eating of shellfish may be remembered after a 2- to 7-week period more distinctly than the eating of many other foods (Khalifa et al, 1986).

In addition to shellfish, vegetables may be contaminated if fields are irrigated with polluted water or fertilized with human excrement. Milk can also be a source of the virus. Foods that require much handling and are not subsequently cooked such as sandwiches, salads, and pastries have also been implicated. Poor personal hygiene is a contributing factor in a large percentage of hepatitis outbreaks (Bean and Griffin, 1990).

The disease may be spread by persons with the infection as early as 7 days before the onset of symptoms; others may transmit the virus without ever becoming perceptibly ill. Flies and cockroaches may also be vectors.

Symptoms. The name *hepatitis* denotes inflammation of the liver, and the liver does become enlarged and tender. Fever is usually present. The initial phase can be serious, with severe vomiting; hospitalization is often necessary. Symptoms usually subside after 2–4 weeks.

Norwalk-Type Viruses

Incidence. The Norwalk agent has been responsible for over 150 acute outbreaks of nonbacterial gastroenteritis in the United States since 1982. Several of these were quite large. Over 2,000 cases were reported in Minnesota when infected persons continued as salad makers and thereby contaminated potato salad, coleslaw, and tossed salad for nine different banquets (Anonymous, 1983; Fleissner et al, 1989). Also, about 4,000 people were infected by this agent in frostings from a bakery. The baker was thought to have shed the virus while making the frosting (Anonymous, 1982).

In the Northeast, several large outbreaks could be tied to the consumption of raw oysters and clams or steamed clams. Steaming just until the clams open is inadequate to kill the virus (Truman et al, 1987).

Two Caribbean cruise ships also were responsible for large outbreaks of Norwalk gastroenteritis. In these cases, the ships were found to have the dismal sanitation score of 16 out of 100 (Kaplan et al, 1982).

Symptoms. Onset of symptoms is 24–48 hours after ingestion of the infected food. Vomiting and diarrhea are characteristic symptoms. Fever, if present, is low grade. The illness appears to be mild and self-limiting. Not everyone is equally susceptible or develops the same degree of immunity. Some people seem to develop neither illness nor antibodies after exposure to the virus, whereas others do develop immunity. However, the immunity seems to be very short-lived, lasting less than a year (Anonymous, 1986).

Human Rotavirus

Studies in developed countries worldwide have found rotaviruses to be the most common enteropathogens identified in young children hospitalized with diarrhea and dehydration (Le Baron et al, 1990). Immunity to the virus can be transmitted by antibodies in breast milk (Bartlett et al, 1987). Cases of severe diarrhea in adults have also been linked to this virus (Gerba, 1988).

TOXINS FROM THE SEA

Illness can result from the ingestion of the flesh organs and eggs of over 500 species of salt water fish. Several of the more common toxins are discussed below and the incidence of disease is shown in Table 8.2.

Ciguatera Toxin

This marine toxin is considered the most injurious from a public health and economic standpoint. First encountered and named by the Spanish explorers in the Caribbean, ciguatera poisoning is most

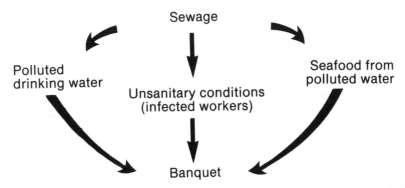

Figure 8.3. Method by which the Norwalk type of virus can infect many people in one outbreak of disease.

frequently associated with tropical fish that inhabit reefs or shallow water (200 feet or less). The toxin is derived from a single-celled marine plant, *Gamberdiscus toxicus*. Herbivorous fish consume this dinoflagellate while foraging through macroalgae (Anonymous, 1980).

Incidence. Humans ingest the toxin by consuming either herbivorous fish that have eaten contaminated plants or carnivorous fish that have eaten contaminated herbivorous fish. The larger the fish, the greater the chance it has picked up the toxin. Likely species include snapper, grouper, sea bass, morays, surgeonfish, jacks, and barracudas. A 3-year study in the Miami area showed that snapper and grouper were the most common offenders. Barracuda has been associated with ciguatera poisoning so often that its sale is prohibited in Miami (Liston, 1989).

Symptoms. Ingestion of this neurotoxin can cause acute sickness. Death can result from paralysis, although the mortality rate is low. Symptoms may develop immediately after a single bite or up to 30 hours later. The first symptom is a tingling of the lips, tongue, and throat; numbness follows. Sensory reversal of hot-to-cold is a symptom specific to the syndrome. Other symptoms include nausea, cramps, diarrhea, visual disturbances, headache, muscle pains, and progressive weakness. Severely affected patients may take many months to fully recover.

Control. It is difficult to know whether a fish is safe to eat because a species and a fishing spot can change whimsically from safe to toxic. There is some evidence that the toxin is more likely to be acquired when the reef ecology has been disturbed by either human intervention or natural events such as hurricanes. More outbreaks occur during the late spring and summer. In the United States, ciguatera poisoning occurs most frequently in Puerto Rico, the Virgin Islands, Hawaii, and Florida.

No easily perceptible changes in the flavor or other characteristics of the fish are consistently noted. However, one report stated that there was a subtle hint of metallic taste in contaminated fish.

The existence of this toxin has been a drawback to the expansion of commercial fishing in tropical areas where fish can and should make a significant contribution to the supply of high-quality available protein. However, the recent development in Hawaii of a test that can quickly, easily, and cheaply determine the toxin's presence in the flesh of fish should minimize the incidence of this poisoning and increase the expansion of fishing in tropical areas (Calif. Dept. of Health, 1986; CDC, 1986b; Jacobson, 1986).

One of the peculiar aspects of ciguatera poisoning is that a marked variability in susceptibility seems to exist. If 10 people eat the same fish, one or two will get severely ill, three or four will be mildly ill, and the rest will not be ill at all. Illness variability does not seem to be related to the amount of toxin ingested (Anonymous, 1986).

Scombroid Poisoning

Scombroid means mackerel-like, and this poisoning is associated with mackerel, sardines, tuna, and bonito. It has also been found in nonscombroid species including yellowtail (amberjack) and dolphin (mahimahi). This poisoning results from fish flesh with incredibly high levels of histamine (Baranowski et al, 1990). Normally, the histamine level of fish is less than 1 milligram percent. In a recent outbreak involving yellowtail, the histamine levels were as high as 430 milligram percent.

Symptoms. These occur 10–90 minutes after the fish is eaten. Classic symptoms are similar to an allergic reaction and include a facial flush, body rash, severe headache, shortness of breath, dizziness, throbbing, thirst, vomiting, and diarrhea. In some cases, shock results, as can occur in any situation where excessive histamine is produced (CDC, 1986c).

Control. Histamine, saurine, and cadaverine are produced by microbial breakdown of the amino acid histidine when freshly caught fish are not kept properly cool. The toxin can be produced in just 3–4 hours with the fish at room temperature. An outbreak involving mahimahi occurred as a result of a quantity of fish that was allowed to undergo alternate freezing and thawing (Anonymous, 1989).

Since this toxin is heat stable, high levels of histamine can be found in fresh, salted, or canned fish. With this type of poisoning, the fish sometimes has a detectable, sharp, pepperlike taste (Settipane, 1986).

Paralytic and Neurotoxic Shellfish Poisoning

Toxic dinoflagellates of the genus *Gonyaulax* cause paralytic shellfish poisoning. When the organism is present in the beds of bivalves, mollusks ingest them and collect the potent, heat-stable neurotoxin. This neurotoxin belongs to the family of toxins known as saxitoxins. They block sodium entry into the nerve cell and can therefore be quite toxic.

Paralytic shellfish poisoning is a sporadic, albeit persistent, problem in many northern coastal waters. States where it is most likely to occur are Alaska, Washington, Oregon, California, Maine, and Massachusetts.

Symptoms. Symptoms occur 30 minutes after infected shellfish are eaten. Lack of coordination, tingling, drowsiness, incoherent speech, numbness, rash, dry throat and skin, and paralysis are some of the neurological symptoms. Fatality ranges from 1 to 10%.

Control. This type of poisoning can be avoided if fish are not gathered from closed areas. The FDA's National Shellfish Sanitation Program surveys coastal waters, closes off certain areas to fishing, and prohibits the sale and distribution of fish from these areas. Shellfish are monitored, and a bed is closed when the toxin level is greater than 80 micrograms of toxin per 100 grams of shellfish meat (Jonas-Davies et al, 1984).

Neurotoxic shellfish poisoning is also caused by a dinoflagellate. This organism is a problem around the Gulf of Mexico and off the coast of Florida. The condition is also known as red tide, red water, and brown water (Sharma and Taylor, 1987). The red tides are easily seen and are associated with fish deaths in the area. In this case, mollusks that feed on the toxic algae are the vehicle. Symptoms are similar to those seen in paralytic shellfish poisoning except that this poisoning is seldom fatal.

Diarrhetic shellfish poisoning has been seen in Europe, Japan, and Latin America. It has not yet been documented in the United States. In this illness, the dinoflagellates produce toxins that cause diarrhea and other gastrointestinal complaints (Taylor, 1988).

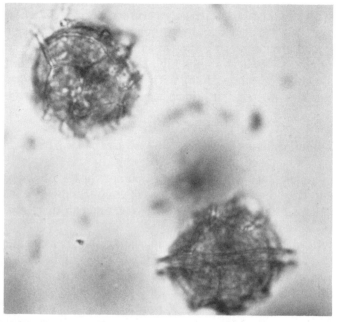

Figure 8.4. Dinoflagellates. This type of microscopic algae can produce toxin that causes paralytic shellfish poisoning. (Courtesy FDA).

Avoiding Seafood Hazards

These foodborne hazards have been documented to occur and are unfortunately occurring with greater frequency (Nightingale, 1987; Snyder and Poland, 1990). Many of them can be prevented if the food is adequately handled and cooked or properly stored, as in the case of scombroid poisoning.

Nearly 90% of all reported foodborne illnesses associated with seafoods are attributable to mollusks and to a very few species of fish. Eighty-one percent of seafood-related illnesses are from nine states and territories: Hawaii, Puerto Rico, Virgin Islands, Guam, New York, California, Washington, Connecticut, and Florida, with 50% coming from the first four areas listed. Hawaii alone accounts for 35%, with major problems being ciguatera and scombroid poisoning (Garrett, 1988).

Seafood hazards also can be partially avoided by surveillance programs that quarantine contaminated waters. Measures to prevent the growth of various toxic algaes are also being tested.

COPING STRATEGIES

The same food safety dilemmas and risk-benefit decisions are posed with parasites, viruses, and fish toxins as were discussed in the chapter on bacterial contamination of food. In the home and in the processing plant, it is crucial to adhere to high standards of sanitation, thus avoiding contamination and cross-contamination of food. Like vegetative bacteria, parasites and viruses are killed by adequate heat treatment. Problems with all these arise when food is eaten raw or does not receive adequate heat treatment.

Toxins in food require a different approach. Extreme care must be taken to ensure that the food is not contaminated in the first place. For instance, fish must not be drawn from contaminated areas. To prevent scombroid, fish must be immediately cooled and not allowed to sit in the hot sun.

With this as with every other issue, we must weigh the risks and the benefits. We must weigh the risk of eating raw meat or fish versus the benefit of a pleasant, somewhat exotic, gastronomic experience. We must weigh the risk of eating a fish that might contain a toxin against the many health benefits attributed to fish. We must weigh the risks of using pesticides to control algaes that cause paralytic shellfish poisoning and the like against the benefit of toxin-free fish. We must weigh the risk of nutrient loss that could occur in the freezing of pork versus the benefit of trichina-free pork. In like manner, we must weigh the risk of additional nutrient loss or decreased nutrient availability that is possibly associated with further cooking. Most important, we must evaluate the risks posed by our own food handling techniques.

REFERENCES

Ahmed, F. E. 1991. Seafood Safety. Committee on Evaluation of the Safety of Fishery Products. National Academy Press, Washington DC.

Anonymous 1980. Poison from the sea. Emer. Med. Dec. 15, 1980. pp. 83-84.

Anonymous. 1982. Norwalk gastroenteristis: A large community outbreak associated with bakery products. Dis. Control Newsl. (Minn. Dept. Health) 9(10):1-2.

Anonymous. 1983. Norwalk gastroenteristis. Dis. Control Newsl. (Minn. Dept. Health) 10(2):6-8.

Anonymous 1986. The truth about raw fish. Reprint from Emergency Med. June 15, 1986.

Anonymous 1989. Scombroid fish poisoning—Illinois, South Carolina. Dairy, Food, Environ. Sanit. 8:520-521.

Baranowski, J. D., Hilmer, A. F., Brust, P. A., Chongsiriwatana, M., and Premaratane, R. J. 1990. Decomposition and histamine content in mahimahi (*Coryphaena hippurus*). J. Food Prot. 53:217-222.

Bartlett, A. V., Bednarz-Prashad, J., DuPont, H. L., Pickering, L. K. 1987. Rotavirus gastroenteritis. Annu. Rev. Med. 38:399-415.

Bean, N. H., and Griffin, P. M. 1990. Foodborne disease outbreaks in the United States, 1973-1987: Pathogens, vehicles, and trends. J. Food Prot. 53:804-817.

Blacklow, N. R., and Cuker, G. 1981. Viral gastroenteritis. New Engl. J. Med. 304:397-406.

Burros. M. 1987. Number of parasites in raw fish reportedly increasing. Mpls. Star Tribune, Oct. 21. p. 3T.

Calif. Dept. of Health. 1986. Ciguatera Fish Poisoning. Calif. Morbidity 18(May 9):1.

Centers for Disease Control (CDC). 1985. Summary: Recommendations for preventing transmission of infection with human T-lymphotrophic virus type III. MMWR 24:681.

Centers for Disease Control (CDC). 1986a. Trichinosis—Maine, Alaska. MMWR 35:33-35.

Centers for Disease Control (CDC). 1986b. Ciguatera fish poisoning. MMWR 35:263-264.

Centers for Disease Control (CDC). 1986c. Restaurant-associated scombroid fish poisoning. MMWR 35:264-265.

Doyle, M. P. 1985. Food-borne pathogens of recent concern. Annu. Rev. Nutr. 5:25-41.

Fayer, R., and Dubey, J. P. 1985. Methods for controlling transmission of protozoan parasites from meat to man. Food Technol. (Chicago) 39(3):57-60.

Fleissner, M. L., Herrmann, J. E., Booth, J. W., Blacklow, N. R., and Nowak, N. A. 1989. Role of Norwalk virus in two foodborne outbreaks of gastroenteritis: Definitive virus association. Am. J. Epidemiol. 129:165.

Garrett, E. S. 1988. Microbial standards, guidelines, and specifications and inspection of seafood products. Food Technol. (Chicago) 42(3):90-93 ff.

Gerba, C. P. 1988. Viral disease transmission of seafoods. Food Technol. (Chicago) 42(3):99-103.

Jacobson, A. S. 1986. Ciguatoxin poisoning. Medical Tribune, June 4.

Jeliffe, E. F. P., and Jeliffe, D. B. 1982. Adverse Reactions to Food. Plenum Press, New York.

Jonas-Davies, J., Sullivan, J. J., Kentala, L. L., Liston, J., Iwaoka, W. T., and Wu, L. 1984. Semiautomated method for the analysis of PSP toxins in shellfish. J. Food Sci.

49:1506-1508.

Kaplan, et al. 1982. Epidemiology of Norwalk gastroenteritis. Annu. Intern. Med. 96:756-761.

Khalifa, K. I., Werner, B., and Timperi, R. 1986. Non-detection of enteroviruses in shellfish. J. Food Prot. 49:971-973.

Le Baron, C. W., Lew, J., Glass, R. I., Weber, J. M., and Ruiz-Palacios, G. M. 1990. Annual rotavirus epidemic patterns in North America. JAMA 264:983-988.

Liston, J. 1989. Current issues in food safety—Especially seafoods. J. Am. Diet. Assoc. 89:911-913.

Liston, J. 1990. Microbial hazards of seafood consumption. Food Technol. (Chicago) 44(12):56-62.

Murrell, K. D. 1985. Strategies for the control of human trichinosis transmitted by pork. Food Technol. (Chicago) 39(3):65-72.

N.Y. State Dept. of Health. 1986. How heat resistant are foodborne viruses? Food Prot. Bull. 2:1-2.

Nightingale, S. L. 1987. Foodborne disease: An increasing problem. Am. Family Phys. 35:353-354.

Roderick, G. E., and Cheng, T. C. 1989. Parasites: Occurrence and significance in marine animals. Food Technol. (Chicago) 43(11):98-102.

Settipane, G. A. 1986. The restaurant syndromes. Arch. Intern. Med. 146:2129-2130.

Sharma, R. P., and Taylor, M. J. 1987. Animal toxins. Pages 456-462 in: Handbook of Toxicology. T. J. Haley and W. O. Berndt, eds. Hemisphere, New York.

Snyder, O. P., and Poland, D. M. 1991. America's "safe" food. Dairy, Food Environ. Sanit. 11:14-20.

Taylor S. 1988. Microbial toxins of marine origin. Food Technol. (Chicago) 42(3):94-98.

Truman, B. I., Madore, H. P., Menegus, M. A., Nitzkin, J. L., and Dolin, R. 1987. Snow mountain agent gastroenteritis from clams. Am. J. Epidemiol. 126:516-525.

How Food Processing Affects Nutritional Quality and Food Safety

"Processed foods." What images do consumers see when they think of these words? New, novel, and nutrient-dead? Technology toying with nature in the name of profit? Food technologists who strip the vitamins and fiber and take them home in their pockets, replacing them with chemicals and worthless fillers? These fragmentary images are like unicorns in that facets of them are true but the image as a whole is formed by mythology. This chapter tries to dispel the myths about food processing and deal with its effects on the safety and nutrient quality of food.

HISTORY OF FOOD PROCESSING

One myth is that food processing and processed foods are new. Although some foods and processes are new, the largest part of food processing is part of ancient history. From the beginning of time, populations have been trying to invent ways to preserve food after harvest and slaughter and to make food more palatable and readily used. Both the use of fire (heat) and of caves and root cellars (cold) to preserve food were probably accidental discoveries of early humans, but both fulfilled the ultimate objective of making the readily available food supply last longer.

Fire not only changed the taste of food but also killed most of the microorganisms, thereby improving the safety of the food and extending its storage life. Caves and root cellars were used both to keep items cool in the summer and to prevent freezing in the winter.

Drying, fermentation, and separation of natural products are all ancient practices. Evidence from Egyptian tombs of the First Dynasty reveals that food was cooked, salted, and dried; bread was baked; and beer was fermented. Fermented products like beer and wine were crucial

171

to many ancient cultures as they provided a way to store the energy of food products in a palatable form without refrigeration. The Bible mentions the use of leavening and fermentation as well as the separating of chaff from wheat.

The ancient Chinese went beyond these traditional preservation methods to techniques that enabled them to alter nature in their flavor. They discovered that by placing fruit near a burning kerosene lamp, they could artificially cause the fruit to ripen. The Chinese did not know the chemistry involved, but they knew that the process worked. Today we understand that the ethylene produced during the burning of kerosene is the same chemical as that released by plants to cause natural ripening (Whelan and Stare, 1976).

Drying and salting were techniques widely used by nearly every culture. Raisins and other dried fruits rank as some of the oldest foods, along with pickled or salted meat and fish products. The use of salt to both flavor and preserve food made it so highly valued that it was used as a form of payment.

Other chemical additives have been used for centuries. Vinegar, nitrate, potash (lye), sugar, and spices were all used to flavor and preserve food. Many of these commodities were so important that they had more value than gold. Exploration of the globe and early trade routes were spurred by the desire not only for the exotic flavors offered by spices but also for their highly important preservative qualities.

The Beginning of Progress

Until the 1800s there were only modifications and improvements of existing preservation techniques—no new food processing methods were added. In 1810, Napoleon, driven by concerns about feeding an army far from its base, offered a prize of 12,000 francs for someone who could develop stable rations for his troops. The French chef Nicholas Appert was awarded the prize for his trial-and-error discovery of how to preserve meat, vegetables, and fruits. His method involved filling bottles with food, sealing them with corks, and heating them in boiling water for certain periods of time. Originally this was called appertization; today, it's called canning. In the more than 170 years of this industry's existence, many advances have been made to improve the quality and safety of the canning process (Goldblith, 1971, 1972).

Also in the 1800s, cooling advanced beyond the use of root cellars, caves, and cooling houses located over streams, with the development of a commercial ice industry. Large blocks of ice were cut from rivers and lakes and stored in underground ice houses. Mechanical refrigeration began during the U.S. Civil War and provided more reliable temperature control than ice boxes (Leveille and Ubersax, 1980a).

Freezing and freezers were the next advance on the scene. Meats were the first items to be commercially frozen and sold, due to both their extreme perishability and their high cost. Some fruits were sugared and frozen. Early attempts to freeze vegetables were unsuccessful, as the products looked and tasted dreadful. Successful freezing of vegetables could not occur until it was discovered that enzymes in the product needed to be inactivated to maintain product quality. In the 1930s, Clarence Birdseye spearheaded the use of blanching to inactivate enzymes and maintain product quality in frozen food (Leveille and Ubersax, 1980a).

Major changes also occurred in the milling industry. Instead of merely crushing the grain, various fractions of the grain were separated. Mills changed from using massive millstones to using steel rollers (Leveille and Ubersax, 1980b).

Where We Are Now

Improvements are constantly being made in the basic processes of canning, freezing, and drying. Furthermore, processors have separated and recombined constituents of food in unusual and creative ways. The desire for more healthy products has spawned products such as high-fiber baked goods and fat-free cakes and salad dressings. The desire for more convenient products has directed the market to produce many food products that are ready or nearly ready to eat or heat. This is a fairly recent phenomenon, launched by the explosion in the number of women working outside the home.

New processing methods are constantly being sought because currently much of the world's harvest is lost. Reduction of these losses translates directly into more food and lower prices for commodities. One method, food irradiation, was generated out of our increasing knowledge of nuclear physics. Another method, biotechnology, came from our increased understanding of molecular genetics. Despite its new name, biotechnology is a very old technique long practiced in the form of animal husbandry and plant breeding but currently using more efficient means of genetic manipulation. Even fermentation is really biotechnology. Irradiation and products of biotechnology created by gene splicing (Vettorazzi, 1988) are discussed or mentioned in subsequent chapters.

ADVANTAGES AND DISADVANTAGES OF FOOD PROCESSING

Advantages

Food processing's basic objective is to make food look, taste, and smell as though it had not been preserved but to allow it to be safely eaten at a later date. Other objectives may include the development of totally new food products, experiences, and levels of convenience, but they are subordinate to the basic objective of preservation. To reach the basic objective, attempts must be made stop microbial growth and chemical reactions that cause the destruction of flavor, color, texture, and nutritive value. Success in food processing frees populations from some of their dependence on geography, climate, and other environmental factors; reduces waste; and makes urbanization practical. Most of us live far from the sites where our food is grown or manufactured. For the food to arrive with optimum quality, it typically must be processed or treated with additives.

In addition, processing helps to convert agricultural products into forms more suitable or desirable for human consumption and to extract commodities that are not readily available in the food as harvested. Examples include the hydrogenation of oils into margarines for use as substitutes for butter, the separation of cereal bran, and the extraction of oil or meal from soybeans.

Processing may indeed make food safer, not only by removing, inhibiting, or killing pathogenic microorganisms, but also by inactivating toxic factors such as those discussed in previous chapters. It may help food retain its nutrients by destroying enzymes like thiaminase, which destroys thiamin (vitamin B-1) in food.

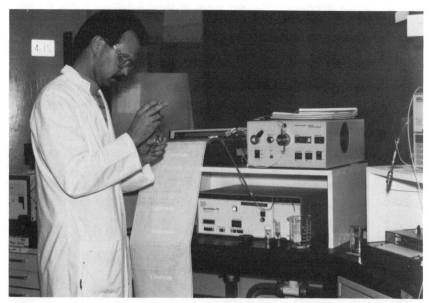

Figure 9.1. Testing for vitamins in a processed food. This liquid chromatograph can measure many food constituents. (Courtesy Hazleton Labs, Madison, WI)

Processing may actually enhance the nutrient value of the food supply. A missing nutrient may be added, such as iodine to salt. Fat may be skimmed from milk and other natural products to reduce the quantities of certain nutrients that contribute to chronic disease. Nutrients that may be lost during processing can be returned through enrichment or fortification.

Processing advances allow choice. We now can choose between regular and reduced-fat, reduced-cholesterol, and fat-free mayonnaise, salad dressings, and ice cream. It is great that nutrients thought to lead to chronic disease can be reduced, but this improvement may mean a tradeoff. Food that is modified with respect to calories, fat, salt, sugar, or other dietary attributes may need increased processing or additives.

Sensory properties of food may be enhanced through processing. On the surface this may seem frivolous, but indeed it is not. Food must taste good to be eaten. Nutrients in food with poor sensory qualities will never be utilized as the food won't be eaten.

Processing and convenience go hand in hand. Convenience built into food products is crucial to the way our society is structured today. With large numbers of women working outside the home, there is simply less time available for lengthy periods of food preparation and preservation. The availability of easy-to-prepare food items may add to the variety and availability of nutrients in the diet.

Fast food and processing are also necessary partners. Society's desire to have food fast means that a lot of food is preprepared, frozen, or otherwise preserved and stored for later use. This often necessitates the use of processing additives.

Processing permits variety. Our mobile society, which eats out and travels widely, has increased the demand for unusual and ethnic foods. Today's grocery store exemplifies this. It may carry canned jalapeno peppers, salsa, a variety of Oriental sauces, and over 120 different kinds of produce. The frozen food sections of many supermarkets provide a culinary way to traverse the globe, with items like Mexican burritos, Greek gyros and phyllo, Indian dal, Italian lasagna, Middle Eastern pita, and Vietnamese egg rolls.

And Some Drawbacks

As can be seen from the previous examples, food processing meets many needs and provides many positive benefits. It also has some drawbacks. First and foremost, it must be stated emphatically that all forms of processing cause nutrient loss. For a fair comparison, however, these losses must be weighed against losses that would occur during normal cooking or storage. In many cases, the processing losses are in place of those that would occur with normal cooking. At times unfair comparisons are reported—the processed product is pitted against its raw counterpart even if the product is rarely eaten raw. Such comparisons exaggerate losses that occur with processing. Accurate comparisons are especially important, as the major loss of vitamins and minerals in foods occurs during final preparation in the home or institution (Erdman, 1979; Karmas and Harris, 1988).

Another drawback of food processing is that some products provide little in the way of nutritional benefit to the consumer. Some are constituted from substitute ingredients of low nutritional value, for example, flavoring and water to replace real juice (Anonymous, 1982). Some products are high in constituents such as salt, fat, or cholesterol, which should be limited for optimum health. Processed foods may also be low in certain important components of the diet, such as fiber or fruits and vegetables.

Total reliance on processed foods may indeed have another outcome. It may foster heavy dependence on them because the skills of food preparation are not learned in a family where food is warmed, not cooked.

Processed foods also tend to use a great deal of packaging to keep the food at optimum quality, creating a possible environmental concern. A food that requires low energy for storage, such as a potato, may be processed into a frozen French fry, which requires high energy for storage. Thus, environmental concerns about solid waste, energy use, and the like add to the list of concerns about certain processed foods.

This chapter focuses on various processing methods, describing what they are, their safety, and their nutritional effects. In addition, it compares the nutritional value of unprocessed products and those having undergone various processing techniques.

PROCESSING METHODS

Stability of Nutrients in Various Processes

Understanding the nutritional effects of processing necessitates having a basic understanding of nutrient stability.

Vitamins. Vitamin C is the least stable of all the nutrients, so it has been studied the most. Other heat-labile vitamins, such as thiamin and B-6, have also been followed in some foods and some processes. Table 9.1 gives the relative stability of the vitamins, considering all possible causes of destruction. Table 9.2 shows the average losses that occur in normal cooking and factors that cause them. Vitamins that are highly water-soluble and affected by heat are the ones that usually show the greatest cooking and processing losses.

In general, cooking methods that use little time or little water (such as microwaving or stir-frying) result in the best nutrient retention in most products (Eheart and Gott, 1964, 1965; Gerster, 1989; Gordon and Noble, 1959; Gould and Golledge, 1989; Klein, 1989; Klein et al, 1981). However, nutrient retention in vegetables cooked in a microwave oven varies dramatically with the amount of water, cooking time, oven power, and other variables (Dietz and Erdman, 1989). Table 9.3 gives an example of this for vitamin C. Losses in vegetable products are greater if the product is trimmed excessively, if it undergoes extensive washing or soaking, if it is chopped into very small pieces, if it is cooked in too much water or for too long a time, or if it is held on

Table 9.1. Overall Vitamin Stability

Highly Unstable	Somewhat Unstable	Fairly Stable
Vitamin C	Folate	Niacin
Thiamin	Pyridoxine	Vitamin K
	Riboflavin	
	Vitamin B-12	
	Vitamin A	
	Vitamin E	
	Vitamin D	

Table 9.2. Vitamin Losses During Cooking[a]

Vitamin	Cooking Losses	Primary Factors
C	0–100%	Oxygen, heat, water, alkaline pH, metals
Thiamin	30–70% vegetables 0–80% meat 0–60% baking	Water, heat, alkaline pH
Riboflavin	10–30% plant 9–39% animal	Light, water, alkaline pH
Niacin	3–27%	Water (but good recovery if liquids are used)
Pantothenic acid	7–56%	Water, heat
B-6 (pyridoxine)	30–82%	Water, temperature > 100°C
Folate	0–50% vegetables 33–95% variety meats 46–95% fish or pork 29–70% egg yolk, chicken	Water, heat, loss of protection from ascorbate
B-12	0–20%	Heat, oxygen
Biotin	0–50%	Oxygen
A (as carotene)	0–60%	Light, heat, oxidation
D	0–40%	Light
E	0–60% oxidized fat	Oxygen, UV light

[a] Adapted from VNIS (1984).

a steam table. Reheating also causes nutrient loss. Table 9.4 shows that in all vegetables the amount of loss increases if they are reheated; however, the actual amount of loss is very dependent on the particular vegetable. In animal products, losses are greater if temperatures fluctuate during frozen storage, if the drippings from meat are discarded, and if the product is held on a steam table.

Minerals. Minerals are less subject to the effects of processing than vitamins (Lachance and Fisher, 1988). Minerals are lost primarily into the cooking liquid or the drip. Use of the drip or the cooking water in sauces, stews, or gravies means that the mineral nutrients are not lost.

Protein. Protein products can either increase or decrease their bioavailability during processing. Normal cooking does not reduce the biological availability of meat (Lachance and Fisher, 1988). However, if the protein undergoes browning reactions that tie up the essential amino acid lysine, it loses nutritional quality. Legume products that contain various protein enzyme inhibitors (Chapter 5) show increased protein availability after some heating. For example, navy beans with a biological value of 37 (measured as the percentage of absorbed protein nitrogen retained in the body) increased in biological value to 71 after boiling for 5 minutes and to 80 after 10 minutes, but they had begun to drop back to 77 after boiling for 60 minutes (Lachance and Fisher, 1988).

Table 9.3. Retention of Ascorbic Acid in Broccoli with Various Cooking Methods

Cooking Method	Percent Retention
Stir-fry	76
Microwave, little water	75
Boil, little water	74
Microwave, much water	56
Boil, much water	44

[a] Adapted from Armbruster (1978).

Table 9.4. Effect of Reheating on Percent Ascorbic Acid Retained in Vegetables[a]

Vegetable	Cooked	Refrigerated 1 Day	Reheated, After Refrigeration
Asparagus	86	82	66
Broccoli	88	68	60
Peas	88	52	43
Cabbage, shredded	73	44	33
Spinach	52	48	32
Green beans	83	41	29

[a] Adapted from Charles and Van Duyne (1958).

Fats. Fats oxidize or isomerize during processing and storage, which could decrease their nutritional value and could actually have a toxic effect. Under some circumstances, oxidized fats can also destroy certain vitamins and alter proteins, thus making other components in the food less nutritious.

Carbohydrates. The nutritional properties of carbohydrates are not greatly affected by processing, although some recent data (Jenkins et al, 1987; Wolever et al, 1986) show that processing may alter the rate at which glucose is absorbed and thus affect insulin and blood sugar. The particle size and organization of fibers are affected by processing and extrusion, and these changes appear to alter the physiological effects of certain fibers.

The effects of all types of processing on nutrient stability in fruits and vegetables are well covered by Salunke et al (1991).

Thermal Processing

Heat or thermal processing of food includes blanching, pasteurizing, and canning, as well as cooking. Blanching is often used as a treatment before some other type of food processing such as canning, freezing, or drying (Leveille and Uebersax, 1979).

Blanching. Blanching's main purpose is to inactivate plant enzymes that destroy desirable color, texture, flavor, and nutrients. It also expels gases that are in the tissue and kills some surface microorganisms. It can also shorten the final cooking time. Blanching before canning may make the packing operation easier (Lund, 1988).

Since blanching is done in water or steam, water-soluble constituents (including vitamins and minerals) may be lost. The particular blanching method used greatly affects the degree of nutrient loss (Fennema, 1988). Although reported nutrient retention values vary dramatically from study to study and may be contradictory, steam blanching usually results in the greatest nutrient retention. This is especially true if the products being blanched have a large surface area or surface-mass ratio and therefore have large losses due to leaching (Fennema, 1988). If the product is to be cooled before subsequently being processed, air cooling results in the greatest nutrient retention (Table 9.5) (Fennema, 1988; IFT, 1986; Poulsen, 1986).

Pasteurization. Pasteurization is a thermal process that inactivates some, but not all, of the microorganisms in the food. Since the food is not sterile, this process must be used with other preservation techniques such as chilling. It aims to kill those microorganisms that either hasten spoilage or are health hazards (Leveille and Uebersax, 1979).

The pasteurization of milk or egg products does little to the nutritional quality of these products (Lund, 1988). Some authors advocating the use of raw milk point out that during pasteurization 100% of the vitamin C is destroyed. While what they say is true, the statement is very misleading as milk is an incredibly poor source of this vitamin. Since the amount of vitamin C in milk is nutritionally inconsequential, the loss of 100% of it is insignificant. The small loss of vitamin B-12 that occurs with pasteurization would be of consequence only if milk were the only animal product in the diet and therefore the only source of this vitamin.

Canning. Canning is a sterilization process in which food in sealed containers is heated at temperatures sufficiently high to kill practically all microorganisms and their spores. For acid foods (pH < 4.6), like most fruits, which do not allow botulism spores to survive, sterilization can be achieved at boiling temperatures. For low-acid foods (pH > 4.6) such as vegetables, milk, meat, and grain products, temperatures above the boiling point are required to kill all the spores and prevent botulism. Temperatures above the boiling point are achieved in a pressure canner and, industrially, in a batch or continuous automatic retort.

Vitamin losses in the canning process are dependent on the length of the heating and cooling as well as on the temperature of the heat treatment. Obviously, the more drastic the heat process, the more severe the potential nutrient loss. Under some conditions, up to 75% of the vitamin B-6 may be lost in canning (Sauberlich, 1987). Nutrient losses also depend on the rate of heat transfer in the food itself. For liquids or foods that remain in distinct pieces while in liquid, heat transfer occurs by the relatively fast process of convection. Such foods require less processing time and hence retain more nutrients than foods such as pureed pumpkin or thick soups and entrees in which heat transfer occurs by the relatively slow process of conduction (Lachance and Fisher, 1988).

Any technique that aids heat transfer reduces processing time and improves vitamin retention (Lund, 1988). In commercial canning operations, cans rotate on a conveyor belt as they move through the retort. The movement speeds heat transfer to the center of the can,

Table 9.5. Vitamin Retention as Affected by Blanching and Cooling Method

Blanching Medium	Cooling Medium	Vitamin C in Peas[a]	Folate In Spinach[b]	Folate In Broccoli[b]
Steam	Air	90.2	...	...
	Water	79.8	58	91
Water	Air	80.4	...	...
	Water	69.1	17	30

[a] Data from Cumming and Stark (1980).
[b] Data from DeSouza and Eitenmiller (1986).

thus reducing the required processing time. Rapid cooling of the cans after they go through the retort also minimizes nutrient loss. Commercially canned food often has more nutrients than home-canned food because of reduced heating and cooling times.

Even the container material for canning affects the amount of processing time required and therefore can change the resultant nutrient value. Glass jars have slower heat transfer and therefore require slightly more processing time than metal cans. Flexible "brick" paperboard packages or retort pouches may allow reduction of the processing time by one third to one half (IFT, 1986; Rizvi and Acton, 1982), which reduces losses of heat-labile nutrients, as shown in Table 9.6. Unfortunately, these new packaging materials are more costly and the process more labor-intensive than traditional canning processes, and strong competition from frozen vegetables has not made them economically competitive for all product categories as yet.

Another advance in thermal sterilization is the increased use of high-temperature short-time processing. The shorter processing time results in less loss of heat-labile vitamins (Burton, 1988; Ulrich, 1990). When combined with aseptic packaging techniques (in which the product and package are sterilized separately and then combined under aseptic conditions), the result is a sterile product without the extremely high temperatures required in traditional canning procedures but with increased retention of labile vitamins such as thiamin (IFT, 1986; Lund, 1988).

After processing, vitamin losses in canned food continue. These losses are very much dependent on the specific food and the length and temperature of storage. If the cans are stored in a cool, dry place, losses may be quite small. In a canned meat product, 15% of the thiamin was lost in 6 months of storage at 21°C and 45% in 24 months (Burger, 1983). Storage at lower temperatures minimizes the losses (Somers and Hagen, 1981). In vegetables stored up to 2 years at temperatures

Table 9.6. Effect of Packaging Material on Nutrient Retention[a]

| | Percent Retention | | | | | | | |
| | Thiamin | | Vitamin B-6 | | Riboflavin | | Ascorbic Acid | |
Food	Retort Pouch	Can	Retort Pouch	Can	Retort Pouch	Can	Aseptic Pack	Can
Sweet potatoes	77.0	60.4	···	···	102.8	88.6	···	···
Lima beans, strained	84.2	59.3	90.5	89.9	···	···	···	···
Beef, strained	90.5	78.4	95.9	97.1	···	···	···	···
Tomato soup	···	···	···	···	···	···	91	59

[a] Data from Rizvi and Acton (1982); Everson et al (1964a, 1964b); Ulrich (1990).

as high as 80°F, losses ranged from 0 to 26% for vitamin C, 6–26% for carotene, 0–20% for niacin, 8–30% for riboflavin, and 5–40% for thiamin. In general, retention in fruits was better than in vegetables during storage (Lund, 1988).

Special attention needs to be paid to vitamin B-6 and pantothenic acid. Losses of these vitamins in canned food might exceed 90%. Schroeder (1971) concluded that the U.S. required daily allowance for these two vitamins could not be met from a diet of highly refined and/or canned foods.

Mineral losses during canning are different from losses of heat-labile vitamins. Many of the minerals leach into the cooking water, and losses may be as much as 50% (IFT, 1986). If the cooking water is used, the nutrients are still available. However, in many cases, the cooking water is thrown out. Factors such as maturity of the vegetable and the specific mineral in question affect the amount of nutrient loss during processing (Rincon et al, 1990). The presence of organic acids such as citric acid, either natural or as food additives, may increase the availability of iron in canned fruits and vegetables (Johnson and Hazell, 1988).

Chill Processing and Modified Atmospheres

Chilling food inhibits microbial growth and destructive enzymatic and chemical reactions. Unlike many other processing methods, chilling allows most foods to maintain their fresh or raw condition and does not change their eating quality. Furthermore, most pathogens do not grow at these temperatures. Therefore, it is a preferred method of short-term food preservation.

Extending shelf life. Refrigerators should maintain temperature between 32 and 40°F. The general rule of thumb is: the lower the temperature, the longer the shelf life of the food. In fact, careful study has shown that, in most cases, rates of microbial growth and of chemical reactions decrease by half for each 10°C that the temperature drops. Some products that will not freeze at 32°F are held at temperatures slightly below the freezing point to further increase their keeping qualities.

Used in combination with other processing methods, including pasteurization, humidity control, and modified atmosphere storage, refrigeration can dramatically extend shelf life. Some of these combinations have been used in the food supply for some time. Pasteurization and refrigeration of dairy products can increase shelf life by weeks, and in certain products, by months. Controlled humidity together with modified atmosphere is now commonly used for the storage of fresh produce such as apples, pears, and cabbage. These practices ensure harvest quality produce well after the harvest has passed.

Eating quality as well as nutrient quality must be maintained. Losses of ascorbic acid (vitamin C) in fresh fruits and vegetables are less with controlled atmosphere storage than with storage in air. For instance, controlled atmosphere storage of Chinese cabbage extended the shelf life to over 5 months from under 3 months in air. During this storage, ascorbic acid in the cabbage was retained (Kader, 1986).

Combination with modified atmosphere. New uses and new combinations of processing techniques for use with chilling are emerging, fueled by consumer demand for fresh products. Modified atmosphere together with refrigeration is now being used for many products besides produce, including fresh and cured meats, plated dinners, pasta and pasta sauces, and bakery items. Using a combination of techniques for bakery products can increase their shelf life by over 400%. Fresh pasta and pasta sauces can last over 3 weeks.

Some products use a complete vacuum. In vacuum-packed products, air may be completely removed. In other types of modified atmosphere packaging, another gas may replace the air. For instance, a product may be flushed with nitrogen or have the carbon dioxide raised and the oxygen reduced to 1-3%. The reduction of air improves product life for many reasons (Huxsoll and Bolin, 1989). It reduces or destroys spoilage organisms and protects against product shrinkage, oxidation, and color deterioration. The addition of carbon dioxide inhibits the growth of undesirable microorganisms including *Staphylococcus aureus, Salmonella, Yersinia enterocolitica*, and *Escherichia coli*. Unfortunately, it has little effect on the growth of *Clostridium* species or their toxins (King and Bolin, 1989; Shewfelt, 1986; Watada, 1986).

While refrigeration combined with modified atmosphere packaging appears to offer great promise for delivering "fresh," convenient food products, it has one major drawback. It may allow the growth of facultative anaerobes (microorganisms able to grow in the absence of air) such as *Clostridium botulinum*, especially if the product is not held at the proper temperature at any point between the processing plant and the diner. With no visible spoilage organisms, the product could be thought safe when, in fact, it contains botulinum toxin (Hintlian and Hotchkiss, 1986). One possible solution is some type of indicator that changes color if the product has not been kept sufficiently cold, since this would greatly increase the likelihood that the botulinum organism might begin to grow. The other precaution being taken by this industry is making certain that no microorganisms are introduced into the food in the first place. Some firms are utilizing the same degree of sterility for production of these refrigerated foods as is found in surgical suites. Nutrient stability data in many different foods produced with these new techniques is emerging.

In some commodities, the vitamins are retained more readily with modified atmosphere storage, whereas in other commodities and other conditions the losses incurred are greater than under normal conditions. In a review, Shewfelt (1990) concluded that it was difficult to draw any general conclusions at this time about the effect of modified atmospheric conditions on vitamin retention.

Frozen Foods

Freezing is considered the best medium-term food storage available because in most cases both the eating quality and the nutritional quality are maintained (Fennema, 1988). Freezing kills some microorganisms, but many simply lie dormant at low temperatures because water is not available to them. Chemical and enzymatic reactions are inhibited, so food deterioration is slowed.

Nutrient retention. For certain products, eating and nutrient quality may not be maintained unless enzymatic reactions are stopped completely. This is done by blanching. Adequate blanching time must be used to inactivate even the most heat resistant of the enzymes. Although the blanching process stops nutrient loss in the freezer, it has its own set of nutrient losses, which must be accounted for in the total nutrient losses of a frozen food. Table 9.7 shows the successive loss of several vitamins during the processing steps for peas. As expected, the water-soluble vitamins, which are highly labile, experience the greatest loss.

In addition to the blanching itself, other factors can have an effect on nutrient retention. Rapid cooling and freezing after blanching are crucial for optimum nutrient retention and product quality. Fast freezing allows the formation of many small ice crystals rather than a few large ones, which means that less cellular disruption and less drip loss will occur on thawing. Both of these factors enable more vitamins and minerals to be left in the product.

Freezing methods. Several methods are used commercially to ensure that foods freeze rapidly. Blast freezers move cold air and cause very rapid freezing of products. Plate freezers have shallow plates

Table 9.7. Vitamin Retention (%) in Peas after Various Processing Steps[a]

Step	Vitamin C	Thiamin	Riboflavin	Niacin	Carotene
Newly harvested	100	100	100	100	100
Blanched	67	95	81	90	102
Blanched, frozen	55	94	78	76	102
Blanched, frozen, heated	38	63	72	79	103

[a] Adapted from Poulsen (1986).

containing refrigerant and quickly freeze the food placed on them. Sometimes food is frozen by placing it directly in a refrigerant such as liquid nitrogen or liquid carbon dioxide. This is the fastest and most expensive of the freezing methods (Leveille and Ubersax, 1980a).

The temperature of frozen storage is a crucial determinant of nutrient retention and overall quality. In general, the lower the temperature of storage, the better the maintenance of all aspects of quality, including flavor, texture, and nutrient retention. Table 9.8 shows that thiamin retention in salisbury steak increases as the storage temperature decreases. The table also shows that greater thiamin losses occur when the temperature fluctuates than when it is kept constant (Kramer et al, 1976). For maximum nutrient retention and quality, the temperature of the product should not be allowed to fluctuate.

Product characteristics also partially determine vitamin stability under frozen storage. During 12 months of frozen storage at $-20°C$, frozen asparagus, green beans, lima beans, and peas all retained their sensory quality, as well as 90% of their original ascorbic acid. Under the very same conditions, broccoli, spinach, cauliflower, and peaches retained their sensory quality but lost anywhere from 20 to 50% of their ascorbic acid. On the other hand, some products such as citrus concentrates lost their sensory quality long before they lost their vitamin C (Kramer, 1979).

Effect of packaging. The packing medium affects nutrient retention during frozen storage. For instance, frozen orange juice concentrate lost 1% of its ascorbic acid after 9 months of frozen storage at $-18°C$, while diluted orange juice lost 32% in just 6 months. During 9 months of frozen storage, cantaloupe packed in syrup lost up to 44% of its vitamin C and without syrup (plain) lost up to 85%. Peaches in syrup with added antioxidant lost an average of 23% of their vitamin C and

Table 9.8. Thiamin Content (mg/100 g) of Salisbury Steak after Varying Conditions of Frozen Storage[a]

Storage Conditions	Temperature	
	Constant	Fluctuating
Initial	3.0	3.0
Three months		
−10°C	2.8	1.8
−20°C	2.9	2.1
−30°C	3.2	2.6
Six months		
−10°C	1.9	1.8
−20°C	2.7	1.7
−30°C	2.7	2.6

[a] Adapted from Kramer et al (1976).

without it lost an average of 69% during 8 months of storage (Fennema, 1977; Hagen and Schweigert, 1983). In meat products, the loss during the freezing process was dependent on the length of storage, the temperature in the freezer, and the cut and type of meat (Fennema, 1977). Table 9.9 shows examples of these losses.

The packaging of frozen foods also affects nutrient retention. If the package allows loss of water from the product, the result is "freezer burn," which is characterized by marked loss of product quality as well as of nutrients. Packaging materials that are impermeable to light and oxygen enhance nutrient retention. The package should also keep out odors, which cause deterioration in the flavor, especially of porous products.

Thawing procedures also affect nutrient retention (Fennema, 1988). Thawing of food at room temperature results in greater drip and nutrient loss than slow thawing in the refrigerator. Also, the growth of microorganisms is slow in the refrigerator, which keeps the product safer than products thawed at room temperature.

Drying

Drying is an ancient food preservation method. Water is either removed or tied up so that it is not available for microbial growth. The dried food has quite a long shelf life. However, losses of nutrients for some products from drying, especially home drying, can be substantial. Home-dried green beans and tomato or tomato products lost over 90% of both vitamin C and carotene; zucchini lost 66% of the vitamin C; while raspberries lost only 30% (Holmes et al, 1979). In contrast, boysenberries lost 80%. Better nutrient retention was seen in commercial drying operations, although retention of vitamins C and A was still a problem. Retention of both these vitamins was better when the controversial additive sulfite was added (Sakunkhe et al, 1991).

Table 9.9. B Vitamin Loss (%) from Meat Products[a]

| Meat | Storage | | Loss[b] | | |
	Months	Temperature (°C)	Thiamin	Riboflavin	Niacin
Liver, beef	2	−20	32	35	0
Beef steak	10	−18	2	43	4
Pork loins	12	−18	11	+44	+14
Poultry					
Light	8	−18	12	3	+10
Dark	8	−18	42	11	0

[a] Adapted from Fennema (1977).
[b] Plus indicates gain.

In some cases the process of drying may actually increase the nutrient in the food. Such is the case with fruits dried on iron grates. Raisins and other fruits dried on iron grates leach iron into the fruit, so that raisins used to have a higher iron content than the grapes themselves. Unfortunately, progress in the raisin industry is causing raisin-makers to switch from iron grates to plastic ones, so there will no longer be a net gain in the iron content.

Drying of milk is done by spraying it either into a rising current of warm dry air or onto a heated drum. All of the vitamin C is lost by these methods. However, since milk is a poor source of vitamin C to begin with, any loss is inconsequential. Other nutrients in milk are retained in the drying process.

Drying also may have an effect on protein utilization, as amino acids such as lysine may become less available through browning reactions that can occur during drying and storage (Table 9.10). This reduction of protein quality is important only for persons who rely heavily on dried milk powder or legumes for their protein.

Comparison of Canning, Freezing, and Drying

Consumers would frequently like a comparison of nutrient losses that occur in foods processed by different methods. Unfortunately, the making of such comparisons is far less than an absolute science. There is no gold standard against which nutrient losses should be compared. It is not clear whether the nutritive values of fruits and vegetables should be compared to values of garden-fresh or of market-fresh produce. Or perhaps the comparison should be made to the average values published in tables such as those in the USDA Handbook 8 series. Further confounding the issue is the fact that the nutrient levels in natural products vary with the variety, the degree of ripeness when picked and when consumed, and certain production factors such as climatic and soil conditions. One study on grapes showed a 35-fold

Table 9.10. Browning Reaction with Loss of Lysine in Milk[a]

Milk Form	Percent Lysine Lost to Browning Reaction
Raw milk	0
Pasteurized milk, 74°C, 40 sec	0–2
HTST[b] pasteurized milk, 135–150°C, a few seconds	0–3
Spray-dried milk powder	0–3
Sweetened condensed milk	0–3
Evaporated milk	10–15
Roller-dried milk	20–30

[a] Adapted from Mauron (1989).
[b] High-temperature short-time.

variation in their ascorbate (vitamin C) levels (Adams and Erdman, 1988).

Handling after harvest has a large impact on the nutrient value in the product. For instance, spinach left at room temperature lost 50% of its vitamin C in 24 hours, and in the refrigerator lost 25% (Fennema, 1977). When coupled with the normal losses that occur during correct preparation, the "fresh" product might have less vitamin C than the frozen or the canned. Some modern processing operations have mobile units that actually go into the field, while others are located adjacent to where the crops are grown and harvested. This means that there is very little to virtually no time between picking and processing, hence no time for nutrient loss due to handling. It is thus possible for processed produce to have greater nutrient values than "market fresh" produce, which may have lost nutrients in the long journey from the field to the store.

Basis of nutritional comparisons. It is crucial that nutritional comparisons be made on the product as served, in that the amount of time the product is stored "fresh" in the refrigerator and the degree of product paring or trimming can affect the amount of vitamins and minerals remaining. In most cases, losses in home-processed and commercially processed foods would be similar. In some commercial processing operations, losses may be more precisely controlled than in the home, with a resultant increase in both product yield and nutrient availability (Adams and Erdman, 1988).

Table 9.11. Average Vitamin Losses (%, Compared to Fresh-Cooked) from Fruits and Vegetables During Processing[a]

Food	Method	Vitamin A	Thiamin	Riboflavin	Niacin	Vitamin C
Vegetables[b]	Frozen, cooked					
	Mean	12	20	24	24	26
	Range	0–50	0–61	0–45	0–45	0–78
	Canned					
	Mean	10	67	42	49	51
	Range	0–32	56–83	14–50	31–65	28–67
Fruits[c]	Frozen					
	Mean	37	29	17	16	18
	Range	0–78	56–83	14–50	31–65	28–67
	Canned					
	Mean	39	47	57	42	56
	Range	0–68	22–67	33–83	25–60	11–86

[a] Adapted from Fennema (1977).
[b] Mean of 10 vegetables: asparagus, lima beans, green beans, broccoli, brussels sprouts, cauliflower, corn, green peas, mashed potatoes, and leaf spinach. Seven vegetables were used for canning: same list as above without broccoli, brussels sprouts, and cauliflower.
[c] Mean of eight fruits: apples, apricots, blueberries, sour cherries, orange juice, peaches, raspberries, and strawberries.

Comparing nutrient losses. Comparison of fresh and processed vegetables using the various handbooks of nutrient data always raises questions about losses that occur in processing, as different varieties are raised for processing than for marketing fresh. Some data from this type of study are found in Table 9.11. As was expected, losses were greatest for the most heat-labile vitamins, thiamin and vitamin C. The average losses due to canning and freezing varied significantly, and for some nutrients the range of loss was wide. In some products no loss occurred; in others the loss was near 80%. These data confirm that the high processing temperatures of canning create more nutrient loss than do blanching and frozen storage.

Data collected for vegetables used in food service showed that more vitamin C was lost in frozen vegetables than in "fresh" vegetables cooked by food-service methods, although if the food was held before serving the losses became similar. The type of vegetable and the exact treatment, cooking method, and holding conditions all had an impact on the amount of vitamin C in the vegetable at serving time (Carlson and Tabacchi, 1988).

For many years, data such as those assessing vitamin C loss in peas during various types of processing (Table 9.12) were lacking for many vegetables and nutrients. A USDA study compared the nutrient retention of vegetables and fish—fresh cooked, frozen heated, and canned heated (Dudek, et al, 1980). This study used the same batch of vegetables or fish in cooking and processing to enable precise comparison of vitamin and mineral losses (Tables 9.13 and 9.14).

As can be seen from Tables 9.12-9.14, the degree of nutrient loss was dependent on the food, the nutrient, and the processing method. For peas (Table 9.12), each successive processing step reduced the vitamin C value. Drying caused the greatest loss, followed by canning. For vitamin A, the processing method did not have a significant effect (Table 9.13). In fact, vitamin A was relatively stable in all the vegetables studied. Vitamin B-6 loss was greater after canning and freezing than after the microwaving or boiling of fresh vegetables. On average, vitamin losses were greater for canned than for frozen products (Tables

Table 9.12. Percentage Retention of Vitamin C in Peas with Processing and Cooking[a]

Process Stage	Form of Peas			
	Fresh	Frozen	Canned	Dried
Blanching	...	75	70	75
Blanching	...	71	63	45
Cooked	44	39	36	25

[a] Adapted from Mapson (1956).

9.12–9.14). However, losses must be considered in the context of the diet. Most of the population would not be at risk just because peas supplied only 13% of the U.S. RDA for vitamin C and the fresh product supplied 26%. For most, the risk comes because they fail to include an adequate number of servings of fruits and vegetables in their diet, not because they used processed fruits or vegetables.

For fish products, only the loss of thiamin in the canned product is worth mentioning (Table 9.14). Again, canned salmon is only one

Table 9.13. Percent of U.S. RDA[a] (per 100 g of product) for Several Vitamins and Minerals in Vegetables Processed in Different Ways

Vegetable	Processing Method	Vitamin[b]			Mineral[c]		
		C	A	B-6	Calcium	Iron	Magnesium
Peas	Boiled	26	12	10.8	3	9	10
(~ 3/5 cup)	Microwaved	22	12	10.7	3	9	10
	Frozen, heated	15	15	4.7	3	7	8
	Canned, heated	13	15	2.4	2	5	5
Carrots	Boiled	3.6	473	11.7	3	3	3
(~ 2/3 cup)	Microwaved	4.2	509	12.9	3	4	3
	Frozen, heated	3.5	588	6.4	3	3	2.5
	Canned, heated	4.0	567	6.0	2	4	2
Spinach	Boiled	22.2	165	11.8	13	20	20
(~ 1/2 cup)	Microwaved	10.6	163	12.4	13	20	20
	Frozen, heated	15.5	158	7.0	17	7	17
	Canned, heated	18.7	204	4.8	19	15	19
Sweet potatoes	Boiled	49	383	11.0	4	3	5
(~ 2/3 cup)	Microwaved	43	411	12.0	3	2.5	5
	Frozen, heated	15	328	9.3	4	3	5
	Canned, heated	20	325	11.7	2	4	5

[a] Recommended daily allowance.
[b] Adapted from Dudek et al (1980).
[c] Adapted from Dudek et al (1981).

Table 9.14. Percent of U.S. RDA[a] (per 100 g of Product) for Several Vitamins and Minerals in Fish Processed in Different Ways

Fish	Processing Method	Vitamin[b]			Mineral[c]	
		Thiamin	Riboflavin	Niacin	Zinc	Iron
Salmon	Broiled	12	8	29	5	7
	Microwaved	12	8	27	4	5
	Frozen, baked	10	8	26	6	4
	Canned, heated	2	8	22	7	6
Shrimp	Broiled	...	...	8	...	...
	Microwaved	...	...	12	...	...
	Frozen, baked	...	...	11	...	...
	Canned	...	...	4	...	...

[a] Recommended daily allowance.
[b] Adapted from Dudek et al (1980).
[c] Adapted from Dudek et al (1981).

of many thiamin sources in the diet, so the loss for most people would not create any risk.

Mineral losses were very similar for the various processing methods and for fresh-cooked products (Tables 9.13 and 9.14). Iron levels in only one vegetable, spinach, seemed to be affected slightly by freezing.

The USDA study (Dudek et al, 1980, 1981) also reported that the major difference in the nutritive value of products was more likely to be due to handling in the home than to the effects of processing. A few examples mentioned included the soaking of potatoes in water for a long time, causing them to lose vitamin C, and the chopping of cabbage ahead of time, which allows the vitamin C to oxidize through contact with air and with enzymes activated when the cabbage is cut. In other situations, market-fresh vegetables are not refrigerated properly, allowing loss of nutrients.

Other important comparisons. In addition to nutrient losses, another comparison needs to be made: how do processing methods compare with respect to salt, sugar, and fat? Fresh produce is free of fat, sugar, and salt until these are added in preparation. Some canned vegetable items have significantly increased sodium levels. For instance, fresh and frozen green beans have under 5 milligrams of sodium per serving, but canned beans have 160 milligrams. Fresh peas have under 5 milligrams of sodium, frozen peas around 100 milligrams, and canned peas 200 milligrams per serving. Frozen peas with a sauce have over 400 milligrams per serving! Obviously, products with sauces such as butter sauces increase the fat content of the final product. However, the increased palatability of vegetables with sauce may mean that more vegetables are eaten. If so, this could be a good use of part of the day's fat allotment!

Milling and Cereal Processing

Milling has evolved from mere crushing of grain to separating and refining of various grain constituents. The grain undergoes a series of steps that remove extraneous matter such as chaff, metal, sticks,

Table 9.15. Effect of Milling on Vitamins from Wheat (Representative Value)[a]

	Extraction of Flour, %			
	100 Whole Wheat	80	70 All purpose	<50 Cake Flour
Thiamin, μg/g	3.8	2.4	0.7	0.5
Riboflavin, μg/g	1.7	0.8	0.6	0.5
Niacin, μg/g	50	15	10	6
Iron, mg/100 g	3.3	1.9	1.0	1.0

[a] Adapted from Mattern (1991) and Zeigler and Greer (1971).

stones, and foreign and unsound kernels. The product is cleaned by dry scrubbing of the kernel and sometimes also by a water wash. After being cleaned, it is tempered so that the water content becomes optimal, and then it is ready to be milled.

Wheat may be stone-ground by simply placing it between two stones and crushing it until the majority of the particles reach the desired size. However, in all large flour mills, the grain is first lightly milled between corrugated steel rolls. The resulting large and small fragments are separated by size with the aid of bolting cloth sifters. The large bran-containing fragments are routed to additional sets of corrugated steel rolls and sifters. The smallest (flour-sized) fragments are combined with other similar flour-sized streams to make flour, while the intermediate-sized fragments are purified of bran by aspiration and then ground to flour size between smooth steel rolls. These steps are repeated in additional sets of rolls, sifters, and aspirating purifiers from five to seven times in a process of gradual reduction of the wheat to various flour streams, bran, and germ.

Whole wheat flour recombines all these fractions. Straight grade flour, containing about 75% of the kernel, consists primarily of the starchy portion (endosperm) of the kernel with the germ and bran removed, resulting in a loss of fiber and micronutrients. These losses are shown in Table 9.15.

Some B vitamins lost in milling are added back through enrichment. Table 9.16 shows the nutrient values of enriched and unenriched flour. Micronutrients not added back that may be of significance in the U.S. diet are vitamin E, vitamin B-6, and magnesium. Some nutritionists are campaigning to change the enrichment standard to include these nutrients.

Thus, the losses of fiber and trace minerals are significant. However, many of the trace minerals in the bran and germ are not readily absorbed by the body because they are tightly bound by the fiber matrix or by the insoluble constituent of bran, phytate.

Table 9.16. Vitamin and Mineral Content (milligrams per cup) of Various Flour Types[a]

	Flour			
Nutrient	Enriched	Straight	Whole Wheat	Cake
Calcium	17.6	26.6	54.5	1.2
Iron	3.2	1.5	4.4	0.5
Thiamin	0.5	0.1	0.7	0.02
Riboflavin	0.3	0.07	0.3	0.02
Niacin	3.9	1.6	2.2	0.8

[a] Data from Ensminger et al (1983).

Production of Ready-to-Eat Cereals

Ready-to-eat cereals provide constant examples of ways that technology can add variety and nutrients to the diet by diversifying ways to eat common grains and fiber sources. The production of breakfast cereals has been covered in great detail by Fast and Caldwell (1990). In the making of such cereals, grains can be directly flaked, shredded, or puffed. In some cases, a dough is formed and then forced through an extruder. After extrusion it is sometimes further formed or shaped, such as by puffing or flaking, and then toasted.

There is little information on nutrient losses due to extrusion. Losses of vitamin C may be less than in processes involving boiling water; losses of thiamin appear to be dependent on the process variables used; losses of riboflavin are minimal (Cubadda, 1988). Losses that occur in one type of extruder under one set of conditions may or may not occur with other extruders or other conditions (Harper, 1988). Since most extruded cereals are fortified after extrusion, the degree of nutrient loss may have little impact on consumers.

To fortify cereals with nutrients that are not heat-stable, most ready-to-eat cereals are sprayed with a vitamin-mineral mixture after being toasted. Studies on the vitamin-mineral mixtures used on the cereal products show them to be relatively stable.

High temperatures used in toasting may cause browning and the loss of some biological value of the protein due to the browning reaction. In the United States, where adequate to excessive levels of protein are ingested, this decrease in biological value is not significant. Furthermore, these losses are similar to those that occur in many baking processes using grain products. However, in some areas where high-quality protein is limited, loss of lysine in the browning reaction can have a detrimental effect on already limiting supplies of lysine (Quattrucci, 1988).

The nutrients from breakfast cereals make a significant contribution to the total diet (Morgan et al, 1986a, 1986b). The actual contribution of the cereal depends on the ingredients, level of fortification, and heat treatment. Cooked hot cereals can be excellent nutritional choices as they are minimally processed and can be prepared with no additional fat, sugar, or salt.

High levels of sugar and salt in ready-to-eat breakfast cereals are of some concern. While cereals in their unprocessed or minimally processed form such as quick oats have less than 5 milligrams of sodium per serving, many ready-to-eat cereals have over 300 milligrams. The labels of cereal cartons must give the salt content per serving, so an alert consumer can monitor sodium levels. Many cereal labels voluntarily give the sugar content per serving.

For those concerned about fiber, the key is to look at the ingredient statement. Cereals with such major ingredients as whole grain, bran, and wheat germ provide more fiber than those formulated with refined flours. Many labels also provide information about the dietary fiber content, although it may be hard to predict fiber's physiological effect from simply knowing the fiber content, because processing or alteration of the matrix may alter the physiological effect. Furthermore, different types of fibers appear to have different physiological effects (Dreher, 1987), and processing may alter the amount and solubility of fiber in certain situations (Chang and Morris, 1990; Marlett, 1991).

Processing of Fat

Fat is pressed or solvent-extracted from nuts and seeds or rendered from tissue. The pressing or solvent-extracting may be done with or without heat. There are claims that cold extraction of oil allows more vitamin E to be retained. However tests comparing the vitamin E content of oils have not been able to substantiate this claim.

After extraction, oils may be winterized and deodorized. In winterization, the fat is cooled so that fatty acids that solidify with chilling are removed. Deodorization removes fatty acids and other residues that have an unusual flavor or smell. This may slightly decrease the nutritional value of the oil, as constituents like the omega-3 fatty acids may be removed (Institute of Shortening and Edible Oils, 1988).

Antioxidants are added to many oils. These help to keep the fat from going rancid. Rancidity not only reduces eating quality but also produces products that are toxic (Tomassi, 1988).

Hydrogenation. Edible oil is hydrogenated to raise its melting point and solidify it. With the aid of a catalyst, hydrogen is added across the double bonds of an unsaturated oil (Diplock, 1991). There are both advantages and disadvantages to an oil or fat that has undergone saturation. The resultant product has the advantage of being spreadable and having properties that make it more useful in a wide variety of baked products. It is also less likely to go rancid and produce substances that are toxic to the body and ruin the taste of the food, and the need for synthetic antioxidants such as BHA and BHT is reduced.

However, hydrogenation causes the fat to be more saturated. This could be a nutritional disadvantage if the saturated fat increases serum cholesterol. Also hydrogenation produces a small percentage of fatty acids that have the *trans* configuration, which does not occur in nature, and there is some concern about the presence of these in margarines and other food products. Shortenings contain 14–60% *trans* fatty acids, margarines 16–70%, and salad oils 8–17%. These fatty acids are metabolized in the same way that saturated fats are and may affect body prostaglandin levels. The Federation of American Societies for Experimental Biology in the United States studied the issue of *trans* fatty acids and came to the conclusion that they posed no problem at current levels of use. The Committee on Medical Aspects of Food Policy of the Department of Health and Social Security in the United Kingdom analyzed the same data and came to the conclusion that *trans* fatty acids might pose some slight risk and should be labeled on food packages. More information is needed on the metabolism and safety of *trans* fatty acids (Grundy, 1990; Gurr, 1988; Mesnick and Katan, 1990).

Other Processes

Fruit ripening. Ethylene produced endogenously by all plants has hormonelike qualities (Nagy and Wardowski, 1988). It can cause growth, ripening, and senescence. Commercially it is used to regulate the ripening of apples, cherries, blueberries, figs, grapes, walnuts, and tomatoes. This is one process that actually increases the vitamin C content of treated tomatoes by as much as 16% over that of untreated fruit. The vitamin A content is unchanged by the process.

The ethylene treatment does make some products change their texture or become more bitter. Carrots, for example, produce bitter isocoumarin when treated this way (Watada, 1986).

Chemical sprays or dips. These are applied to fruits to regulate growth and development, prevent insect attack, and control microorganisms. Plant hormones and many of the pesticides discussed in Chapter 13 are also used. (For a discussion of the subject, see Shewfelt, 1986.)

Fermentations. Fermentation is a form of biotechnology that was harnessed in antiquity. Many types of food products are preserved by fermentation, including sauerkraut, pickles, soy and other oriental sauces, cheeses, yogurt, wine and beer, some sausages, tempeh, miso, some breads, and vinegar.

The storage of these products decreases the nutrient value of the product, but microbial activity may increase some of the micronutrients in the food. The B vitamins and particularly vitamin B-12 can increase with fermentation. Protein bioavailability also increases in some cases (McFeeters, 1988), and phytic acid may be reduced.

Foods that are fermented may of necessity be highly salted to keep them safe. High levels of salt are a concern for the segment of the population whose blood pressure increases when salt is ingested.

Products of fermentation may be beneficial; for instance, yogurt cultures can recolonize the gut after antibiotic therapy. On the other hand, some of the products of fermentation in wine, beer, or soy sauce may be mutagenic (Gammack, 1988) (see also Chapter 5).

ROLE OF FORMULATED PREPARED FOODS

Highly processed foods are available today. The problem with many of these foods stems not from the fact that they lost nutrients during processing but from the fact they increase items in the diet that most people wish to limit, such as salt, fat, saturated fat, and sugar. Sodium levels for some of the meals may be very high. A single meal item may provide 900–1200 milligrams of sodium. That is practically half of the recommended intake of sodium. Unfortunately, some snack items add little to the diet except salt and fat. Consumers should choose these items infrequently. Within a single product, such as popcorn, the range of salt and fat levels can be incredible, a trace to 380 milligrams for sodium, 1–15 grams for fat. Popcorn can be air popped and have no added sodium or fat. In many of the microwave varieties, the levels of salt and saturated fat can vary dramatically.

Other highly formulated foods may use refined ingredients that probably are not used in the home kitchen. Refined ingredients are chosen because they are uniform, stable, predictable, and have no off-flavors. Some of these make the product have lower nutrient levels than its home-produced counterparts; others such as whey, guar gum, or barley bran may increase the nutritional value.

Perceptions of the usefulness of a formulated product may change. When filled milk, cream, and ice cream products were first introduced, they were viewed as cheap, inferior imitations made by removing milk fat and replacing it with other fats. Currently these items are popular because they can be formulated to reduce the saturated fat, fat, and cholesterol contents of these foods.

SUMMARY

In thinking about processed foods, it is useful to put them into the following framework:

Relatively nonformulated—frozen vegetables, fruit juices, canned foods;

Partially formulated—unrefined ingredients make up the bulk of the product (e.g., processed entrees);

Highly formulated—refined ingredients make up more than 50% of the product (e.g., drinks, desserts, some breakfast cereals, snack foods); or

Foods designed to achieve a special nutritional objective (e.g., low-cholesterol, low-fat cheese).

Selections of food that are processed but relatively nonformulated or partially formulated probably comprise the best nutritional bet and still afford some measure of convenience. Convenience need not be synonymous with high formulation. For persons with special dietary needs, some highly formulated foods may offer the very requirements that they are looking for, such as low fat and low cholesterol. Other foods in the highly formulated category may fulfill the consumer's demand for good taste but do not add to overall nutritional well being. Although many consumers blame the processors for producing products that lack nutritional attributes, consumers cannot entirely wash their hands of the blame if they continue to purchase these products.

As with most issues in food safety and nutrition, the issue of food processing causes several risk-benefit questions to surface and shows again the need for moderation and variety. The risk-benefit questions can be summarized as follows:

1) Do gains in the increased availability and variety of food offset the attendant nutrient losses?

2) Are the effects of no processing better or worse than the effects of processing (i.e., does "market fresh" produce actually have higher levels of nutrients than its processed counterparts)?

3) If the food is not processed, what are the effects of not being able to have that food and its nutrients all year long?

4) If the palatability or availability of the food is increased, are the nutrients from the food more likely to be eaten?

5) Does processing decrease perishability, and therefore cost, making the food available to a wider portion of the population?

6) Does the extent of nutrient loss due to processing have any significant impact on the population eating the food?

The notion of moderation and variety has importance when thinking about processed food. If all food that an individual consumes is highly processed, it is questionable that all the needed nutrients and fiber can be obtained without getting too many nutritional negatives—salt, sugar, fat, or calories. It is clear that one can choose processed foods that yield adequate levels of micronutrients without high levels of salt, fat, and sugar. It is also possible to select a diet that has few micronutrients and too many nutritional negatives. Wise selection requires choosing a wide variety of items to ensure intake of all the micronutrients and using foods that are high in nutritional negatives only on an occasional basis.

REFERENCES

Adams, C. E., and Erdman, J. W. 1988. Effects of home food preparation practices on nutrient content of foods. Pages 557-606 in: Nutritional Evaluation of Food Processing, 3rd ed. E. Karmas and R. S. Harris, eds. AVI (Van Nostrand Reinhold Co.), New York.

Anonymous. 1982. "Fruit" drinks. Consumer Rep. 47(Sept.):456-457.

Armbruster, G. 1978. Comparison of reduced ascorbic acid content of microwave- and conventional-cooked fruits and vegetables. Pages 103-104 in: Proc. Microwave Power Inst., 13th, Ottawa.

Burger, I. H. 1983. Effect of processing on nutritive value of food: Meat and meat products. Pages 323-336 in: Handbook of Nutritive Value of Processed Food. Vol I: Food for Human Use. M. Recheigl, ed. CRC Press, Boca Raton, FL.

Burton, H. 1988. Ultra-High Temperature Processing of Milk and Milk Products. Elsevier Applied Science, New York.

Carlson, B. L., and Tabacchi, M. H. 1988. Loss of vitamin C in vegetables during the foodservice cycle. J. Am. Diet. Assoc. 88(7):65-67.

Chang, M.-C., and Morris, W. C. 1990. Effect of heat treatments on chemical analysis of dietary fiber. J. Food Sci. 55:1647-1649 ff.

Charles, V. R., and Van Duyne, F. O. 1958. Effect of holding and reheating on the ascorbic acid content of cooked vegetables. J. Home Econ. 50:159-164.

Cubadda, R. 1988. Nutritional properties and safety of food processed by extrusion cooking. Pages 157-170 in: Nutritional and Toxicological Aspects of Food Processing. R. Walker and E. Quattrucci, eds. Taylor and Francis, New York.

Cummings, D. B., and Starck, R. 1980. The development of a new blanching system. J. Can. Diet. Assoc. 41(1):39-43.

DeSouza, S. C., and Eitenmiller, R. R. 1986. Effects of processing and storage on the folate content of spinach and broccoli. J. Food Sci. 51:626-628.

Dietz, J. M., and Erdman, J. W. 1989. Effects of thermal processing upon vitamins and protein foods. Nutr. Today 24(4):6-14.

Diplock, A. T. 1991. Antioxidant nutrients and disease prevention: An overview. Am. J. Clin. Nutr. 53:189S-193S.

Dreher, M. 1987. Handbook of Dietary Fiber. An Applied Approach. Marcel Dekker, New York.

Dudek, J. A., Elkins, E. R., Chin, H., and Hagen, R. 1980. Investigations to Determine Nutrient Content of Selected Fruits and Vegetables—Raw, Processed, and Prepared. National Food Processors Association, Washington, DC.

Dudek, J. A., Behl, B. A., and Elkins, E. R. 1981. Determination of Effects of Processing and Cooking on the Nutrient Composition of Selected Seafoods. National Food Processors Association, Washington, DC.

Eheart, N. S., and Gott, C. 1964. Conventional and microwave cooking of vegetables. J. Am. Diet. Assoc. 44:116-119.

Eheart, M. S., and Gott, C. 1965. Chlorophyll, ascorbic acid and pH changes in green vegetables cooked by stir-fry, microwave, and conventional methods and a comparison of chlorophyll methods. Food Technol. (Chicago) 19(5):185-188.

Ensminger A. H., Ensminger M. E., Kolande, J. E., and Robson, J. R. K. 1983. Food compositions. In: Food and Nutrition Encyclopedia, 1st ed. Pegus Press, Clovis, CA.

Erdman, J. W. 1979. Effect of preparation and service of food on the nutrient value. Food Technol. (Chicago) 33(2):38-48.

Everson, G. J., Chang, J., Leonard, S., Luh, B. S., and Simone, M. 1964a. Aseptic canning of foods. II. Thiamine retention as influenced by processing method, storage time, and temperature, and type of container. Food Technol. (Chicago) 18(1):84-86.

Everson, G. J., Chang, J., Leonard, S., Luh, B. S., and Simone, M. 1964b. Aseptic canning of foods. II. Pyridoxine retention as influenced by processing method, storage time, and temperature, and type of container. Food Technol. (Chicago) 18(1):87-89.

Fast, R. B., and Caldwell, E. F., eds. 1990. Breakfast Cereals and How They Are Made. American Association of Cereal Chemists, St. Paul, MN.

Fennema, O. 1977. Loss of vitamins in fresh and frozen foods. Food Technol. (Chicago) 31(12):32-33, 38.

Fennema, O. 1988. Effects of freeze preservation on nutrients. Pages 269-318 in:

Nutritional Evaluation of Food Processing, 3rd ed. E. Karmas and R. S. Harris, eds. AVI (Van Nostrand Reinhold Co.), New York.

Gammack, D. B. 1988. Nutrition and toxicology of fermented foods. Pages 231-239 in: Nutritional and Toxicological Aspects of Food Processing. R. Walker and E. Quattrucci, eds. Taylor and Francis, New York.

Gerster, H. 1989. Vitamin losses with microwave cooking. Food Sci. Nutr. 42F:173-181.

Goldblith, S. A. 1971. A condensed history of the science and technology of thermal processing. Part 1. Food Technol. (Chicago) 25(12):56-62.

Goldblith, S. A. 1972. A condensed history of the science and technology of thermal processing. Part 2. Food Technol. (Chicago) 26(1):64-68.

Gordon, J., and Noble, I. 1959. Comparison of electronic vs. conventional cooking of vegetables. J. Am. Diet. Assoc. 35:241-244.

Gould, M. F., and Golledge, D. 1989. Ascorbic acid levels in conventionally cooked versus microwave oven cooked frozen vegetables. Food Sci. Nutr. 42F:145-152.

Grundy, S. M. 1990. Trans-mononusaturated fatty acids and serum cholesterol levels. New Engl. J. Med. 323:439-445.

Gurr, M. I. 1988. Lipids: Products of industrial hydrogenation, oxidation and heating. Pages 139-156 in: Nutritional and Toxicological Aspects of Food Processing. R. Walker and E. Quattrucci, eds. Taylor and Francis, New York.

Hagen, R. E., and Schweigert, B. S. 1983. Nutrient contents of table-ready foods: Cooked, processed and stored. Contemp. Nutr. 8(2):1-2.

Harper, J. M. 1988. Effects of extrusion processing on nutrients. Pages 365-391 in: Nutritional Evaluation of Food Processing, 3rd ed. E. Karmas and R. S. Harris, eds. AVI (Van Nostrand Reinhold Co.), New York.

Hintlian, C. B., and Hotchkiss, J. H. 1986. The safety of modified atmosphere packaging: A review. Food Technol. (Chicago) 40(12):70-76.

Holmes, Z. A., Miller, L., Eewards, M., and Benson, E. 1979. Vitamin retention during home drying of vegetables and fruits. Home Econ. Res. J. 7(4):258-264.

Huxsoll, C. C., and Bolin, H. R. 1989. Processing and distribution alternatives for minimally processed fruits and vegetables. Food Technol. (Chicago) 43(2):124-128.

Institute of Food Technologists (IFT). 1986. Effect of food processing on nutritive values: A scientific status summary by the Institute of Food Technologists' Expert Panel on Food Safety. Food Technol. (Chicago) 40(12):109-116.

Institute of Shortening and Edible Oils. 1988. Food Fats and Oils. The Institute, Washington, DC.

Jenkins, D. J. A., Wolever, T. M. S., Collier, G. R., Ocana, A., Rao, A. V., Buckley, G., Lam, Y., Mayer, A., and Thompson, L. V. 1987. The metabolic effects of a low glycemic index diet. Am. J. Clin. Nutr. 46:968-975.

Johnson, I., and Hazell, T. 1988. Iron in processed foods. Nutr. Food Sci. 9:2-4.

Kader, A. A. 1986. Biochemical and physiological basis for effects of controlled and modified atmospheres on fruits and vegetables. Food Technol. (Chicago) 40(5):99-104.

Karmas, E., and Harris, R. S, eds. 1988. Nutritional Evaluation of Food Processing, 3rd ed. AVI (Van Nostrand Reinhold Co.), New York.

King, A. D., and Bolin, H. R. 1989. Physiological and microbiological storage stability of minimally processed fruits and vegetables. Food Technol. (Chicago) 43(2):132-135.

Klein, B. P. 1989. Retention of nutrients in microwave-cooked foods. Contemp. Nutr. 14(2):1-2

Klein, B., Luo, C., and Boyd, G. 1981. Folacin and ascorbate retention in fresh, raw, microwave, and conventionally cooked spinach. J. Food Sci. 46:640-641.

Kramer, A. 1979. Effects of freezing and frozen storage on nutrient retention of fruits and vegetables. Food Technol. (Chicago) 33(2):58-61 ff.

Kramer, A., King, R. L., and Westhoff, D. C. 1976. Effects of frozen storage on prepared foods containing protein concentrates. Food Technol. (Chicago) 30(1):56-62.

Lachance, P. A., and Fisher, M. C. 1988. Effects of food preparation procedures in nutrient retention with emphasis on foodservice practices. Pages 505-506 in: Nutritional Evaluation of Food Processing, 3rd ed. E. Karmas and R. S. Harris, eds. AVI (Van Nostrand Reinhold Co.), New York.

Leveille, G. A., and Uebersax, M. A. 1979. Fundamentals of food science for the dietitian: Thermal processing. Diet. Curr. 6(3):11-18.

Leveille, G. A., and Uebersax, M. A. 1980a. Fundamentals of food science for the dietitian: Frozen foods. Diet. Curr. 7(4):19-24.

Leveille, G. A., and Uebersax, M. A. 1980b. Fundamentals of food science for the dietitian. Diet. Curr. 7(1):1-8.

Lund, D. 1988. Effects of heat processing on nutrients. Pages 319-354 in: Nutritional Evaluation of Food Processing, 3rd ed. E. Karmas and R. S. Harris, eds. AVI (Van Nostrand Reinhold Co.), New York.

Mapson, L. W. 1956. Effect of processing on the vitamin content of foods. Br. Med. Bull. 12(1):73-77.

Marlett, J. A. 1991. Dietary fiber content and effect of processing on two barley varieties. Cereal Foods World 36:576.

Mattern, P. J. 1991. Wheat. Pages 1-54 in: Handbook of Cereal Science and Technology. K. J. Lorenz and K. Kulp, eds. Marcel Dekker, New York.

Mauron, J. 1989. Browning reaction products and food quality. Pages 667-683 in: Trends in Food Science and Technology. M. R. Rao, N. Chandrasekhara, and K. A. Ranganath, eds. Food Science and Technology Assoc., Mysore, India.

McFeeters, R. F. 1988. Effects of fermentation on the nutritional properties of food. Pages 423-446 in: Nutritional Evaluation of Food Processing, 3rd ed. E. Karmas and R. S. Harris, eds. AVI (Van Nostrand Reinhold Co.), New York.

Mesnick, R. P., and Katan, M. B. 1990. Effect of dietary trans fatty acids on high-density and low-density lipoprotein cholesterol level in healthy subjects. New Engl. J. Med. 323:439-445.

Morgan, K. J., Zabik, M. E., and Stampley, G. L. 1986a. Breakfast consumption patterns of U.S. children and adolescents. Nutr. Res. 6:635-646.

Morgan, K. J., Zabik, M. E., and Stampley, G. L. 1986b. Breakfast consumption patterns of older Americans. J. Nutr. Elderly 5(4):19-44.

Nagy, S., and Wardowski, W. F. 1988. Effects of agricultural practices, handling, processing, and storage on fruits. Pages 73-100 in: Nutritional Evaluation of Food Processing, 3rd ed. E. Karmas. and R. S. Harris, eds. AVI (Van Nostrand Reinhold Co.), New York.

Poulsen, K. P. 1986. Optimization of vegetable blanching. Food Technol. (Chicago) 40(6):122-129.

Quattrucci, E. 1988. Heat treatments and nutritional significance of Maillard reaction products. Pages 113-124 in: Nutritional and Toxicological Aspects of Food Processing. R. Walker and E. Quattrucci, eds. Taylor and Francis, New York.

Rincon, F., Zurera, G., Moreno, R., and Ros, G. 1990. Some mineral concentration modifications during pea canning. J. Food Sci. 55:751-754.

Rizvi, S. S. H., and Acton, J. C. 1982. Nutrient enhancement of thermostabilized foods in retort pouches. Food Technol. (Chicago) 36(4):105-109.

Salunke, D. K., Bolin, H. R., and Reddy, N. R. 1991. Storage, Processing, and Nutritional Quality of Fruits and Vegetables, 2nd ed. Vol. 2. Processed Fruits and Vegetables. CRC Press, Boca Raton, FL.

Sauberlich, H. E. 1987. Vitamins—How much is for keeps? Nutr. Today 22(1):20-28.

Schroeder, H. A. 1971. Losses of vitamins and trace minerals resulting from processing and preservation of foods. Am. J. Clin. Nutr. 24:562-573.

Shewfelt, R. L. 1986. Postharvest treatment for extending the shelf life of fruits and vegetables. Food Technol. (Chicago) 40(5):70-81.

Shewfelt, R. L. 1990. Sources of variation in the nutrient content of agricultural commodities from the farm to the consumer. J. Food Qual. 13:37-54.

Somers, I. I., and Hagen, R. E. 1981. Effects of Processing, Cooking and Storage on Nutrients in Meats. National Livestock and Meat Board, Chicago.

Tomassi, G. 1988. Nutritional and safety aspects of natural antioxidants. Pages 1-14 in: Nutritional and Toxicological Aspects of Food Processing. R. Walker and E. Quattrucci, eds. Taylor and Francis, New York.

Ulrich, P. 1990. Aseptic processing and packaging of food. Pages 1.166-1.170 in: Processing and Quality of Foods, Vol. 1. P. Zeuthen, J. C. Cheftel, C. Eriksson, T. R. Gormley, P. Linko, and K. Paulus, eds. Elsevier Applied Science Publ., New York.

Vetorazzi, G. 1988. Safety aspects of genetically engineered food products. Pages 215-230 in: Nutritional and Toxicological Aspects of Food Processing. R. Walker and E. Quattrucci, eds. Taylor and Francis, New York.

Vitamin Nutrition Information Service (VNIS). 1984. Vitamin losses in food preparation. VNIS, Hoffman-La Roche Inc., Nutley, NJ.

Watada, A. E. 1986. Effects of ethylene on the quality of fruits and vegetables. Food Technol. (Chicago) 40(5):82-85.

Whelan, E. M., and Stare, F. J. 1976. Panic in the Pantry: Food Facts, Fads and Fallacies. Antheneum, New York.

Wolever, T. M. S., Jenkins, D. J. A., Kalmusky, J., Jenkins, A. L., Giordano, C., Giudici, S., Josse, R. G., and Wong, G. S. 1986. Comparison of regular and parboiled rices: Explanation of discrepancies between reported glycemic responses to rice. Nutr. Res. 6:349-357.

Ziegler, E., and Greer, E. N. 1971. Principles of milling. Pages 115-200 in: Wheat: Chemistry and Technology, 2nd ed. Y. Pomeranz, ed. American Assoc. Cereal Chem., St. Paul, MN.

Food Additives

CHAPTER 10

Additives have been used in food processing since antiquity. Ancient historical records indicate that salt was used to cure meat in 3,000 B.C. By 900 B.C., during the time of Homer, the use of salt and smoke were already old practices. However, even Homer, with his ageless wisdom, was unaware of the chemicals produced by hardwood smoke that brought about food preservation. Starting in the Middle Ages, nitrate in the form of saltpeter was added to meat to increase the preservative action of salt and smoke as well as to change its color and flavor and prevent botulism.

Not all additives have been employed with beneficial aims. Many early accounts indicated that flour, tea, wine, and beer were widely adulterated. Food laws had to be enacted to eliminate additives used as harmful or cheap filler agents.

Addition of artificial colors has had an even more checkered history. Disreputable practices using actual poisons such as compounds of mercury, arsenic, and lead to color food have occurred (see Chapter 11). Thus, the history of food additives has been shaped by two forces—one associated with clear technological advances that brought about the preservation or improved preparation of a food, and the other associated with duping the consumer to make him or her think that a food was of higher quality than it actually was. With this two-pronged past, it is easy to understand why consumers are wary of food additives. This wariness, together with a general mistrust of science and technology, makes it easy for a consumer to be persuaded by an activist decrying food additives. The fear is understandable when the activist points out that the average consumer in the United States annually consumes his or her *weight* in additives (Pim, 1981). That is, we each consume an average of 139 pounds of additives a year.

WHAT IS AN ADDITIVE?

The thought that we consume our own weight in additives is indeed stunning, although that fact alone without additional information gives a distorted picture.

Part of the problem comes from the fact that most consumers think they know what additives are—chemical names on the ingredient statement. However, many do not recognize the chemical name as something that they themselves use in food products. For instance, potassium tartrate in the list of ingredients may not be identified as the cream of tartar commonly required in a recipe. In like manner, sodium aluminum sulfate phosphate in an ingredient list is rarely recognized as the principal leavening acid in the most widely sold brands of baking powder. Some nutrients added to enrich or fortify might also scare consumers. For instance, thiamin hydrochloride sounds too much like hydrochloric acid for some consumers to recognize that it is simply thiamin, or vitamin B-1.

Even more astonishing is the fact that many ingredients on the label statement that aren't considered additives by most consumers are legally additives and are thus included in the 139-pound figure. A few examples include spices, yeast, and sweeteners (including honey, sugar, and corn syrup). What, then, is the legal definition of a food additive? The 1958 Food Additives Amendment of the Food, Drug, and Cosmetic Act defines a food additive as

> any substance the intended use of which results or may reasonably be expected to result, directly or indirectly, in its becoming a component or otherwise affecting the characteristics of any food.

Translated, this definition means that food additives are components of food. If one analyzes the recipe for a meatloaf, it is difficult to decide whether the bread crumbs that help to bind the product together are an additive or an ingredient. We may be left wondering whether the ketchup that is placed on the top of the meatloaf is a flavor additive or an ingredient. Because of the difficulty in classifying ingredients in a recipe, the legal definition is written very broadly.

With a firm understanding of the definition of a food additive, let's take another look at the per capita consumption. Although we supposedly consume our weight in food additives, the lion's share of that weight comes from things normally not considered additives by consumers, including sucrose (table sugar) corn sweeteners, dextrose (glucose), and salt. If we exclude those from our calculation, the per capita additive consumption is 5 pounds per year (Leading Edge, 1989). Figure 10.1 shows the breakdown of the 5 pounds of food additives in 1988.

If we further subdivide the remaining 5 pounds and don't include as additives black pepper, mustard, and other spices; carbon dioxide;

citric acid; starches; and sodium bicarbonate and other leavenings, the per capita consumption dwindles still further, to around a couple of pounds per year. Thus the over 1,800 chemicals with foreign sounding names account for about 2 pounds per year. The median level of use for these 1,800 chemicals is 0.5 milligram (0.014 ounce) per year—i.e., half are used in levels greater than this amount and half are used at less. As a benchmark for comparison, one ordinary 5-grain aspirin tablet contains 325 milligrams of aspirin.

Seventy-five percent of the 1,800 additives are used as flavors. This fact is just another indication that they must be used in very small amounts. Large doses of flavor compounds in a food are more objectionable than overuse of perfume or room deodorant. In other words, the use of most flavor additives is self-limiting. Attesting to this is the fact that, although flavoring agents comprise the largest number of additives used in food, they amount to only slightly more than 5% by weight of the 5 pounds per year per capita consumption of additives in the United States.

A flavor additive is just one example of an additive function. All additives can be classified according to function, a list of which would

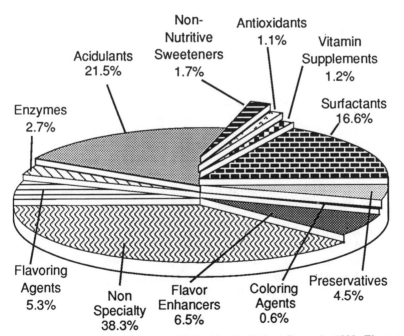

Figure 10.1. Consumption of food additives in the United States in 1988. The total was 1.4 billion pounds. (Reprinted, with permission, from Leading Edge, 1989)

include leavening agents, maturing and bleaching agents, sequester-ants, humectants, anticaking agents, clarifying agents, curing agents, emulsifiers, stabilizers and thickeners, acids and bases, foaming agents and foam inhibitors, nutrients, colors, flavors, sweeteners, preservatives, and antioxidants.

Box 10.1

Why Are Food Additives Used?

Food additives *should* be used to:
- Improve or maintain nutritional quality. Examples include the addition of vitamin D to milk, vitamins A and D to margarine, and B vitamins to white flour.
- Improve keeping quality and reduce wastage. Examples include calcium propionate to keep bread from molding, nitrate to prevent botulism in cured meats, BHA or BHT to keep oils from going rancid, and sodium benzoate to preserve soy sauce.
- Make food more readily available throughout the year and through-out the country. Examples include lemon juice or citric acid to keep fruit from browning and salt to preserve pickles and sauerkraut.
- Maintain food quality characteristics. Examples include corn-starch added to powdered sugar to help prevent lumping, emulsi-fiers added to salad dressings to prevent oil separation, or leaven-ings to make baked products rise.
- Facilitate the fast and convenient preparation of the food. Exam-ples include use of phosphate additives in products like instant oatmeal or instant pudding.
- Make food more appealing. This is the most controversial of the reasons. Examples include the coloring of strawberry ice cream to make it pink or, in general, the use of artificial or synthetic colors and flavors.

Food additives *should not* be used if they:
- Disguise inferior or damaged food, or are used to deceive the consumer.
- Reduce nutrient quality of the product.
- Are in excess of the technical amount needed to do the desired function.
- Are unneeded because of alternate or equally acceptable processing or packaging techniques.

(Adapted from Graham, 1980)

CLASSIFICATION OF FOOD ADDITIVES

Additives are classified into two groups—direct (intentional) and indirect (unintentional). Direct additives are chemicals intentionally added to food during processing to perform a specific function such as preserving or improving the quality of the product. Direct additives are the focus of this chapter.

Indirect additives are chemicals that unintentionally make their way into foods sometime during the production, processing, storage, or packaging. Examples include pesticide residues, animal feed adjuvants and drugs, processing aids such as iodine cleaners, plant growth regulators, and components of packaging materials that migrate into foods. The law reads that indirect additives are permitted only if they cannot be avoided by "good agricultural and manufacturing practices" and then only if the amounts present are considered so insignificant as to be safe (Gilchrist, 1981). These are discussed in subsequent chapters.

Generally Recognized As Safe

Another additive-related term that needs to be defined is GRAS. GRAS stands for the words "generally recognized as safe," but it implies a longer definition from the Food Additives Amendment of 1958 to the United States Food, Drug, and Cosmetic Act. The amendment, in effect, allowed the use of any substance already used in food that is

> generally recognized, among experts qualified by scientific training and experience to evaluate its safety, as having been adequately shown through scientific procedures (or, in the case of a substance used in food prior to January 1, 1958, through either scientific procedures or experience based on common use in food) to be safe under the conditions of its intended use.

Subsequently, the general recognition of safety (including official recognition of the GRAS acronym) was further dealt with in food regulations (21 CFR 170.30), in which the eligibility for classification as GRAS was defined in detail.

GRAS status thus means that the safety of the additives referred to has been proven through their use in the food supply and/or by other scientific procedures. Substances have either GRAS status or regulated food additive status. Any substance not in use before the passage of the Food Additives Amendment must have its safety tested before it is approved for use in the food supply, and it must be recognized as either GRAS or as an approved and regulated food additive. The required toxicity testing procedures were outlined in Chapter 3.

Currently there is little difference between regulated food additives and GRAS substances in what is known about their safety. This was not always the case. In the late 1960s and early 1970s, questions of

---

Box 10.2

Excerpts from the FDA's Affirmation of GRAS Status for Mannitol and Sorbitol

Mannitol (1,2,3,4,5,6-hexanehexol) and its sterioisomer sorbitol are both solid hexahydric alcohols prepared commercially by catalytic reduction of glucose. Both occur naturally in small amounts in a variety of foods. . . . Sorbitol is a normal constituent of such fruits as cherries, plums, pears, apples, and many berries. . . .

The scientific literature review yielded the following information as summarized in the report of the Select Committee on GRAS substances:

Orally administered sorbitol is absorbed and metabolized rapidly by man through normal glycolytic pathways, ultimately to carbon to carbon dioxide and water. After a 35 g dose (equivalent to 583 mg per kg) in normal and diabetic adults, for example, less than 3 percent of the sorbitol was excreted in the urine in any case and the concentration of sorbitol in the blood was found to be immeasurably small. No evidence of toxicity was reported.

The oral LD_{50} of sorbitol in male and female mice is reported to be 23,200 and 25,700 mg per kg, respectively; in male and female rats, 17,500 and 15,900 per kg, respectively. The oral LD_{100} in the male rat is separately reported as 26,000 mg per kg.

The following short-term studies of the oral administration of sorbitol are relevant:

In male rats, fed 5 percent sorbitol in a balanced diet, no toxic effects were observed during the 3 months of feeding. Feed consumption is not reported, but estimates based on other data presented indicate that sorbitol was being fed at levels of approximately 5 g per kg per day.

Rhesus monkeys fed sorbitol at a level of 8 g per kg per day for 3 months remained unaffected.

Man, consuming 10 g of sorbitol each day (equivalent to 167 mg per kg) for one month remained unaffected.

Normal children, 5–6 years old, and normal infants, 20–35 months old, fed 9.3 g of sorbitol (equivalent to 500 or more mg per kg) remained unaffected except for the appearance of diarrhea in the younger group.

The laxative threshold for sorbitol, established in 12 normal adults, has been reported to be 50 g (equivalent to 833 mg per kg). It is also reported, in a study involving 86 volunteers, that a dosage level of 25 g per day in two doses does not cause laxation.

The following long-term studies of the oral administration of sorbitol

safety began to arise about additives already assumed to be "on the GRAS list," such as cyclamate and saccharin. Following the banning of cyclamate, President Richard Nixon directed the FDA to reevaluate

---

are relevant:

Rats fed 5 percent sorbitol (equivalent to 5 g per kg per day) through three generations showed no deleterious effects on growth rate or liver glycogen storage capacity. There were no gross histological abnormalities in kidney, liver, spleen, pancreas, or duodenum attributable to sorbitol. A subsequent report has indicated that weanling rats, given sorbitol at levels of 10 to 15 percent in the diet for 17 months and observed over 4 successive generations, showed no evidence of deleterious effects on weight gain, reproduction, lactation, or histological appearance of the main organs.

Rats fed 16 percent sorbitol for 19 months showed a tendency to become hypercalcemic after one year, with the appearance in some animals of bladder concretions and a generalized thickening of the skeleton. No feed consumption or animal weight figures were reported, but sorbitol level was estimated to be of the order of 16 g per kg.

No oral studies of the carcinogenic activity of sorbitol have been reported. However, studies in rats revealed that injected sorbitol, in the form of an iron-sorbitol citric acid product, produced no injection site tumors.

Sorbitol, at dose levels of 5 g per kg did not produce any measurable mutagenic response in the host-mediated assay in mice, in the metaphase chromosomes of rat bone marrow, or in the dominant lethal test in the rat. A slight increase was noted in the mitotic recombination frequency for *Saccharomyces cerevisiae* in the host-mediated assay, and a moderate, dose-related adverse effect was exhibited by human embryonic lung cells scored at anaphase.

Sorbitol elicited no teratogenic response in pregnant mice or rats fed a daily dose of 1,600 mg per kg for 10 days, or in hamsters fed 1,200 mg per day for 5 days.

The Joint Food and Agriculture Organization/World Health Organization Committee on Food Additives indicates the acceptable daily intake of sorbitol for man as follows: "conditional acceptance (as a food additive or as a food) not limited."

. . . The label and labeling of food whose reasonably foreseeable consumption may result in a daily ingestion of 50 grams of sorbitol shall bear the statement: "Excess consumption may have a laxative effect."

Reprinted from the Federal Register (1979).

all substances on the GRAS list. The FDA in turn called upon the expertise of a prestigious society of research scientists, the Federation of American Societies for Experimental Biology (FASEB) to provide expert opinions on the safety of the 415 additives (Gilchrist, 1981; Siu et al, 1977).

The GRAS Review

The review was exhaustive. It involved five major steps and took 10 years. First, scientific literature as far back as 1920 was combed for references with any mention of the substance. For just one substance, over 16,000 studies were collated. From the literature search, a complete bibliography, along with abstracts, was prepared for each substance. Second, monographs were prepared that summarized the bulk of the existing data and provided reprints and translations of the important articles. Thus any information on metabolism, chemical toxicity, occupational hazards, degradation, carcinogenicity, mutagenicity, teratogenicity, dose response, histology, behavioral effects, detection methods, processing, and reproductive effects was collected. Third, the National Research Council of the National Academy of Sciences surveyed the food industry to determine the level of use in food of each of the substances on the list. These data, together with data on food intake from the USDA and the Market Research Corporation of America, were used to project the estimated daily human intake (Elton, 1981). Fourth, testing for mutagenicity and teratogenicity was undertaken. Fifth, all the data from the first four segments were evaluated by the Select Committee on GRAS Substances of FASEB.

Box 10.2 shows an excerpt from the FDA proposal as published in the *U.S. Federal Register* to affirm the GRAS status of mannitol and sorbitol. It gives an idea of the type of information and the exhaustive amount of information collected. All of the GRAS proposals are a matter of public record and are available in the *Federal Register*.

From evaluations such as those in Box 2, a safety status for each additive was issued (Irving, 1978; Smith and Rulis, 1981). GRAS substances were placed in one of five categories (Table 10.1).

After completion of the GRAS review, the same type of exhaustive procedure of safety assessment was extended to potentially all chemical substances that are added to food. The impetus for this review resulted from two major factors. One was dramatic changes in the exposure of the population to additives due to economic factors, advances in food processing technology, and societal trends in the consumption of processed foods. Another was the marked increase in the sophistication of toxicological testing and criteria used to judge food additive safety (Smith and Rulis, 1981).

Consumer Perceptions

Despite the government's efforts to obtain data about the safety of GRAS substances and additives, consumers' reactions to the use of these substances are at best mixed, and more negative than positive. A survey done before 1980 found that 59% of the women surveyed favored the banning of "food additives used only to improve appearance of food even if there was no positive evidence of harm." In another survey of over 370 households in the Washington, D.C. area, 42% had a negative response to food additives—i.e., they either believed them dangerous or disapproved for some other reason. In a survey done by *Good Housekeeping*, 75% said they were extremely concerned or very concerned about chemicals in food. About the same number of consumers said that they always or frequently used label information about additives when making purchasing decisions. About 60% said that they tried to avoid preservatives, artificial flavors, and artificial colors (Sloan et al, 1986). This response is interesting, as only 6% could actually name a preservative when asked. Food Marketing Institute surveys of shoppers reveal that over 40% are concerned about food additives and preservatives and class them as a serious hazard to their health (FMI, 1990).

It is also interesting that, although the use of preservatives does not vary with the type of bread made, more than 90% of consumers thought that white bread contained preservatives whereas variety

Table 10.1. FDA GRAS^a Status Evaluation Following Review

Class	Status	Number of Compounds	Typical Examples
I	Safe at current and future use levels in accordance with Good Manufacturing Practices	305	Vegetable oils, casein, benzoates, tartrates, phosphates
II	Safe for use at current level; need further research to see if higher levels in the dietary would be a hazard	68	Alginates, some zinc, iron compounds, tannins, vitamins A and D
III	Additional studies recommended because of unresolved questions	19	Caffeine, BHA, BHT
IV	FDA urged to establish safer conditions or prohibit addition of ingredient to foods	5	Salt and four modified starches
V	Insufficient data on which to make a recommendation	18	Glycerides, certain iron salts, carnauba wax

^a Generally recognized as safe.

breads did not (Martinsen and McCullough, 1977).

In a large consumer survey in Canada, 68% said that the government had inadequate control over food additives (Martinson and McCullough, 1977). In that same survey, 60% said they were prepared to pay more for additive-free food. In a smaller survey in Britain, 87% were willing to pay more for certain types of additive-free food (Drew, 1986).

The anxieties felt by consumers stem from some very necessary survival instincts. The giving and receiving of food in all cultures is characteristic of nurturing relationships. The giving of food implies caring; the receiving and eating of it implies trust. With large, profit-driven corporations as the givers, the relationship is not there. Lost with the relationship is the trust. For many, the safety of the additive may be just a symptom of a deeper issue—loss of trust. The desire for additive-free food is an attempt to have a product like Grandmother made—one to be trusted.

CANCER, DELANEY, AND *DE MINIMUS*

The big fear with food additives, as with other chemicals in our environment, is that they may cause cancer. In a *Good Housekeeping* consumer survey, 83% agreed that chemicals can cause cancer. Over half the respondents felt that stories about other chemicals in the environment affected their feelings about chemicals in food. Over half the respondents feared that cancer might result from the use of artificial ingredients (Sloan et al, 1986). The fact that the chemicals in food were used in very small quantities did not reduce the fear of chemicals in foods for over 45% of the respondents. Other fears were related to allergy or to hyperactivity.

The fear regarding cancer is certainly not new. The so-called "Delaney clause" of the Food Additives Amendment shows that some concern existed in 1958. This clause, sponsored by Rep. James J. Delaney of New York and others, banned from U.S. food and beverages any food additive found to be carcinogenic at any level in humans or animals. It reads as follows:

> No additive shall be deemed to be safe if it is found to induce cancer when ingested by man or animals, or if it is found, after tests which are appropriate for the evaluation of the safety of food additives, to induce cancer in man or animal.

Much controversy has been generated over this clause since its passage. It is considered by some to be scientifically untenable because it is a zero tolerance law. That is, if cancer is shown in any species when the chemical is given in any amount by any means, not even necessarily fed, then the substance must be banned. The way the law is currently written means that a substance that causes cancer in animals but is known not to be harmful to humans must nevertheless be banned.

There is no consensus on what should be done with the Delaney clause (GAO, 1981). Eight out of nine of the former FDA commissioners favored amending the law—also 12 out of 15 biomedical researchers, but none of six consumer group representatives. Those in favor of changing the law say it leaves no room for scientific judgment and refuses to admit that zero risk is an impossibility. Further, they feel that the resources devoted to upholding the clause are not in line with the very slight risk that is posed. This group feels that the clause is a political, not a scientific, issue. Those in favor of leaving the law the way it is feel that it protects even the most susceptible individual, which they argue should be the case for chemicals added to the food supply since everyone is exposed to them. Further, this group believes that since science cannot extrapolate risk well, the zero tolerance must be kept (GAO, 1981).

Several attempts have been made to amend the Delaney clause. In 1985, FDA Commissioner Frank Young interpreted the law in a new way using the concept of *de minimus*. In this instance, two cosmetic color additives were added to the permanent listings. The decision was based on the fact that, although these color additives did cause cancer in laboratory animals, the risk to humans was so trivial that it need not be considered under the law, hence the *de minimus* status (Smith, 1987).

Legislation written to replace the Delaney clause should economize regulatory resources, recognize feasibility in the marketplace, establish priorities for public health relevance, be consistent for diverse sub-

---

Box 10.3

De minimus

This concept is currently being used by several regulatory agencies as an exception to the Delaney clause. *De minimus* is used when a substance is known to cause cancer in laboratory animals but the dietary risk is deemed negligible. The substance is then allowed to be used. For regulatory purposes, negligible means that the lifetime risk of cancer is increased by no more than 1 in a million (Federal Register, Oct. 19, 1988, 53 FR 41104 and 41125). For many, this interpretation is necessary because analytical capabilities have increased so dramatically.

stances, and be independent of everchanging analytical capabilities. Further, it should be a prototype applicable to other concerns such as teratogenicity (Hayes and Campbell, 1986)—a tall order.

The safety of many food additives has come under scrutiny. Some have been banned by invoking of the Delaney clause, others simply by using other provisions of the Food, Drug, and Cosmetic Act. A list of banned additives is given in Table 10.2.

Only one of these additives, cobalt salts, was banned because approved usage was associated with toxic effects in humans. In the mid-1960s, 28 cases and 11 deaths were associated with the cardiotoxic action of cobalt used in beer as a foam stabilizer. However, the people involved were not ordinary beer drinkers. They drank 5–12 quarts of beer daily for periods of 5–40 years! Cobalt may not have been the only cause of cardiotoxicity.

Some additives under scrutiny in the 1990s by one or more groups for various reasons include the substances listed in Box 10.4. In the remainder of this chapter, we will look at the functional additives that have had some safety concerns raised about them. Chapter 11 is devoted to flavor and color additives and the possible effects of additives on behavior and hyperactivity.

SWEETENERS

Noncaloric sweeteners lead all other food additives in dollar sales, and their use appears to be growing (Leading Edge, 1989). Concerns about sweeteners seem to grow along with their acceptance. The banning of cyclamate, the subsequent controversy about saccharin, and spurious reports about aspartame (Roberts, 1990) have raised consumer concern about the safety of sweeteners and other additives. The cyclamate banning sparked a systematic review of all additives. Interest in artificial sweeteners continues because of the strong interest in

Table 10.2. Banned Food Additives

Additive	Function	Year Banned
Dulcin	Sweetener	1950
P-4000	Sweetener	1950
Safrole	Flavoring	1960
Cobalt salts	Foam stabilizer	1966
Calamus	Flavoring	1968
Nordihydroguaretic acid	Preservative	1968
Cyclamate	Sweetener	1969
Diethylpyrocarbonate	Ferment inhibitor	1972
Chlorofluorocarbons	Propellant	1978
Colorant	Color	Various[a]

[a] See Table 11.2.

dieting and because saccharin, aspartame, and other sweeteners are frequently in the news (Lecos, 1985; Stamp, 1990).

Cyclamate

Discovered in 1937 by a graduate student at the University of Illinois, cyclamate is 30 times sweeter than sugar. Its sweetness coupled with a less bitter aftertaste than saccharin made it very viable as an artificial sweetener. Furthermore, it was stable under conditions of high heat, which allowed it to be readily incorporated into a wide variety of food products. Cyclamate was first marketed in 1950 as a dietetic aid in the United States by Abbott Laboratories. Since it was 20 times less expensive than saccharin, although less sweet, the use of cyclamate-saccharin mixtures soared.

Initial approval in the United States. In 1955 the National Academy of Sciences (NAS) reported cyclamate safe for human consumption. Since it was used in the food supply before 1958, cyclamate was classified as GRAS with the passage of the Food Additives Amendment. In the early 1960s, because of its popularity and increasing consumption, especially in carbonated beverages, the FDA again requested the NAS to assess the safety of cyclamate at the current levels of use. The NAS reported back to the FDA that, although reasonable quantities of cyclamate posed no hazard to humans, additional studies were needed to resolve some questions about cyclamate's

---

Box 10.4

Some Food Additives Under Safety Scrutiny

Sweeteners: sugar, xylitol, mannitol, aspartame, other intense sweeteners.

Functional aids: phosphate, brominated vegetable oil, modified starches, carrageenan, bleaching agents.

Antioxidants and preservatives: nitrate, nitrite, BHA, BHT, sulfites,[a,b] benzoates, propionates, sorbates, acidulants (including phosphate)

Colors and flavors or flavor enhancers: FC&C Red. No 3, FD&C Red No. 40, FD&C Yellow No. 5,[a] caffeine, synthetic flavors, monosodium glutamate

[a]With label requirements.
[b]Banned for use in salad bars.

---

safety. Two studies suggested it might be carcinogenic. One study at the highly respected Wisconsin Alumni Research Foundation indicated that high concentrations of a mixture of cyclamate and saccharin were possibly carcinogenic to rats (Price et al, 1970). A European study in which mice were fed cyclamate in their drinking water also indicated carcinogenicity (Rudali et al, 1969).

Banned in 1970. Despite the many unanswered questions left by the study, on October 18, 1968, cyclamate was removed from the GRAS list. Abbott was asked to submit a food additive petition for cyclamate, but this petition was denied based on data implicating cyclamate as a possible carcinogen when surgically implanted on the bladders of a cancer-susceptible strain of rats. Therefore, in 1970, the FDA imposed a total ban on cyclamate and decided that it would not be allowed even as a drug under medical supervision.

Further studies. In 1973, Abbott sought FDA's permission to remarket the sweetener in foods designed for special dietary purposes. In the intervening years, there have been several attempts to reinstate cyclamate. One new petition included nearly 500 new toxicological assessments attesting to cyclamate's safety. However, even after a series of petitions and court fights, cyclamate was still not permitted in the food supply. One FDA commissioner, Jere Goyan, based his position on the fact that cyclamate had not been proven not to cause cancer. In 1985 the NAS reviewed all existing data and concluded that cyclamate and its incriminated metabolite, cyclohexamine, were not themselves carcinogens but could be cancer promoters (NAS/NRC, 1985). The Society of Toxicology published an article stating that the FDA's cyclamate decision was an example of how not to do and interpret animal studies (Munro, 1987).

In 1984, another petition to reinstate cyclamate was submitted to the FDA. This included over two dozen studies indicating that high doses of cyclamate throughout the lives of laboratory animals did not cause cancer. Further, 15 epidemiological studies showed no significant increase in the relationship between bladder cancer risk and the use of artificial sweeteners, both saccharin and cyclamate (Calorie Control Council, 1985; Morris and Przybyla, 1985). The FDA's Cancer Assessment Committee has exonerated cyclamate as a carcinogen (Newberne and Conner, 1986), as did a 1985 report issued by the NAS.

Despite these extensive studies, cyclamate was still not an allowed food additive in the United States when this chapter was written (1991). The position of the United States was clearly different from that of over 40 other countries that allow it as a food additive. It is also considered a safe additive by the World Health Organization (WHO) and the European Community (Malaspina, 1987). In fact, in 1982 WHO raised its estimate of acceptable daily intake from 4 to 10 milligrams per kilogram of body weight. The decision to prevent its use in the food supply thus may be more a political than a scientific one.

Saccharin

As with other artificial sweeteners, the discovery of saccharin was accidental. It was originally used as an antibacterial agent and food preservative. In the early 1900s, its property of being 300 times sweeter than sugar was recognized, and it was added to foods for diabetics.

Many products could be sweetened with saccharin because it was stable over a wide range of temperatures and conditions. The only drawback to its use was a slightly bitter aftertaste. This flavor quality was overcome somewhat in the 1950s when a 1:10 mixture of saccharin and cyclamate gained popularity because the sweetness imparted by the mixture was greater than the sum of the sweetnesses of each component, while the bitterness of saccharin was held below its threshold of perception for most consumers.

Safety testing. Saccharin's safety testing began not long after its discovery in 1879 and continued for over 100 years, with highly controversial results. As far back as 1907, Dr. Harvey Wiley, chief of the USDA's Bureau of Chemistry, wanted saccharin removed from use in canned foods. President Theodore Roosevelt, a diabetic, voiced an irate reaction to Wiley's proposal by saying, "You tell me that saccharin is injurious to health? My doctor gives it to me every day. Anybody who says saccharin is injurious to health is an idiot." Subsequently, a group of Wiley's scientific advisors suggested that a saccharin dose of 0.3 gram per day was safe but that levels above 1 gram per day would cause digestive disturbances. Ultimately, saccharin was banned, but only for a short time, as it was reinstated during World War I due to the shortage of sugar (Hutt and Sloan, 1979; Lecos, 1981; Wightman, 1977). It was also used a great deal during World War II and then continued as a sweetener for diabetics.

Long-term animal experiments with saccharin have been made since the 1950s in a variety of species. A seven-generation chronic feeding study in rats, a three-generation chronic feeding study in mice, and chronic feeding studies on hamsters and monkeys all suggested that the ingestion of saccharin would not represent a health threat to humans. In 1958 it was given GRAS status due to its long use in the food supply.

GRAS status was removed in 1972 when a study by the Wisconsin Alumni Research Foundation suggested a possible link between saccharin (or impurities in the saccharin) and bladder cancer in rats. A subsequent Canadian study showed that bladder cancer incidence was high in second-generation rats fed high doses of saccharin.

Ban in the United States prevented by Congressional moratorium. As a result of these studies, the FDA in 1977 proposed a total ban of saccharin, invoking both the Delaney clause and the general safety clause of the Food, Drug, and Cosmetic Act. Since saccharin was at that time the only available artificial sweetener, public response to the ban was overwhelming. The FDA received over 100,000 letters and Congress over 1 million letters. Not only was the response overwhelming, it was also polarized. The diametric opposition was characterized by two letters that the FDA Commissioner received. One writer said that he hoped 1,000 fat people would come sit on the commissioner's head, while the other said that he hoped that the commissioner would be reincarnated as a white rat for studies with the carcinogen saccharin.

Because of this public outcry, the United States Congress passed the Saccharin Study and Labeling Act (1977), which declared a moratorium on the ban until 1979 so that additional research on the safety could be conducted. The act also required that saccharin-containing products exhibit a warning label reading as follows: "Use of this product may be hazardous to your health. This product contains saccharin, which has been determined to cause cancer in laboratory animals." Congress has continued to enact extensions of the moratorium on the ban.

Further studies. Since 1977 many further studies have been done to assess long-term hazards of using saccharin. Metabolic studies showed it to be metabolically unchanged after being slowly absorbed and rapidly excreted. This is important evidence against its carcinogenicity, as no known carcinogens are excreted unchanged. A 1983 study (IFT Expert Panel, 1986b) involving 2,500 second-generation male rats revealed that high doses of saccharin caused changes in rat bladder tissue if the rat was exposed to saccharin during the suckling period but not if exposure was during the fetal period or after the suckling period. The incidence of tumors was clearly a function of dose, with numbers of tumors declining sharply as the dose decreased. From this experiment, the risk of consumption of two cans of diet soda daily was extrapolated, with the increased risk of human bladder cancer calculated at less than one in a million. Variation in risk using different methods of extrapolation based on rodent experiments ranged from 0.2 cancers to 144,000 cancers in the next 70 years in the United States (Munro, 1987).

Epidemiological studies in Scandinavia, Japan, England, and the United States have failed to reveal any overall association between saccharin ingestion and bladder cancer. Data from these studies include people with exposure to artificial sweeteners beginning decades ago (Concon, 1988; Newberne and Conner, 1986). One study found no elevated risk for the population in general, but they did find a positive association for several subgroups (Hoover, 1980). These included white males who were heavy smokers and nonwhite females with no known exposure to bladder carcinogens. The American Medical Association's Council on Scientific Affairs recommends a moratorium on the saccharin ban, since the evidence on the carcinogenicity of saccharin in humans has not been forthcoming. It also recommends careful monitoring of any adverse effects of saccharin and warns that young children and pregnant women should carefully consider the use of saccharin (AMA, 1985, 1986).

Legally, saccharin is now classified as a cocarcinogen (tumor promoter) with very low potency. Extrapolations suggest that saccharin at 30-300 milligrams per day (0.43-4.3 milligrams per kilogram per day) does not increase human cancer risk (Byard, 1984). It is allowed in the United States under the Congressional moratorium on banning its use. It is approved for use in 80 countries and has been determined as safe by both the FAO/WHO Joint Expert Committee on Food Additives (JECFA) and the Scientific Committee for Foods of the European Economic Community (Arnold and Clayson, 1985; Arnold and Munro, 1983).

Aspartame

Aspartame has a much shorter history than either cyclamate or saccharin. It, too, was discovered by accident. A chemist at G. D. Searle Co. who was trying to synthesize gastric peptide from its constituent amino acids spilled some of the mixture, inadvertently tasted it, and noticed that it was sweet. The sweetness was subsequently shown to be due to a dipeptide formed from the two naturally occurring amino acids, phenylalanine and aspartic acid. Interestingly, neither of the constituent amino acids tastes sweet alone, but they do so when joined together as a dipeptide.

Since this sweetener is comprised of amino acids, it yields the same number of calories as amino acids and proteins—4 Calories per gram. However, as it is 180 times sweeter than sucrose, significant caloric reductions may be achieved (Table 10.3). Its sweet taste without the bitter aftertaste often associated with artificial sweeteners makes aspartame advantageous from a sensory standpoint. Its sweetness varies with pH and temperature. In some products, particularly those with acid fruit flavors, aspartame enhances the existing flavor.

Stability. The stability of aspartame is dependent on pH, moisture, and length of storage. Sweetness can be maintained in a soft drink held at room temperature for 6 months. However, aspartame is not stable to high heat, so its use in certain types of food is not feasible (Homler, 1984).

Safety. Since this sweetener is comprised of amino acids found in common foods, there was an initial feeling that it was going to be a very safe sweetener. The two constituent amino acids are found commonly in meats, vegetables, fruits, dairy products, and cereal grains. Table 10.4 compares levels of these constituents in common servings of food to the amount in a serving of diet soda sweetened solely with aspartame.

A food additive petition for approval of aspartame was filed in 1973 by G. D. Searle. The petition included 113 studies. The petition, submitted to the FDA and in Canada to the Health Protection Branch, was the most comprehensive ever submitted.

As required, the petition included tests on the metabolism and potential toxicity of the sweetener. Metabolic studies showed that during digestion it is hydrolyzed, as are all other dipeptides, to its constituent amino acids. Like amino acids from any food, these could be used for energy or could be incorporated into body proteins. Minor amounts might be used in the production of neurotransmitters.

Some methanol is also produced during digestion because aspartame contains a methyl ester. However, blood levels of methanol were far below those considered to have any adverse effect even when aspartame

Table 10.3. Sweetness of Artificial Sweeteners[a]

Sweetener	Sweetness Relative to Sucrose (Sucrose = 1)
Acesulfame K	200
Alitame	2,000
Aspartame	180
Cyclamate	30
Saccharin	300
Sucralose	600

[a] Adapted from Munro (1989).

Table 10.4. Calories in Sweetened Food[a]

Food	Sweetener		Calorie Savings from Aspartame
	Sugar	Aspartame	
Soft drink, 12 oz.	156	2	154
Lemonade, 8 oz.	86	5	81
Gelatin dessert	81	10	71
Hot chocolate mix, 6 oz.	110	50	60

[a] Data from NutraSweet Co. (1987).

was fed at six times the estimated daily intake (Stegnick and Filer, 1984). Further support for the lack of toxicity of methanol resulting from aspartame ingestion comes from looking at levels of formate in the body. Methanol toxicity results from its metabolism to formate and the buildup of formates. No formates were detected after aspartame doses of 200 milligrams per kilogram (Horwitz and Bauer-Nehrling, 1983). Other tests on carcinogenicity, teratogenicity, and other toxicity were negative.

Initial approval stayed. In 1974 the FDA approved the use of aspartame as a sweetener in certain foods. However, the approval was stayed after formal objections were filed. Objections were based on allegations that aspartame might lead to brain damage and might cause mental retardation and endocrine dysfunction. Searle agreed to withhold the

---

Box 10.5

Outline of Aspartame Metabolism and Breakdown and Fate of the Products

Aspartame

Aspartylphenylalanine   Diketopiperazine

Aspartic acid + Phenylalanine

Methanol

On a per weight basis, metabolism of aspartame yields approximately 50% phenylalanine, 40% aspartic acid, and 10% methanol. All three of these compounds are metabolized. Even if all sucrose were to be replaced with this sweetener, the levels of the metabolic breakdown products would not create any problems, as the normal intake of phenylalanine is 3.6 grams and of aspartic acid, 6.8 grams (Salminen and Hallikainen, 1990).

Diketopiperazine is formed with the chemical breakdown of the compounds in foods that are heated or stored. With its formation comes a perceptible loss of sweetness. This product has been reviewed extensively in the toxicological evaluation of aspartame and found to have no toxic or carcinogenic effects. JEFCA has given this compound an acceptable daily intake of 7.5 milligrams per kilogram of body weight (Salminen and Hallikainen, 1990).

marketing of aspartame until the safety issues had been resolved. A board of scientific inquiry was established to evaluate the charges.

One concern focused on the neurotoxic potential of high plasma levels of phenylalanine. Subsequent investigation showed that doses of aspartame equivalent to 24 liters of diet soda or 600 sweetener tablets doubled the plasma phenylalanine level. Even with the level doubled, it was 5–10 times below the level required to see any toxic symptoms in all groups studied, including infants, pregnant women, and people with phenylketonuria (PKU). However, it was recommended that aspartame contain a warning label for PKU, as individuals with this disease need to control their phenylalanine intake from all sources (AMA, 1985). Fifteen thousand people in the United States suffer from PKU.

With respect to the aspartic acid component, the concerns focused on possible focal brain lesions or endocrine disorders from aspartic acid alone or in combination with glutamate from the diet. A loading dose of aspartame did produce higher plasma levels, but they were 15 times less than those required to show any adverse effect. The board of inquiry and the FDA concluded that there was no evidence that aspartame either alone or in combination with glutamate contributes to brain damage, mental retardation, or endocrine dysfunction, and in 1981 its use both as a tabletop sweetener and in foods in the United States was again approved.

Further safety considerations. Some fear has been voiced that aspartame has the potential to change brain neurotransmitters, resulting in behavior change and possibly seizures. Neurobehavioral changes and changes in brain neurotransmitter levels have occurred in young adult male rats fed 4.5 or 9% aspartame in the diet. Recent studies utilizing levels relevant to human exposure showed no aspartame-induced changes in neurotransmitters. Furthermore, studies of monkeys given high levels of aspartame have not shown any behavioral abnormalities, abnormalities in electroencephalograms, or seizure activity (Alfin-Slater and Pi-Sunyer, 1987). Dr. Richard Reuben, Chairman of Pediatric Neurology at New York University Medical Center and Chairman of the Epilepsy Institute, reviewed data from over 5,000 patients, many of whom consumed aspartame regularly, and reported no change in seizure incidence (Epilepsy Institute, 1986; Brody, 1987).

Some food intolerance to aspartame may exist. Some reports link its use with the onset of migraine headaches. These have not been causally linked. It is possible that some individuals may be sensitive to aspartame, just as some individuals are sensitive to components in red wine or other foods that give them headaches. There is also some disputable evidence that aspartame is a source of urticaria (hives) in some hypersensitive individuals (Kulczycki, 1986; Roberts, 1990).

The FDA has received over 3,000 complaints regarding aspartame. The Centers for Disease Control (CDC) followed up on these by interviewing a sample of 517 complainants. The CDC concluded that certain individuals may have a sensitivity to aspartame but that the complaints were mild and did not provide evidence of a serious health hazard (Bradstock et al, 1986). Reported side effects belonged to three groups: 1) central nervous system effects such as mood changes, insomnia, and seizures, 2) gastrointestinal effects including abdominal pain, nausea, and diarrhea, and 3) gynecological effects including irregular menses.

An international meeting on dietary phenylalanine and brain function was held in 1987. The double-blind studies reported at this meeting failed to establish any adverse clinical or behavioral effects on children or adults. Some studies did suggest that, although aspartame does not cause neurological abnormalities, it may promote them in animals at risk. In general, the empirical studies did not find a causal relationship between aspartame and seizures or behavioral defects in normal children or adults (Caldwell, 1987).

There is some concern about the heavy use of aspartame (Maher, 1986). Heavy consumption of soft drinks puts some individuals at risk because they consume levels near the acceptable daily intake (ADI). In particular, a young child consuming a liter of diet soda a day would receive 26 milligrams of aspartame per kilogram, nearly twice the ADI for children. This concern is set in a regulatory environment that is enabling more and more different uses for aspartame. Roberts (1990) also reports cases where adults abuse aspartame by drinking several liters of diet soft drinks per day as well as using other saccharin-containing products, and he describes some of their complaints.

The carcinogenic potential of aspartame was evaluated by the FDA's Center for Food Safety and Applied Nutrition. The FDA commissioner concluded that aspartame and its decomposition product diketopiperazine did not contribute to tumor formation in rats and mice (Federal Register, 1981; Reno et al, 1975).

The use of aspartame during pregnancy and lactation has also been questioned. Initial studies in rats fed aspartame at 6% of the diet showed an increase in infant mortality of 6%, a reduction of maternal weights during lactation by 4–6%, and delayed eye opening, righting, swimming, and startle response in 6% of the offspring. However, in primates fed doses at the 99th percentile of projected intake (e.g. replacing all sweeteners in the diet with aspartame), there was no evidence of risk to the developing fetus. Since even excessive levels do not appear to affect the fetus, and since aspartame does not readily cross the placenta, it is considered safe during pregnancy when used in moderation (Franz, 1986; Sturtevant, 1985).

Both the American Medical Association (AMA, 1986) and the American Academy of Pediatrics (AAP, 1985) have concluded that aspartame use is safe for humans who do not suffer from PKU and is safe to the fetus during pregnancy at levels currently being used. The American Diabetes Association has also endorsed the safety of aspartame (ADA, 1985). Furthermore, studies indicate that aspartame had no effect on blood glucose levels (Nehrling et al, 1985). The American Dental Association has also issued a positive statement about the safety of aspartame and has endorsed its use because it does not promote tooth decay (Goodman et al, 1987). Aspartame has been given an ADI of 40 milligrams per kilogram by the JECFA of the World Health Organization (Wells, 1989). It is currently allowed in over 50 countries.

Acesulfame-K

Like many other sweeteners, the 1967 discovery in Germany of acesulfame-K was accidental. Acesulfame-K is a derivative of acetoacetic acid and is about 130 times sweeter than sucrose (Wells 1989). It has some structural similarity to saccharin (Newberne and Conner, 1986).

Acesulfame-K has several advantages over aspartame. It withstands processing at higher temperatures and therefore can be used to sweeten baked goods (Hood and Schoor, 1990). When aspartame and acesulfame-K are used together, the sweetness quality is better than with either sweetener alone. It eliminates shelf-life problems seen in soft drinks sweetened with aspartame and costs less but it may cause a lingering bitter taste at high concentrations (Wells, 1989).

Box 10.6

Acesulfame's Approval Process

Acesulfame K was assessed for the first time by the JEFCA in 1981. At that time no acceptable daily intake was set, as there were some questions about some of the data. One mouse experiment indicated carcinogenicity, but the studies did not meet the requirements required by the JEFCA. Subsequent studies showed that acesulfame was neither carcinogenic nor mutagenic. With more complete toxicological information, the safety of the compound was established. However once any questions have been raised about the safety of a product, some consumers and consumer groups will fear it (Salminen and Hallikainen, 1990).

A large number of pharmacological and toxicological tests have initially given acesulfame-K a clean slate. Metabolic studies on acesulfame-K show it to be excreted unchanged (O'Sullivan, 1983). The JECFA has approved its use, and it is being used in 20 countries. It received approval for use in the United States in the summer of 1988. Packets of this sweetener are available on the consumer market.

Sugar Alcohols

Sorbitol, mannitol, maltitol, and xylitol are all sugar alcohols (polyols) that occur naturally in small amounts in fruits and vegetables. Five to 15 grams of xylitol is produced per day in the body during normal carbohydrate metabolism.

Theoretically, these sugar alcohols yield the same number of calories as sugar, but in actuality they are incompletely metabolized, so the net energy yield is lower (Wursch and Anantharaman, 1989). While this might seem to be advantageous to a dieter, it is not, because the lower energy yield is offset by a sweetening power that is much lower than that of sucrose for all polyols except xylitol. Thus, more sweetener must be used to achieve the effect of sucrose. Most are of little use for the lucrative diet market, even though some consumers believe that candies and foods labeled "dietetic" are lower in calories.

While they might not save any calories, they may have an effect on appetite. Mice fed diets containing xylitol or sorbitol were less likely to consume excess food and become overweight than those fed diets containing glucose or sucrose (Potezny et al, 1986).

Diabetics can use these sweetening agents in limited amounts as they do not require insulin to enter the cell. Amounts must be limited (not over 50–60 grams per day), as any excess goes to the liver and is converted to glucose (ADA, 1980). The fact that they are absorbed more slowly and help reduce the formation of toxic ketone bodies also is advantageous to the diabetic.

These sweeteners have been used for products like sugarless gums and mints and as sweeteners for medicines, since they are believed to be less cariogenic (cavity-producing) than other sugars. This is due to slow fermentation by mouth bacteria (Linke, 1986). Studies have shown that there were 30% fewer cavities in rats fed sorbitol and mannitol, and xylitol virtually eliminated dental caries (Makinen, 1989). Conflicting reports suggest that studies indicating that sugar alcohols were safe for teeth were done for too short a time span. One study showed that the population of the decay-producing bacteria *Streptococcus mutans* increased with the consumption of sorbitol (Hirsch, 1985). Some evidence suggests that over longer time spans the mouth bacteria evolve to utilize the sugar alcohols. If this proves true, sugar alcohols may become as cariogenic as the sugars they are designed to replace (Linke, 1986).

Other safety questions have arisen about sugar alcohols. Chronic consumption of sugar alcohols in rats has reduced weight gain, as well as producing diarrhea, abdominal pain, bloating, cecal enlargement, excessive calcium excretion, calcium deposits in the bladder, and neoplasms when fed at high doses (Bar, 1985; Jain et al, 1985). These naturally occurring compounds have all yielded negative results with respect to mutagenicity tests (JECFA, 1986), but one study indicated that xylitol may be tumorigenic (ADA, 1980).

One 13-year-old girl was treated for acute abdominal pain after eating 16 pieces of candy containing sorbitol (Lipin, 1984). Thus, high doses of this compound may be of concern (Bar, 1985).

Lycasin is a new polyol developed from glucose syrup that contains a sorbitol and other oligosaccharides. It is about 75% as sweet as sucrose but tastes and eats like it. Technically, it is advantageous because it resists browning. This polyol does not promote tooth decay (Linke, 1986; Rugg-Gunn, 1989) and is less likely to promote diarrhea than other polyols. This substance was approved for use as an additive in 1987 in the United States (Roquette Freres, 1990).

"New" Sweeteners

A wide variety of other sweeteners not presently approved for use in the United States is being used in other countries (Nichol, 1991). Some of these may be petitioned and approved for use as food additives in the United States as well.

Glycyrrhizin. One GRAS additive already in use has some applications as a sweetener. This is the substance glycyrrhizin from the licorice plant. Although it is primarily used for its flavor characteristics, it could be used as a sweetener, as it is 50 times sweeter than sugar and is anticariogenic. It is currently being used in candies, root beer, liqueurs, chocolate, and vanilla. It is also used in a variety of ways in Japan (IFT Expert Panel, 1986b; Sela and Steinberg, 1989).

Thaumatin (trademark name: Talin). This sweetener is found in the fruit of a West African plant called *katemfe*. Sweetness in this case comes from a protein consisting of 207 amino acids. The protein is up to 5,000 times sweeter than sucrose, but the sweetness develops slowly and leaves a licoricelike aftertaste. It not only imparts sweetness, but also extends the flavor and aroma of other food constituents. This protein's sweetening effect is stable over a wide range of pH and temperature. It is approved for use in the United Kingdom, Mexico, Japan, and Australia (except in baby food). In the United States, the only allowed use is in chewing gum, but petitions for other uses have been filed. In 1985 JECFA agreed it was safe for use in food but did not specify an ADI (Wells, 1989).

Miraculin. This is a basic glycoprotein that comes from the berry of the West African miracle fruit. It has a taste-modifying effect that causes sour foods to taste sweet and is used to sweeten sour vegetables, fruits, or yogurt. One company attempted to start a plantation to grow the plant in the Caribbean and introduce it into countries of the Caribbean basin and Brazil. However, the company went bankrupt trying to do the testing required for FDA approval. Food additive status was never established in the United States.

Stevioside. Leaves of a Paraguayan shrub yield the sweetener stevioside. It has long been used by the natives of Paraguay to sweeten bitter drinks. The glycoside is extracted and crystallized. This glycoside is 300 times sweeter than sugar. It is said to have a delectable flavor in small quantities but a bitter flavor in high quantities. The shrub is cultivated in Japan, where this sweetener is used as a food additive. In Japan the compound is not subjected to testing because of its natural origin. Stevioside is not approved for use in the United States, as some preliminary studies indicated that it may be toxic to rats. Some research has been directed toward making analogues of this compound that are less absorbable in the human gastrointestinal tract (DuBois et al, 1984; IFT Expert Panel, 1986b; Inglett, 1981). It is used in Paraguay and Brazil as well as Japan, with approval in the European Community applied for (Wells, 1989).

Monellin. This is a protein 2,000 times sweeter than sucrose. It has long been used in Africa, where natives were familiar with the sweetening imparted by the red serendipity berry. Research in the United States has been done in part at Monell Chemical Senses Center. Studies found that it loses much of its sweetness when heated to temperatures above 70–80°C. No FDA approval has been sought because of the high cost involved and some technical limitations to the use of this sweetener.

Left-handed sugars. A left-handed sugar (L-sugar) that has been patented for use in food (trade name: Lev-O-Cal) could become available as a sweetener. This molecule is the mirror image of ordinary sugar. L-sugars occur naturally in minute amounts in foods such as sugar beets, plantains, seaweed, and some algae. L-sugars look, taste, and bake like sugar but yield no calories to mammals. Mammals have enzymes that metabolize right-handed sugars but none to metabolize left-handed ones (Holmes, 1986). Some safety testing of L-sugars has been done (Gorton, 1983), but much more will be required before submission of a petition for approval. No countries currently allow use of this sugar.

Sucralose. Other substitutes for sucrose are the chloroderivatives of sucrose, one of which is called sucralose (Anonymous, 1988; Jenner, 1989). The various derivatives may be five to 2,000 times sweeter than sucrose. In 1987, the McNeil Specialty Products Company petitioned the FDA to approve marketing of sucralose as a tabletop sweetener as well as for use in 13 product categories. Sucralose has the technical advantage of being stable at baking temperatures (Hood and Campbell, 1990). According to the material from the company, studies in experimental animals and humans show its safety, even at levels greatly exceeding the maximum possible human intake. Further clinical studies on sucralose show that it does not affect either blood glucose or insulin levels and thus would be safe for use by diabetics (Goldsmith, 1988).

Neosugars. Another approach is the manufacture of neosugars. These are made by attaching several fructose molecules onto sucrose. They occur in nature and are easily manufactured in the laboratory. These products are not digested by mammals, so they do not yield calories. They provide roughly 40–60% of the sweetening power of sucrose (Mayer and Goldberg, 1985).

Alitame. A 4,500-page petition has been accepted for study by the FDA for another dipeptide sweetener comprised of aspartic acid and alanine, called alitame. It has a clean, sweet taste like sugar but is 2,000 times sweeter than sucrose. It has stability that would allow it to be used in a wide variety of foods, including baked goods as well as beverages (Freeman, 1989; Wells, 1989).

D,L-**Amino malonyl-D-alanine isopropyl ester.** This is another peptide sweetener. It is 58 times sweeter than sucrose and has a sucroselike taste with no aftertaste. The sweetener is more stable than aspartame. Other peptide sweeteners are being developed, some by genetic engineering (Janusz, 1989).

Dihydrochalcones. These are derived from bioflavonoids, natural constituents of fruits and citrus. The sweetening abilities of this family of compounds vary considerably; they are 300–2,000 times sweeter than sugar. The perception of sweetness is delayed, and a lingering aftertaste of menthol or licorice follows. Both these factors will limit their use to products such as mouthwash or fruit juices, which would not be adversely affected by either factor. They are currently permitted for use in Belgium, Zimbabwe, and Spain but not in Canada, the United States, or the United Kingdom (Wells, 1989).

Palatinose (Palatinit). This is the disaccharide isomaltulose found in honey and cane sugar. It is being commercially produced in Japan from sucrose. This sweetener is not an allowed additive in the United States. The Japanese Dental Science Association has found that replacing part of the sucrose in foods with palatinose would control the cariogenicity of sucrose (Anonymous, 1985a). It is lower in calories only because it is not readily absorbed, but it behaves in other respects like sugar (Anonymous, 1990b).

Hernandulcin. Another substance that may have some use as a sweetener comes from the petals and leaves of the Mexican plant *Lippa dulcis* and was known to the Aztecs. The active component, hernandulcin, rediscovered and synthesized by researchers at the University of Illinois, is more than 1,000 times sweeter than sucrose. It does have some drawbacks, not being as pleasant as sucrose, with perceptible off- and aftertastes as well as some bitterness. Its initial toxicity tests show it to be nonmutagenic and of low toxicity (Compadre et al, 1985).

Summary. Sweeteners, it can be argued, are nonessential food additives except in the case of diabetics (Finer, 1987). However, one interesting argument suggests that if all the diet products that we ingest were replaced by calorie-bearing ones, the risk from overweight would far outweigh any slight increase in cancer risk that would possibly occur from the use of artificial sweeteners (Cohen, 1978; Rogers and Blundell, 1989). Although some studies support the use of artificial sweeteners as diet aids (Tordoff and Alleva, 1990), others do not (Anonymous, 1990).

BHA, BHT, AND OTHER PHENOLIC ANTIOXIDANTS

Unsaturated fats are easily oxidized. The products of oxidation, peroxides and hydroperoxides as well as their breakdown products, are detrimental not only to the taste and smell of the product, but also to the optimum health of the person ingesting the product. Every effort must be made to inhibit fat oxidation reactions.

Strategies that inhibit oxidation in foods containing fats are: 1) production and consumption of the food quickly, 2) storage of food under vacuum or modified atmosphere, 3) hydrogenation of fat to reduce the degree of unsaturation, 4) avoidance of prooxidants such as certain metal ions, and 5) use of natural or synthetic antioxidants such as butylated hydroxyanisole (BHA) or butylated hydroxytoluene (BHT).

Many oils naturally contain vitamin E compounds (tocopherols) that can act as antioxidants (Dougherty, 1988; Megremis, 1990). Unfortunately, such antioxidant capacity is readily overpowered. Furthermore, natural antioxidants such as vitamin E are destroyed during processing steps such as frying or baking and thus may not be viable alternatives as antioxidant additives (Buck, 1985).

Phenolic antioxidants do survive processing steps and prevent fat oxidation (Coulter, 1988). The antioxidant selected depends on the food requiring protection, as the oxygen-trapping ability varies with the components of the food. In some cases, antioxidants are used in combination because the protection is greater than with either antioxidant alone. The allowed amounts of phenolic antioxidants like BHA and BHT are limited. In the United States, a maximum of 0.02% of the fat content as BHA and/or BHT is GRAS, but in certain rancidity-prone but low-fat foods (e.g., dry breakfast cereals, instant potatoes), BHA and/or BHT at up to 50 ppm of the food itself is permissible as a regulated food additive.

In addition to preventing fat rancidity, phenolic antioxidants such as BHT have antimicrobial activity and may help retain vitamin E in the food. BHT exhibits antimicrobial activity against several common food pathogens and viruses (Branen et al, 1980).

Since the introduction of phenolic antioxidants such as BHA and BHT into the food supply in the 1940s, these substances have been widely used. They are found in various prepared foods containing fat such as breakfast cereals, mixes, dried soups, processed meat and fish products, potato flakes, crackers, and snack foods. Their wide use reflects both their effectiveness and their ability to increase the availability of low-cost, convenient food. For instance, the shelf life of instant cereals increases from 2 to 50 days when an antioxidant is added. Without antioxidants, it is doubtful that such products would be marketed at all. If they were, they undoubtedly would be costly, since the prices of most widely marketed products are in large part determined by their perishability or the type of packaging system required.

In today's market a fat-containing food that boasts no antioxidants has consumer appeal. It may also be giving some subtle messages that,

Table 10.5. Food Additive Intakes in Japan, Finland, and the United States[a,b]

Additive	Average Daily Intake, milligrams per person		
	Japan	United States	Finland
BHA	0.001	1.8	...
BHT	0.023	5.6	0.72
Cyclamate	0.91	banned	...
Nitrate	35.5	1.5	1.5
Nitrites	0.0018	2.2	0.2
Proprionates	36.3	260.0	10.4
Saccharin	0.91	7.1	1.1
Sorbates	36.3	72.0	27.8
Sulfur dioxide	21.0	7.0	4.6

[a] Data are estimate from food consumption surveys and market basket surveys.
[b] Adapted from Louekari et al (1990).

Table 10.6. Comparative Activity of Antioxidants in Pork Fat[a,b]

Antioxidant	Percent	Days to Reach Rancidity
None	0	3
BHA	0.01	14
BHA	0.02	28
BHT	0.02	18
Plus tocopherol	0.02	15
	0.05	15
	0.2	15

[a] Adapted from Bauerfeind (1975).
[b] Pork fat was at 45°C. Rancidity is defined as peroxide at a level of 20 millequivalents).

if known, would decrease consumer appeal. One is that the product has a very short shelf life, as the fat will go rancid very quickly and produce oxidation products that are unhealthful and bad-tasting. Another is that the food is in a package sealed without oxygen, a process that usually requires extensive— and expensive—packaging. Another is that the fat is saturated enough to not require an antioxidant and is therefore high in saturated fats, which tend to elevate serum cholesterol.

Safety of Phenolic Antioxidants

In terms of metabolism, phenolic antioxidants are absorbed along with fats in the small intestine. Thus, intestinal cells are exposed to greater levels of BHA and BHT than other body tissues. From the intestine they are taken to the liver, where they are temporarily stored. Liver function may be altered by BHA and BHT, and some experimental animals have shown evidence of liver enlargement. Since BHA and BHT are not readily excreted, they accumulate in adipose tissue (Slaga, 1981).

Two of these aspects of metabolism create some concern: the liver enlargement that results when the animal is required to detoxify compounds and the fact the these compounds may be stored in the body rather than excreted.

Changes at the cellular level also cause some concern. Cell membrane permeability is changed in vitro, although it is not clear whether this effect occurs in vivo. High levels slow the rate of synthesis of DNA and RNA, which in turn reduce rates of cell division and growth (Daniel, 1981). In tissue culture, antioxidants decreased cell survival and caused chromosomal damage (Ito and Hirose, 1989).

Rat feeding studies appear to document some of the changes that occur both at the cellular level and in metabolism. Levels of BHT greater than 0.02% of the diet reduced body weight, impaired clotting, caused changes in the cells of the intestinal mucosa, and increased liver weight (Fulton et al, 1980). Hemorrhagic effects of massive doses of BHT in rodents (but not dogs) are of concern but are reversed by the administration of vitamin K (JECFA, 1987). Another concern is that BHT may cross the placenta and have some adverse developmental effects on offspring (Vorhees et al, 1984). Researchers in Canada reported that, at inordinately high dietary levels of BHA and BHT (2% of the diet, or 20,000 ppm, which is 1,000 or more times the maximum likely human dietary concentration), increased numbers of proliferative lesions were found in the forestomachs of rodents. Some feel these stomach lesions may be specific to the rodent, since other species lack a forestomach, but the findings nevertheless raised a warning flag indicating that these antioxidants should be tested further (Ito et al, 1983; Newberne and Conner, 1986). Subsequent studies on BHA in animals without a forestomach still left JECFA in doubt because dogs showed no abnormalities but monkeys and pigs did show abnormalities in the esophagus (JECFA, 1987). It was subsequently decided (JECFA, 1989) that, although the rat forestomach lesions were of concern, animals not having a forestomach had no problems, and an ADI could be established.

As with many food additives, a small percentage of the population shows hypersensitivity to antioxidants. In this case, reactions have been documented in humans and guinea pigs (Daniel, 1981).

Possible Carcinogenicity

The possible carcinogenicity of BHA and BHT causes both concern and confusion, as studies have shown them to be both cancer-promoting and anticarcinogenic. Studies by the National Cancer Institute involving both rats and mice fed large doses of BHT (6,000 ppm) did not produce even the slightest suggestion that BHT was carcinogenic. In fact, several studies indicate that antioxidants such as BHT are actually anticarcinogenic (Slaga, 1981; Weisburger and Evarts, 1977). A Japanese study involving the feeding of fish (*Rivulus ocellatus*) showed dietary BHA to be carcinogenic to the liver at 0.01% of the diet (Park et al, 1990), but no epidemiological evidence of this exists (NAS/NRC, 1989). BHT and BHA also appear to protect experimental animals from tumors found at a wide variety of sites and induced by a variety of known carcinogens, including nitrosamines and polycyclic hydrocarbons (Newberne and Conner, 1986; Slaga, 1981; Weisburger et al, 1986). Other studies with experimental animals indicated that feeding either 4,000 or 6,000 ppm had no effect on tumor incidence. Furthermore, several tests of chronic toxicity have not revealed any adverse effects at all.

Other studies have indicted antioxidants to be carcinogens and tumor promoters (NAS/NRC, 1989). In one study with rats fed 500 ppm BHT along with a known carcinogen, BHT acted as a tumor promoter (Maeura and Williams, 1984). In another, it caused chromosome damage and acted as a tumor promoter in the lungs and liver (Olsen et al, 1983; Slaga, 1981). In yet another study, liver carcinogenesis was inhibited by BHT, although high levels of BHT increased bladder cancer (Ito et al, 1987; Prochaska et al, 1985; Wattenberg, 1983). Data on both BHA and BHT were compiled by JECFA (1987).

Health Benefits of BHA and BHT

While all these effects are worrying, the ingestion of antioxidants has also been shown to have beneficial effects. Laboratory rats lived longer when fed BHA and BHT (Harman, 1980). Like the natural antioxidant vitamin E, antioxidants also appear to protect the liver against ethanol damage, choline deficiency, and the noxious effects of rancid fat.

The nutrient-antioxidant interaction is indeed an interesting one. For instance, adequate protein in the diet prevented adverse effects of BHT, whereas low protein diets did not (Nikonorow and Karlowski, 1976). If the antioxidants were fed at levels greater than 0.5% of the diet, vitamin A in the liver decreased and excretion of vitamin C in the urine increased (Pascal, et al, 1979). Perhaps some of the contradictory data presented here and in the previous section on carcinogenicity would be clarified if we knew more about the interaction of nutrients and antioxidants.

Weighing Risks and Benefits

Personal and regulatory decisions about the safety and use of these antioxidants are not clear-cut. It is not a simple matter of well-documented concerns about safety, in which case a substance should be banned. There are concerns about safety and carcinogenicity, but also evidence of beneficial effects and cancer inhibition. Further complicating the risk-benefit decision is concern about the well-documented risk that occurs when rancid (oxidized) fat is ingested. Data from the 1920s established that the feeding of oxidized fat caused vitamin E deficiency (Evans, 1922, 1962). Furthermore, peroxide breakdown products in oxidized fat may damage cell membranes, proteins, and DNA. More recent studies have shown that products of oxidized fat are carcinogenic, mutagenic, and cytotoxic (Sims and Fioriti, 1977). Researchers at the Cleveland Clinic believe that antioxidants are partially responsible for the decline in stomach cancer in the United States. In Europe, where antioxidants are not used in cereals, baked goods, and margarine, higher death rates from stomach cancer are found. Thus the benefit from the use of antioxidants may be greater than the risk incurred.

From a regulatory perspective, the existence of contradictory data was the reason that BHA and BHT were in the group of 19 compounds placed in Class III after the FASEB GRAS review. Substances in this class require additional studies because of unresolved questions in the research data. The International Life Sciences Institute stated that these phenolic antioxidants "can exert promoter and inhibitor activity depending on the time of administration . . . , type of carcinogen, and the target organ affected" (Malaspina, 1987). Similar conclusions were reached by JECFA (1989).

The FDA estimates the average daily intake at slightly less than 4 ppm. At this low concentration, no adverse effects are expected. The JECFA (1987) gave these compounds ADIs of 0.3 milligrams per kilogram for BHA and 0.125 milligrams per kilogram for BHT, based on no-effect levels in studies. It also asked that more studies feeding them to monkeys and pigs be undertaken. In one country, India, a much more conservative posture with respect to the use of antioxidants has been taken. Antioxidants are not allowed without special permission from the Indian government (Achaya, 1981).

PRESERVATIVES

Sulfites

Sulfites have been used for centuries in food processing as preservatives and sanitizers. Wines have been treated with sulfur dioxide since Roman times. Sulfites continue to be widely used in food processing to prevent enzymatic browning of fresh fruits and vegetables and nonenzymatic browning of many foods, to reduce microbial spoilage, to condition dough, to bleach, and to control fermentation reactions (Taylor and Bush, 1986; Taylor et al, 1986). Although some of these uses are cosmetic, many help keep food safe by preventing microbial reactions. Some so-called cosmetic reactions actually maintain the nutritional value of the product (IFT Expert Panel, 1986a). For example, dried apricots that look lovely because they have retained their bright orange color also have retained their carotene, a vitamin A precursor.

Sulfites occur naturally in fermented foods like wine and beer. Sulfite additives include sulfur dioxide, sulfite salts, bisulfite salts, and metabisulfite salts (Anonymous, 1985c). The amount of sulfite added to food is limited by taste, nutritive value, and law. Their flavor automatically limits the amount of sulfite compounds added to food to less than 500 ppm. To preserve the nutritional and microbiological safety of the diet, use of sulfite is prohibited in foods that are good sources of thiamin and in fresh meat, where they restore the red color and make the meat appear fresher than it should. Sulfites may be useful in destroying aflatoxin in some foods (Yagen et al, 1989).

The U.S. Bureau of Alcohol, Tobacco, and Firearms requires that wines sold in the United States not contain sulfite at more than 350 ppm. The average wine contains 150 ppm. Any alcoholic beverage with over 10 ppm sulfite must be labeled as containing sulfite. All wines would of necessity carry the label, as about 50 ppm are produced during the fermentation. Even wines marketed as "sulfite-free" contain naturally occurring sulfites.

In addition to foods and alcoholic beverages, drugs (including some drugs used to treat asthmatics) also are a source of sulfites (Riggs et al, 1986). The average sulfite consumption for Americans was calculated by FASEB to be 6 milligrams per day, and for wine and beer drinkers, 10 milligrams per day. Table 10.7 gives the sulfite levels of many common foods.

Safety of sulfite. Several pathways that metabolize other sources of sulfur in the diet (e.g., sulfur amino acids like methionine) can readily metabolize sulfite through oxidation to sulfate, which in turn is readily excreted in the urine. Some sulfite is bound. The toxicity of bound forms is being studied (Hui et al, 1989). Sulfite food additives contribute only a small fraction to the total sulfate excreted.

Sulfite is not found to be carcinogenic or mutagenic in a variety of experimental animals. In certain cases, sulfites have inhibited known mutagens (Taylor et al, 1986), even though sulfites have also induced

Table 10.7. Sulfite Levels in Foods as Consumed[a]

Food	Sulfite (ppm)
Dried fruit, excluding dark raisins and prunes	1,200
Bottled lemon juice, nonfrozen	800
Bottled lime juice, nonfrozen	160
Wine	150
Molasses	125
Dried potatoes	35–90
White or sparkling grape juice	85
Wine vinegar	75
Maraschino cherries	50
Pectin	10–50
Fresh shrimp	10–40
Corn syrup	30
Sauerkraut	30
Corn starch	20
Frozen potatoes	20
Fresh mushrooms	13
Beer, soft drinks	10

[a] Adapted from FASEB (1985).

mutations in vitro. Based on these data, the FASEB GRAS review committee affirmed its GRAS status (Steffy, 1980).

The only toxic effect known for sulfite is that it destroys thiamin. Where the rats had adequate thiamin, sulfite levels up to 300 milligrams per kilogram per day caused no ill effects, but thiamin-deficient rats showed toxic effects from sulfite doses as low as 50 milligrams per kilogram (Yang and Purchase, 1985).

Sulfite sensitivity. Although sulfite appears to be generally safe, it has one major drawback for a very small segment of the population— sulfite sensitivity. This diagnosis emerged in the late 1970s. Onset of symptoms in hypersensitive individuals can be extremely rapid, just minutes after eating a food. Sensitivity may manifest itself as asthma, headaches, dizziness, hypotension, anaphylaxis, hives (urticaria), abdominal pain, and even death (Sullivan and Smith, 1985). Diagnosis is not easy because the degree of sensitivity varies. Individuals with extreme sensitivity may react to doses as low as 3 milligrams of sulfite, while those with high tolerances may not react until they have ingested 120 milligrams. The sensitivity for those with a high tolerance might never be discovered, as 120 milligrams of sulfite in a single meal would be rare indeed. For instance, a meal chosen for its high sulfite levels— sulfited lettuce, dehydrated potatoes, shrimp, and wine—gave a total of 93 milligrams (IFT Expert Panel, 1986a).

The actual number of persons at risk is a source of debate. One estimate is 5–10% of the 10 million asthmatic people in the United States. Another is 5–10% of the steroid-dependent asthmatics, or perhaps 1–2% of the total asthmatic population. Response also appears to vary with the mode of administration (Malaspina, 1987).

Forms of sulfite. Different forms of sulfite have differing metabolic effects. Free sulfites readily liberate sulfur dioxide, which causes the sensitivity reaction. Bound sulfites are unreactive and liberate no sulfur dioxide (Anonymous, 1985a). Reaction from sulfites on lettuce and other fresh fruits and vegetables on the salad bar may be severe, as the sulfite is free, whereas in many other foods sulfites are bound. Studies with shrimp and dehydrated potatoes as the source of sulfite have shown no adverse effects in most individuals known to be sulfite sensitive. The question remains whether the lack of reaction in this case is caused by lower levels of sulfite or because the sulfite is bound and unable to create an adverse reaction (Taylor, 1986).

Limitations on use. By the summer of 1986, sulfites had been linked to 13 deaths and over 300 illnesses among asthmatic people, causing the FDA to revoke its GRAS status for use on fresh fruits and vegetables in August, 1986. The ban was enacted even though under 5% of asthmatics were affected. Potatoes used by restaurants were not covered by the ban, but this remains under consideration by the FDA.

As of January 1987, all packaged foods in the United States containing more than 10 ppm sulfiting agent were required to be so labeled (Langdon, 1987). The labeling was designed to help sensitive individuals avoid foods that contain sulfite. This regulation was enacted despite 10 ppm being at the lower limit for detection of sulfites. Moreover, the measurement method does not distinguish between the two metabolically different forms of sulfite, free and bound, which appear to have differing toxicities (Hui et al, 1989).

Since the ban of sulfite, alternatives have been sought. Usually two additives must be used in tandem to replace the various functions performed by sulfites. One type combines ascorbic acid and polyphosphate to prevent browning (van Pelt, 1987); another uses citric acid or the isomer of ascorbic acid, erythorbic acid (Yang and Purchase, 1985).

The ADI for sulfite is set by WHO/FAO at 0.7 milligrams per kilogram of body weight—e.g., 37 milligrams for a 50-kilogram (120-pound) woman (JECFA, 1987). Thus, the ADI is far in excess of the 6–10 milligram average intake. Over 99% of consumers are not at risk from the use of this additive. The risk-benefit posture of the FDA in banning sulfite from fresh fruits and vegetables was a case in which the risk to an extremely small minority of consumers seemed so severe that the product was banned in situations where high-risk individuals could not be warned through labeling, as with fresh fruits and vegetables. In labeled foods, the benefit of this additive could be afforded to a majority of consumers while minimizing the risk to those few who need to avoid sulfites in their diet.

Benzoic Acid (Benzoates) and Their Derivatives

Benzoate may be the additive most recognized by consumers because it is widely used in the production of canned soda. Over 50% of the benzoate used in the country goes into the manufacture of soft drinks.

In soda and other acid foods (pH <4), benzoates prevent growth of yeasts, molds, and bacteria. Benzoates occur naturally in a variety of food products such as prunes, cinnamon, cloves, and most berries. In fact, benzoate levels found naturally in cranberry juice are magnitudes higher than are allowed to be added to food. Ironically, some cranberry juice manufacturers are able to advertise that their product has no artificial preservatives! Of course, there are no preservatives—the product doesn't need any. The labeling is accurate despite the fact that it may be misleading.

Concerns about benzoates stem not from hints of carcinogenicity or teratogenicity, but from long-term studies with diets of 5% sodium benzoate that show toxicity and disturbed growth. However, at 1% of the diet, benzoates showed no adverse effects even when eaten by young animals during the developmental period (Crane and LaChance, 1985). Humans fed 1,000 milligrams per day for 88 days (20 times the average consumption of benzoate) showed no observable ill effects.

As with sulfites, some individuals show hypersensitivity to benzoates. Although the sensitivity appears to be much milder than the sensitivity that can occur with sulfite and has occurred only after skin applications, it does raise questions about hypersensitivity to ingested benzoate (Miller, 1987).

According to the U.S. Code of Federal Regulations, no more than 0.1% can be used in foods (Lueck, 1980). Many foods use much less than this; for instance, most carbonated beverages contain 0.03–0.05% benzoate.

In recent years, product developers have tried to seek alternatives to benzoates such as sorbates. This is partly because of concerns about toxicity and partly because the flavor of benzoic acid is readily perceptible.

Parabens is a derivative of benzoic acid. It is also active against yeasts, molds, and *Clostridium botulinum*. Other derivatives of benzoic acid are used in other countries in sauces and vinegars. Studies on these showed no difference in tumor incidence in mice between treated animals and controls despite the report that these compounds had mutagenic potential. The dose fed to the mice vastly exceeded the level that a human would ingest, so there is probably little concern about the safety of such additives (Inai et al, 1985) except for dermatitis in sensitive persons (Concon, 1988). In the United States, the level of parabens in food is limited to 0.1%.

Propionates

Since the 1930s, calcium and sodium propionate have been used in the United States on a large scale in the preservation of bread and baked goods and on a smaller scale in processed cheeses. In bread, propionates prevent molds, yeasts, and the bacteria responsible for "ropy" bread, *Bacillus mesentericus*. Interestingly, propionate is produced naturally in some cheeses. In Swiss cheese, the production of propionic acid is said to be responsible for the holes.

Propionate itself is actually a short-chained fatty acid and is metabolized like any odd-chained fatty acid. It is thus not a foreign compound, as it is formed during the metabolism of odd-chained fatty acids from food sources.

Toxicity studies of propionate show that diets of 1–3% calcium or sodium propionate produced no adverse effects. Similar results occurred in a study lasting over 1 year with diets comprised of 75% bread made with 5% sodium propionate. Diets containing 24% propionate did kill young rats after only 5 days and adult rats after 20. These high-dose experiments are testaments to disordered fat metabolism and bear no relationship to the use of propionates as a food additive (Lueck, 1980). However, Germany has chosen to ban its use (Brümmer et al, 1988), and sorbic acid has been proposed as a replacement.

The GRAS status of propionates was affirmed. Calcium propionate is usually used because it adds to the calcium levels of the diet. Because of their low toxicity, no limit exists on the use of propionates except Good Manufacturing Practices and the fact that these preservatives impart a flavor. As a result of their extensive use in bread and baked products, calcium and sodium propionate account for 75% by weight of all chemical preservatives used (King, 1981; Robach, 1980).

Sorbates

Sorbates are one of the few additives permitted in all the countries of the world. They are used primarily to prevent growth of yeasts and molds, although they can inhibit or partially inhibit the growth of some bacteria. They are even effective against aflatoxin-producing molds and are widely used in foods such as cheese products, non-yeast-raised baked products, and dried fruit and fruit drinks.

Sorbate is a naturally occurring polyunsaturated fatty acid that is metabolized to carbon dioxide and water like other ingested fatty acids. Found naturally in the berries of the mountain ash (rowan), it was introduced as a food preservative after extensive testing. Many different investigations employing sorbate levels up to 10% have failed to show any adverse effect. All tests of teratogenicity and mutagenicity have been negative (Lueck, 1980). It was shown to be carcinogenic if it was injected under the skin, but several food additive regulatory agencies have stated that this type of test has no bearing in deciding its use as a food additive. One study that fed 5% sorbic acid even showed an increase in the growth rate and life span of male rats.

Sorbate is GRAS and may be used at 0.1-0.2% levels. Because of its innocuousness, there is a worldwide trend toward selecting it over other food preservatives such as nitrite or propionates (Brümmer et al, 1988; Lueck, 1980). It is often used in combination with other preservatives. The combination of nitrate and sorbate appears to reduce nitrosamine formation but may form a mutagenic combination. More research is needed (Concon, 1988).

Nitrates

Nitrate and nitrite are controversial food additives because nitrite can react with amines in food or in the body to form nitrosamines, which are indeed something to be concerned about because they are potent carcinogens (NRC, 1981).

During the height of the controversy about the use of nitrate in cured meat, attention was focused on the use of nitrate as a food additive. The focus was there because of the Delaney clause requiring the banning of a food additive that is carcinogenic or is metabolized to a carcinogen. This focus created a somewhat distorted picture, as the smallest dietary source of nitrate is from its food additive use. Only 6% of the average daily nitrate intake comes from cured meats. The other 94% is found naturally in foods (especially vegetables) and water, or is produced endogenously in human saliva and gastric juices. Nitrate occurs in vegetables at levels as high as 3,000 ppm. Vegetables such as cabbage, cauliflower, carrots, celery, lettuce, radishes, beets, and spinach account for 86% of the daily nitrate intake. Saliva and well waters account for another 8%. Human saliva has 6-10 ppm nitrate, and concentrations may increase when vegetables high in nitrate are consumed (IFT Expert Panel, 1987).

Role in foods. As food additives, nitrite and nitrate are used to give the characteristic flavor and color of cured meats, and thus they are functional aids. They also prevent botulism and are thus preservatives (Sen, 1980). As yet, there is no safe substitute, but research to lower the required levels of nitrate has been successful, and exposure to this additive in cured meats has been reduced from previous levels. Ninety percent of cured meat samples now contain less than 50 ppm nitrate; only 0.1% contain more than 200 ppm. The risk of botulism is considered greater than the risk from nitrites, so their use is allowed (Tannenbaum, 1984).

Safety. Nitrate poisoning in humans is uncommon. It is caused by the ingestion of 8-15 grams of nitrate. Symptoms include severe gastroenteritis with abdominal pain, blood in the stools and urine, weakness,

Table 10.8. Dietary Sources of Nitrate[a]

	Amount Provided	
Source	Micromoles per Day per Person	Percent
Vegetables	1,050 (4,200)[b]	87 (97)[b]
Fruits and juices	69	6
Water	32 (2,850)[c]	3 (68)[c]
Cured meat	26	2
Bread and cereals	26	2
Others	13	<1

[a] Adapted from NAS/NRC (1981).
[b] Parentheses show results from vegetarian diet.
[c] Parentheses show results from high-nitrate water area.

and collapse. Chronic ingestion of smaller doses causes dyspepsia, mental depression, and headache. Induction of vitamin A deficiency and thyroid depression have also been reported (Magee, 1983).

Nitrite can be produced from nitrate in the intestine of infants, where the gastric pH is near neutrality. Nitrites can also be produced in any disease state where the gastric pH is above 5.5. At these near-neutral pHs, intestinal bacteria convert nitrate to nitrite. The reduction can also occur during the long storage of some vegetables (Mueller et al, 1986; Sen, 1986).

Nitrite's major toxic effect is methemoglobinemia, a condition occasionally seen in infants drinking well water with high levels of nitrate or ingesting vegetables like beets and spinach that are naturally high in nitrate. It results in hemoglobin that is unable to carry oxygen; hence the baby turns blue. In some areas in the United States the water contains as much as 50 milligrams of nitrate per liter. At this level of nitrate, the safety margin is less than two. There is fear that nitrate in the groundwater will increase, leading to increased incidence of methemoglobinemia in infants (O'Donoghue, 1981). In some areas, parents are advised not to mix infant formula with water containing nitrate levels over 10 ppm. There is also concern that nitrite crosses the placenta (Vorhees et al, 1984).

Carcinogenicity of nitroso compounds. The near ban of nitrite was not related to the possible increased incidence of methemoglobinemia. The real concern centered around nitrosamines. These derivatives are formed when nitrate and a secondary amine react to form nitroso compounds—nitrosamines and nitrosamides.

In both humans and experimental animals, simultaneous ingestion of nitrites and amines can lead to the formation of nitrosamines in the stomach. High levels of amine are found in protein foods, especially

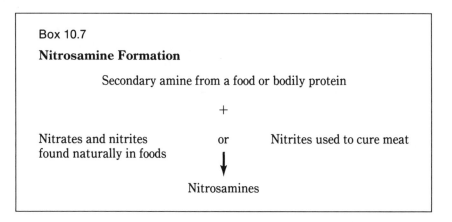

Box 10.7

Nitrosamine Formation

Secondary amine from a food or bodily protein

+

Nitrates and nitrites   or   Nitrites used to cure meat
found naturally in foods

Nitrosamines

fish of marine origin. Amines in fish may actually increase with frozen storage. The nitrosation reaction can be blocked both in vivo and in vitro by adequate levels of ascorbic acid, its isomer (erythorbate), vitamin E, or BHA (Gray et al, 1982).

Nitroso compounds are toxic, carcinogenic, and mutagenic to a wide variety of species. In fact, nitrosamines are among the most potent of all carcinogens, as they bind directly to DNA. They appear to cause tumors in many different organs (Hill, 1988). Ninety percent of all nitroso compounds tested have been shown to be carcinogenic in experimental animals. Humans may not be as susceptible as experimental animals, since human cells contain specific DNA repair activity for removing akyl groups inserted by nitrosamines (Williams and Weisburger, 1986).

Epidemiological data about nitrates and gastric cancer. Correlations have been made between high levels of nitrate in food and drinking water and high levels of gastric cancer (Hill, 1988). It is crucial to point out, however, that the correlation appears to hold true only if the diet is also deficient in fresh fruits and vegetables (Weisburger et al, 1986). In fact, studies from Britain, Japan, and Canada found that persons with high risk for gastric cancer actually had low exposures to nitrate (Tannenbaum and Correa, 1986). In France and Britain, no relationship between nitrate in drinking water and gastric cancer was found (Forman, 1987). Bladder cancer was related to nitrates in the water supply but not to consumption of preserved meats (NAS/NRC, 1989).

The relationship between salivary nitrate and stomach cancer is equally confusing. For instance, epidemiological studies in England found a curious relationship between salivary nitrate and persons at high risk for stomach cancer. Persons living in areas with high risk of stomach cancer had consistently lower levels of nitrate in the saliva (Forman, 1987). The authors say that their findings are "inconsistent with the notion that nitrate exposure is a risk factor for cancer of the stomach." The confusion surrounding these epidemiological studies may be sorted out when other aspects of diet known to influence the nitrosation reaction are considered. Nitrosamines have been implicated in cancer at other sites, but the evidence is even weaker than it is for stomach or bladder cancer (Hill, 1988).

Preformed nitrosamines. Some foods actually contain preformed nitrosamines (Miller and Miller, 1986). They are found in cheeses, beer, cured meats, mushrooms, and premixes of nitrate and spices, as well as in small amounts in foods dried by direct flame such as dried milk, soups, and instant coffees (Hotchkiss, 1984). Diet surveys in Japan indicate that dried fish is a major source of nitrosamines in the Japanese diet (Mohri, 1987; Yamamoto et al, 1984).

Foods are not the only sources of preformed nitrosamines; cigarettes, drugs, and cosmetics are also sources (Tricker and Preussmann, 1988). Rubber products from tires to pacifiers and baby bottle nipples contain nitrosamines (Havery and Fazio, 1985; Sen, 1980; Tannenbaum, 1984). During sterilization of the nipples, nitrosamines are transferred from the nipple to the formula. The United States Consumer Product Safety Commission and the FDA have placed limits on the nitrosamine content of baby nipples and pacifiers (Havery and Fazio, 1985). Elastic rubber nets sometimes used to package hams can also be a source of nitrosamine. Cured meats with cotton nets did not show nitrosamines, so it is recommended that nonrubber nets be used for meat packages (Anonymous, 1986).

Regulation. Since the early 1970s, research and regulation were designed to reduce, eliminate, and monitor preformed nitrosamines in foods and food-contact materials. The levels of nitrate allowed in the curing of bacon have been reduced, and the required levels of ascorbate have been increased. Increased vitamin E together with lecithin is another system used to reduce the levels of nitrosamine in bacon (Wilkens and Gray, 1986).

The USDA began to monitor bacon in 1978, and now bacon with nitrosamine levels at 10 ppb or greater is not allowed to be sold (FSIS, 1986). Other cured meats can achieve reduction in nitrate levels through addition of other botulism-preventing additives such as hypophosphate, lactic acid bacteria, and sulfite, as well as by increased ascorbate and erythorbate. Research in the rubber industry has also been directed toward the elimination of problems with bottle nipples (Havery and Fazio, 1985).

Until recently, beer was a major source of nitrosamines in Western diets. Flame-dried barley malts used in the production of domestic and imported beers and scotch whiskey were incriminated as the source of nitrosamines. The process for the manufacture of malt has been altered so that the resultant beverages would have nitrosamine contents less than 5 ppb (Tricker and Preussmann, 1988).

In humans, the most likely source of nitrosamine is cigarettes (Table 10.9). A person who smokes one pack per week inhales *14-18 ppm* (not ppb) of nitrosamine per week (IARC, 1986). The next most likely source is endogenous production in the body. This source far exceeds nontobacco environmental sources and all food sources including cured meats (Tannenbaum, 1984). The health risk from foods such as bacon with nitrosamine at parts per billion levels remains obscure. It is quite likely that larger amounts were eaten in the past than are consumed now, since consumption of fatty foods has decreased and nitrite levels have been lowered (Roberts and McClure, 1990).

Even practices in the home can influence the nitrosamine levels in foods such as bacon (Skrypec et al, 1985; Vecchio et al, 1986). Protein in older bacon is more likely to break down to form secondary amines. Bacon that is higher in fat is more likely to yield more nitrosamines when cooked. Microwaved bacon has much lower nitrosamine levels than fried bacon (Österdahl and Alriksson, 1990). Less nitrosamine forms in bacon cooked at low temperatures than in bacon fried at high temperatures.

Weighing risks versus benefits. Complete elimination of nitroso derivatives from food and water appears difficult. Consumers can choose to eat meats that are uncured if they are worried about nitrate in cured meat. In fact, that is the recommendation that *Consumer Reports* made in 1982 (Anonymous, 1982a, 1982b). However, consumers must be aware that meat products without nitrate must be handled very carefully due to the risk of botulism (USDA, 1979). Also, uncured hot dogs and sausages are higher priced because of their much shorter shelf life.

Finally, we must weigh two opposing views. One asks if it will do any good to restrict consumption of cured meats if they constitute such a small percentage of the nitrate exposure (Nitrate Safety Council, 1978). The other position, that of the Community Nutrition Institute, is that there is no safe level of human exposure to nitrate or nitrosamines, so

> the only acceptable long-term solution to the health problem of carcinogens which form as a direct consequence of using a certain additive is the prohibition of the use of that additive [CNI, 1979].

Table 10.9. Estimated Exposure to Nitroso Compounds in the United States[a]

Source	Route of Exposure	Exposure[b]
Cigarette smoker	Inhalation	17
New car interiors	Inhalation	0.5
Cosmetics	Dermal contact	0.41
Beer	Ingestion	0.34
Cooked bacon	Ingestion	0.17
Scotch	Ingestion	0.03
Cheeses	Ingestion	0.03

[a] Adapted from IFT Expert Panel (1987). The exposure data was based on 1981 levels of nitrate consumption. Values for some items may actually be less than listed as significant changes have been made in certain products specifically to lower their nitrate content.
[b] In micrograms per person per day.

The current regulatory posture is one of maintaining the lowest possible levels of nitrosamine in food and of promulgating regulations that will lower both the nitrate and nitrosamine levels allowed while still protecting against botulism. Experts in this area suggest that the benefits of these compounds outweigh the risks (Cassens, 1990; Newberne and Conner, 1986; Roberts and McClure, 1990).

Acidulants

Acids prevent the growth of many types of microorganisms. As a result, they are added as preservatives to many foods. For instance, botulism is completely inhibited at acid pH. Acids also help many food products attain better flavor, texture, or color. For instance, the browning reaction of fruits and vegetables is inhibited by acid. Acids may also inactivate enzymes, chelate metals that catalyze reactions, reduce heat processing requirement, and modify flavor (Dziezak, 1990). Many acidulants are acids present naturally in common fruits or are products of fermentation (Beuchat and Golden, 1989).

Citric acid, the characteristic acid found in citrus fruits, is normally produced during human and animal carbohydrate metabolism. It accounts for more than 60% of the acidulants used in foods, yet the citric acid that is a food additive is less than 10% of the daily citrate intake. Citric acid is GRAS and its uses are unrestricted.

Malic acid makes up 90% or more of the acid in apples, plums, watermelons, cherries, peaches, and bananas and is also found in many other fruits. Malic acid is beginning to replace citric acid as a food additive. Like citric acid, it is a metabolite of normal carbohydrate and so is not a health concern. It also has GRAS status.

Acetic acid is the acid in vinegar, pickles, and sauerkraut. Other organic acids include tartaric, lactic, and fumaric. All are found in foods and have GRAS status.

Phosphoric acid is the second most commonly used acid in food. The only inorganic acid allowed, it is familiar to consumers who read the labels on cola-type soft drink cans. Phosphates are also used to aid in the rehydration of certain products, to prevent curdling in products containing milk, to provide some antioxidant properties, to help preserve color, and to act as leavening acids.

Although phosphorus is an essential nutrient, there is concern that excessive phosphorus consumption in the form of soft drinks, processed foods, and diets high in meat can upset the calcium-phosphorus ratio and lead to osteoporosis and other problems associated with mineral imbalance (Massey, 1982; NAS/NRC, 1989). Other minerals such as iron may also be adversely affected (Concon, 1988).

FUNCTIONAL AIDS

Brominated Vegetable Oils

Brominated vegetable oils (BVOs) are used in soft drinks so that flavoring oils will not separate from the liquid mixture in citrus soft drinks and fruit-flavored beverages. In some instances, BVOs are used to give a desirable cloudy appearance to beverages (Achaya, 1981; Freydberg and Gortner, 1982).

Experimental animals fed BVOs showed deposits of brominated fat in the heart and other tissues that apparently were not metabolized. Adverse effects were seen with as little as 20 milligrams per kilogram. Since the estimated daily human intake in countries where BVOs are used in soft drinks is 0.2 milligrams per kilogram, WHO recommended against their use. In two countries where they were used, England and the Netherlands, higher levels of bromine were found in children than in countries where this additive was not in use (Anonymous, 1984). Both countries subsequently banned the use of this additive (Anonymous, 1984).

Modified Food Starch

Starches, long-chain polymers comprised of thousands of glucose units, are used to thicken food. Although these work well in some foods, unmodified starches may lose their ability to thicken under certain conditions such as high or freezing temperatures or high acidity. If the starch chains are linked to one another, the resultant modified starch retains its thickening power in situations where regular starch does not (Light, 1990).

Cross-linking joins adjacent starch chains with a bond about every 1,000 or 2,000 glucose units. The process uses either phosphate or adipic acid. The phosphate or adipate remaining in the starch is very small (Wurzburg, 1986).

Further stabilization can be achieved by changing some of the hydroxyl groups on the starch to acetyl groups. This change helps to prevent the syneresis (weeping) that is characteristic of starch products. (A maximum of 2.5% acetyl groups is allowed.)

Concern about modified starches stems not from worries about their toxicity but rather from the fact they are used in the preparation of commercial baby food. Questions about whether modified starches are digested and provide the energy necessary for the infant have been raised.

Research has attempted to answer these questions. Both in vitro and in vivo studies have shown that the digestability is not impaired and that the glucose response to both regular and modified starch is

the same.

Four-day diet records of over 600 infants showed that slightly more than one fourth ate foods containing modified starches (Filer, 1988). The average intake was 3.6 grams, or 14 calories, per day, with a range of 3-67 calories per day. The average intake was less than 2% of the total caloric intake for infants and is considered to be of no practical importance (Filer, 1988). Both the National Academy of Sciences and the Committee on Nutrition of the American Academy of Pediatrics concluded that modified starches were safe and appropriate for use.

Carrageenan

Carrageenan is a plant gum derived from many varieties of seaweed and used to thicken a wide variety of products. Originally it was called Irish moss because the seaweed was collected in Ireland and made into a pudding. The name actually comes from a town in Ireland, Carragheen.

This particular additive, despite its "natural" heritage and its functional role as dietary fiber (polymer of sulphated D-galactose), is one of those additives that always seems to have a cloud of unresolved safety questions surrounding it. Controversy heated up when a letter was written to the prestigious British medical journal *Lancet* promulgating the idea that carrageenan was carcinogenic. The letter also reported that it caused ulcers. Two studies on rats were cited in support of the letter. Both studies used a semisynthetic diet with very high dietary levels of carrageenan (up to 15% of the diet). At these levels the growth of the rats was inhibited and the large intestine ulcered. Carrageenan alone did not induce tumors. However, when the rats were given a known carcinogen along with the carrageenan, there was a marked increase in the number of tumors over the number that occurred with the carcinogen alone (Concon, 1988).

Studies feeding lower levels of carrageenan showed no statistically significant increase in the number of tumors. An editorial reviewing these data for *Food and Cosmetic Toxicology* stated that carrageenan at dietary levels that do not ulcerate the intestine poses no carcinogenic threat to humans (Hopkins, 1981).

Other unresolved concerns about carrageenan have to do with its effects on the immune system and growth. One report in *Lancet* suggested that carrageenan at dietary levels may impair immune responses (Thomson, 1980, 1981). At 2,000 times the estimated daily intake, carrageenan decreased the number of live births in rats. The ADI is 75 milligrams per kilogram. It is allowed as a food additive but is not a GRAS substance (Concon, 1988).

Bleaching and Oxidizing Agents

Bromate. Potassium bromate is used as an oxidizing agent to improve the baking performance of hard wheat flours. Concern about bromate has been raised because Japanese studies indicate that prolonged administration of relatively large quantities of bromate can increase the number of kidney tumors. The studies were done on rats and involved feeding 500 ppm in drinking water (Fujii et al, 1984).

For many years, 75 ppm was permitted in flour or 50 ppm in bread formulations. The amount of bromate present in bread products after baking was traditionally believed to be zero, as heat reduces bromate to the bromide. However, researchers using increasingly sensitive analytical methods have reported traces of bromate in baked products, and bromate in certain uses in the state of California requires a warning label as a possible carcinogen. In 1989 the JECFA reduced the level allowed in flour to 0-60 milligrams per kilogram. These developments significantly reduced the use of bromate as a flour additive by millers and bakers (Spooner, 1990). It is not permitted in the United Kingdom and European countries.

Chlorine. Gaseous chlorine is used to bleach flour, and in some countries it is a food additive. In Germany, Chloramine T is added to foodstuffs as a preservative to suppress bacterial growth. The use of nitrogen trichloride as a maturing agent in flour was discontinued in the 1940s, as it caused running fits in dogs. Other uses of chlorine and certain chlorine compounds in food and as sanitizers in the food industry have been questioned, as data show that chlorine from swimming pools and drinking water can produce products that have mutagenic effects (Sussmuth, 1982).

SUMMARY

Food additives provide many benefits and perhaps put us at risk in some cases. In each case, it is important to decide whether this is a health issue or a trust issue and what the risks are, if any, from omitting the additive from the foods in which it is found or of omitting the foods.

Some argue that additives serve useful functions for the food industry but that their benefits to the consumer are difficult to determine (Millstone, 1985). Clearly, longer shelf life is advantageous to producers. However, it is my contention that consumers benefit even more from longer shelf life, to the extent that it means lower prices, reduced spoilage and waste, and less frequent trips to the store, as well as inherently safer food in many cases.

Another set of questions about additive use concerns need. Do con-

sumers need convenience, or is convenience created by the industry to sell more products? Do the products that are available fulfill a real need and offer true nutritional equivalency? Do we need salad dressings that don't need to be shaken? Does the presence of additives and the wish for convenience allow us to select a tomatoless pizza with cheese that isn't cheese and tomato flavor that came without the vitamins from the tomato? Do we need foods to be colored or can consumers learn to accept a less pretty product? Is the product safer with the additive present? Does the product have a lot of nutritional negatives? These and other similar questions must be answered when making food selections.

The final questions about additive use are in regard to safety. Is the additive safe to consume at customary dietary levels? Are the risks of using the additive less than the benefits accrued?

In this chapter I have tried to outline what is known about many sweeteners and functional additives relative to their use and safety. Armed with information about safety, consumers may decide that a bread with preservatives is less risky than one that may allow the growth of a possibly aflatoxin-containing mold. They may also decide that they will omit the snacks with antioxidants, not because of concern about the antioxidant but because the product, although tasty, provides little in terms of nutritional benefit and serves up heaping portions of fat, salt, and calories. In like manner, they may select a cereal with an antioxidant because the risk from rancid fat appears to be greater than the risk posed by the antioxidant. They may decide that brominated vegetable oil used in soft drinks is unsafe and is in a product that is not crucial to the diet. The same kind of decision may be made in regard to artificial sweetener use. Some may decide that the risks of sugar are less than the risks of artificial sweeteners; others may decide the opposite. Informed choice in an open marketplace allows us as consumers to exercise the right to vote by our choices.

REFERENCES

AAP. 1985. Report: Committee on Nutrition Task Force. December. American Academy of Pediatrics, Elk Grove Village, IL.

Achaya, K. T. 1981. Regulatory aspects of food additives. Indian Food Packer 35(3):11-14.

ACHS. 1982. Saccharin. American Council on Science and Health, Summit, NJ.

Alfin-Slater, R. B., and Pi-Sunyer, F. X. 1987. Sugar and sugar substitutes. Postgrad. Med. 82(2):46-56.

AMA. 1985. Aspartame: Review of safety issues. JAMA J. Am. Med. Assoc. 254:400-402.

American Diabetes Association (ADA). 1980. Statement on sweeteners. J. Am. Diet. Assoc. 76:549.

American Diabetes Association (ADA). 1985. Statement on use of aspartame. The Association, New York.

American Medical Association (AMA). 1986. Saccharin: Review of safety issues. Conn. Med. 50(3):191-194.

Anonymous 1982a. Experts offer clues to risk of nitrite in cured meats. Consum. Rep. 47:117.

Anonymous 1982b. Nitrite: What is the situation now? Consum. Rep. 47:577.

Anonymous 1984. Brominated oil residues and toxicity. Food Chem. Toxicol. 22:319-320.

Anonymous 1985a. Palatinose. Food Eng. 57(May):75.

Anonymous 1985b. Dining out on sulfites. Emerg. Med. June, pp. 46-65.

Anonymous 1985c. Sulfites in foods: Additives under fire. Anal. Prog. 2(2):1-7. (Medallion Laboratories, Minneapolis, MN)

Anonymous 1986. Package nets can form nitrosamines in meat. Chem. Eng. News 64(May 5):38.

Anonymous 1988. Future ingredients. Food Technol. (Chicago) 42(1):50-64.

Anonymous 1990. Saccharin consumption increases food consumption in rats. Nutr. Rev. 48(3):163-165.

Arnold, D. L., and Clayson, D. B. 1985. Saccharin—A bitter-sweet case. Pages 227-244 in: Toxicological Risk Assessment. D. B. Clayson, D. Krewski, and I. Munro. CRC Press, Boca Raton, FL.

Arnold, D. L., and Munro, I. C. 1983. Artificial sweeteners: Their toxicological etiology is an interesting mix. Pages 211-229 in: Environmental Aspects of Cancer: The Role of Macro and Micro Components of Foods. E. L. Wynder, G. A. Leveille, J. H. Weisburger, and G. E. Livingston, eds. Food and Nutrition Press, Westport, CT.

Bar, A. 1985. Safety assessment of polyol sweeteners—Some aspects of toxicity. Food Chem. 16:231-241.

Barndt, R. L., and Jackson, G. 1990. Stability of sucralose in baked goods. Food Technol. (Chicago) 44(1):62-66.

Bauernfeind, J. C. 1975. Tocopherols. In: Encyclopedia of Food Technology. A. H. Johnson and M. S. Peterson, eds. AVI, Westport, CT.

Beuchat, L. R., and Golden, D. A. 1989. Antimicrobials and their use in foods. Food Technol. (Chicago) 43(1):134-142.

Bradstock, K., Serdula, M. K., Marks, J. S., Barnard, R. J., Crane, N. T., Remington, P. L., and Trowbridge, F. L. 1986. Evaluation of reactions to food additives: The aspartame experience. Am. J. Clin. Nutr. 43:464-469.

Branen, A. L., Davidson, P. M., and Katz, B. 1980. Antimicrobial properties of phenolic antioxidants and lipids. Food Technol. (Chicago) 34(5):42-53.

Brody, I. 1987. Aspartame and seizures. Chem. Eng. News 65(43):4.

Brümmer, J. M., Brack, G., and Seibel, W. 1988. Zur Situation de Konservierung von Backwaren. Getreide Mehl Brot 42(1):17-21.

Buck, D. F. 1985. Antioxidant applications. Reprint from Manuf. Confect. 65(June).

Byard, J. L. 1984. Metabolism of food toxicants: Saccharin and aflatoxin B1, A contrast in metabolism and toxicity. Adv. Exp. Med. Biol. 177:147-151.

Caldwell, E. 1987. Further developments with low- or no-cal sweeteners. Cereal Foods World 32:575.

Calorie Control Countil (CCC). 1985. Alternative Sweeteners. The Council, Atlanta, GA.

Cassens, R. G. 1990. Nitrate-Cured Meat. Food and Nutrition Press, Turnbull, CT.

Cohen, B. L. 1978. Saccharin's benefits outweigh risks in overweight people. Science 199:983-985.

Community Nutrition Institute (CNI). 1979. Nitrite: Why the Debate? Consumer Division, The Institute, Washington, DC.

Compadre, C. M., Pezzato, J. M., Kingborn, A. D., and Kamath, S. K., 1985. Hernandulcin:

An intensely sweet compound discovered by review of ancient literature. Science 227(4685):417-419.

Concon, J. M. 1988. Food Toxicology: Contaminants and Additives, pt. B. Marcel Dekker, New York.

Coulter, R. B. 1988. Extending shelf life by using traditional phenolic antioxidants. Cereal Foods World 33:207.

Crane, S. C., and LaChance, P. A. 1985. The effect of chronic sodium benzoate consumption on brain monoamines and spontaneous activity in rats. Nutr. Rep. Int. 32(1):169-177.

Daniel, J. W. 1981. The safety of antioxidants—Fact or fallacy. Pages 224-236 in: Impact of Toxicology on Food Processing. J. C. Ayres and J. Kirshman, eds. AVI, Westport, CT.

Dougherty, M. E. 1988. Tocopherols as food antioxidants. Cereal Foods World 33:222.

Drew, K. 1986. The role of additives in food distribution. Food Mark. 2(1):3-14.

DuBois, G. G., Bunes, L. A., Dietrich, P. S., and Stephenson, R. A. 1984. Diterpenoid sweeteners. Synthesis and sensory evaluation of biologically stable analogues of stevioside. J. Agric. Food Chem. 32:1321-1325.

Dziezak, J. D. 1990. Acidulants: Ingredients that do more than meet the acid test. Food Technol. (Chicago) 41(1):76-83.

Elton, G. A. H. 1981. Additives and contaminants in the food supply. Food Technol. (Aust.) 33(4):184-188.

Epilepsy Institute. 1986. Communication regarding seizures and epilepsy by Richard Reuben. The Institute, New York.

Evans, H. H. 1922. On the existence of a hitherto unrecognized dietary factor essential for reproduction. Science 56:650-651.

Evans, H. H. 1962. The pioneer history of vitamin E. Vitam. Horm. 20:379-387.

FASEB. 1985. The Re-examination of the GRAS Status of Sulfiting Agents. Federation of the American Societies for Experimental Biology, Bethesda, MD. 96 pp.

Federal Register. 1979. Affirmation of GRAS status for mannitol and sorbitol. Dec. 7, p. 70569.

Federal Register. 1981. Aspartame. Commissioner's final decision. 46:38283-38308. July 24.

FMI. 1990. Trends: Consumer Attitudes and Supermarkets. Food Marketing Institute, Washington, DC.

Filer, L. J. 1988. Modified food starch—An update. J. Am. Diet. Assoc. 88:342-344.

Finer, N. 1989. Are sweeteners really useful to diabetics? Pages 215-239 in: Progress in Sweeteners. T. H. Grenby, ed. Elsevier Applied Science, New York.

Food Safety and Inspection Service (FSIS). 1986. Nitrate levels in bacon. Fed. Regist. 51(115):21731-21736.

Forman, D. 1987. Gastric cancer, diet, and nitrate exposure. Br. Med. J. 294:528-529.

Franz, M. 1986. Is it safe to consume aspartame during pregnancy? A review. Diabetes Educ. 12(2):145-147.

Freeman, T. M. 1989. Sweetening cakes and cake mixes with alitame. Cereal Foods World 34:1013.

Freydberg, N., and Gortner, W. A. 1982. The Food Additives Book. Bantam, New York.

Fujii, M., Oikawa, K., Saito, H., Fukuhara, C., Onosaka, S., and Tanaka, K. 1984. Metabolism of potassium bromate in rats. I. In vivo studies. Chemosphere 13:1207-1212.

Fulton, P. W., Wall, V. J., and Hutton, C. W. 1980. Effects of butylated hydroxytoluene on selected tissues in the rat. J. Food Sci. 45:1446-1448.

GAO. 1981. Regulation of Cancer-Causing Food Additives—Time for a Change. U.S. Government Accounting Office, Washington, DC.

Gilchrist, A. 1981. Foodborne Disease and Food Safety. American Medical Assoc., Monroe, WI.

Goldsmith, L. A. 1988. Sucralose. McNeil Specialty Products, New Brunswick, NJ. April.

Goodman, S. B., Kapica-Cyborski, C., and Suzuki, J. B. 1987. Aspartame: A real sweet story. RDH July/Aug.

Gorton, L. 1983. Mirror-image sugar—Sweetness without calories. Milling Baking News, Mar. 8, pp. 16 ff.

Graham, H. D. 1980. The proper use of food additives. Pages 278-284 in: Safety of Foods. H. D. Graham, ed. AVI, Westport, CT.

Gray, J. I., Reddy, S. K., Price, J. F., Mandagere, A., and Wilkens, W. F. 1982. Inhibition of N-nitrosamines in bacon. Food Technol. (Chicago) 36(6):39-45.

Harman, D. 1980. Free radical theory of aging: Dietary implications. Am. J. Clin. Nutr. 25:839-843.

Havery, D. C., and Fazio, T. 1985. Human exposure to nitrosamines from foods. Food Technol. (Chicago) 39(1):80-84.

Hayes, J. R., and Campbell, T. C. 1986. Food additives and contaminants. In: Casarett and Doull's Toxicology, 3rd ed. C. D. Klaassen, M. O. Amdur, and J. Doull, eds. Macmillan, New York.

Hill, M. J. 1988. N-nitroso compounds and human cancer. Pages 142-162 in: Nitrosamines: Toxicology and Micobiology. M. J. Hill, ed. Ellis Horwood, Chichester, England.

Hirsch, P. 1985. Sugarless cavities. Omni, June, p. 49.

Holmes, J. 1986. Left-handed sugar: Sweet substitute. Insight 7(June 30):48-49.

Homler, B. E. 1984. Properties and stability of aspartame. Food Technol. (Chicago) 38(7):49-55.

Hood, L. L., and Campbell, L. A. 1990. Developing reduced-calorie bakery products with sucralose. Cereal Foods World 35:1171.

Hood, L. L., and Schoor, M. 1990. Evolution, properties, and applications of an approved high-intensity sweetener. Cereal Foods World 35:1184.

Hoover, R. 1980. Saccharin—Bitter aftertaste. New Engl. J. Med. 302:573-575.

Hopkins, J. 1981. Carcinogenicity of carrageenan. Food Cosmet. Toxicol. 19:779-781.

Horwitz, D. L., and Bauer-Nehrling, J. K. 1983. Can aspartame meet our expectations? J. Am. Diet. Assoc. 83:142-145.

Hotchkiss, J. H. 1984. Sources of N-nitrosamine contamination in foods. Adv. Exp. Med. Biol. 177:287-298.

Hui, J. Y., Beery, J. T., Higley, N. A., and Taylor, S. L. 1989. Comparative subchronic oral toxicity of sulphite and acetaldehyde hydroxysulphonate in rats. Food Chem. Toxicol. 27:349-359.

Hutt, P. B., and Sloan, A. E. 1979. NAS issues saccharin report. Nutr. Policy Issues, No. 5. pp. 1-2.

IARC. 1986. Tobacco Smoking. Sci. Publ. 38. Int. Agency for Research on Cancer, Lyon.

IFT Expert Panel on Food Safety and Nutrition. 1986a. Sulfites as food ingredients. Food Technol. (Chicago) 40(6):47-52.

IFT Expert Panel on Food Safety and Nutrition. 1986b. Sweeteners: Nutritive and non-nutritive. Food Technol. (Chicago) 40(8):195-206.

IFT Expert Panel on Food Safety and Nutrition, 1987. Nitrate, nitrite, and nitroso compounds in foods. Food Technol. (Chicago) 41(4):127-136.

Inai, K., Akamizu, Y., Eto, R., Nishida, T., and Tokuoka, S. 1985. Tumorogenicity study of butyl and isobutyl p-hydroxybenzoates administered orally to mice. Food Chem. Toxicol. 23(6):575-578.

Inglett, G. E. 1981. Sweeteners—A review. Food Technol. (Chicago) 35(3):37-41.

Irving, G. W. 1978. Safety evaluation of the food ingredients called GRAS. Nutr. Rev. 36:351.

Ito, N., and Hirose, M. 1989. Antioxidants—Carcinogenic and chemopreventive properties. Adv. Cancer Res. 53:247-302.

Ito, N., Fukushima, S., Hagiwara, A., and Shibata, M., and Ogiso, T. 1983. Carcinogenicity of butylated hydroxyanisole in F344 rats. J. Natl. Can. Inst. 70:343-352.

Ito, N., Fukushima, S., and Hirose, M. 1987. Modification of the carcinogenic response of antioxidants. Pages 253-294 in: Toxicological Aspects of Food. K. Miller, ed. Elsevier Applied Science, New York.

Jain, N. K., Rosenberg, D. B., Ulahannan, M. J., Glasser, M. J., and Pitchumoni, C. S. 1985. Sorbitol intolerance in adults. Am. J. Gastroenterol. 80:678-681.

Janusz, J. M. 1989. Peptide sweeteners beyond aspartame. Page 1046 in: Progress in Sweeteners. T. H. Grenby, ed. Elsevier Applied Science, New York.

Jenner, M. R. 1989. Sucralose: Unveiling its properties and applications. Pages 121-143 in: Progress in Sweeteners. T. H. Grenby, ed. Elsevier Applied Science, New York.

Joint Expert Committee on Food Additives (JECFA). 1986. Toxicological Evaluation of Certain Food Additives and Contaminants: WHO Food Additive Series 20. Cambridge University Press, New York.

Joint Expert Committee on Food Additives (JECFA). 1987. Toxicological Evaluation of Certain Food Additives and Contaminants: WHO Food Additive Series 21. Cambridge University Press, New York.

Joint Expert Committee on Food Additives (JECFA). 1989. Toxicological Evaluation of Certain Food Additives and Contaminants: WHO Food Additive Series 24. Cambridge University Press, New York.

King, B. D. 1981. Microbial inhibition in bakery products—A review. Baker's Dig. 55(10):8-12.

Kulczycki, A. 1986. Aspartame-induced urticaria. Ann. Intern. Med. 104(2):207-208.

Langdon, T. T. 1987. Preventing of browning in fresh prepared potatoes without the use of sulfiting agents. Food Technol. (Chicago) 41(5):64-67.

Leading Edge. 1989. Natural and Synthetic Food Additives. Leading Edge Reports, Cleveland, OH.

Lecos, C. 1981. The sweet and sour history of saccharin, cyclamate, and aspartame. FDA Consumer, HHS Publ. (FDA) (U.S.) 81-2156.

Lecos, C. 1985. Sweetness minus calories. FDA Consumer 19(2):18.

Light, J. M. 1990. Modified food starches: Why, what, where, and how. Cereal Foods World 35:1081.

Linke, H. A. B. 1986. Sugar alcohols and dental health. World Rev. Nutr. Diet. 47:136-162.

Lipin, R. 1984. Outbreak of diarrhea linked to dietetic candies. JAMA J. Am. Med. Assoc. 252:1672.

Louekari, K., Scott, A. O., and Salminen, S. 1990. Estimation of food additive intakes. Pages 9-32 in: Food Additives. A. L. Branen, P. M. Davidson, and S. Salminen, eds. Marcel Dekker, New York.

Lueck, E. 1980. Antimicrobial Food Additives. Springer-Verlag, New York.

Maeura, Y., and Williams, G. M. 1984. Enhancing effect of butylated hydroxytoluene on the development of liver altered foci and neoplasms induced by N-2-fluorenylacetamide in rats. Food Chem. Toxicol. 22:211-215.

Magee, P. N. 1983. Nitrate. Pages 198-210 in: Environmental Aspects of Cancer: The Role of Macro and Micro Components of Foods. E. L. Wynder, G. A. Leveille, J. H. Weisburger, and G. E. Livingston, eds. Food and Nutrition Press: Westport, CT.

Maher, T. J. 1986. Neurotoxicology of food additives. Neurotoxicology 7(2):183-196.

Makinen, K. K. 1989. Latest dental studies on xylitol and mechanism of action of xylitol in caries limitation. Pages 331-362 in: Progress in Sweeteners. T. H. Grenby, ed. Elsevier Applied Science, New York.

Malaspina, A. 1987. Regulatory aspects of food additives. Pages 17-58 in: Toxicological Aspects of Food. K. Miller, ed. Elsevier, New York.

Martinsen, C. S., and McCullough, J. 1977. Are consumers concerned about chemical preservatives in food? Food Technol. (Chicago) 31(9):56-59 ff.

Massey, L. 1982. Soft drink consumption, phosphorus intake and osteoporosis. J. Am. Diet. Assoc. 80(5):581-583.

Mayer, J., and Goldberg, J. 1985. Quest for artificial sweetener proceeds, but at slow pace. Minneapolis Star Tribune, Feb. 27.

Megremis, C. J. 1990. Stabilizing wheat flakes with mixed tocopherols. Cereal Foods World 35:316.

Miller, K. 1987. Food intolerance. Pages 347-372 in: Toxicological Aspects of Food. K. Miller, ed. Elsevier Applied Science, New York.

Miller, E. C., and Miller, J. A. 1986. Carcinogens and mutagens that may occur in foods. Cancer 58:1795-1803.

Millstone, E. 1985. Food additive regulation in the UK. Food Policy 10(3):237-252.

Mohri, T. 1987. Dietary intakes of nitrosamine precursors. Kyushu Yakugakkai Kaiho 41:105-112. (Engl. abstract) (Food Sci. Technol. Abstr. 1990 3A117)

Morris, C. E., and Przybyla, A. 1985. Cyclamate. Special report. Chilton's Food Eng. Int. 57(11):67-75.

Mueller, R. L., Hagel, H.-J., Wild, H., Ruppin, H., and Domschke, W. 1986. Nitrate and nitrite in normal gastric juice. Oncology 43(1):50-53.

Munro, I. C. 1987. International perspectives on animal selection and extrapolation. In: Human Risk Assessment: The Role of Animal Selection and Extrapolation. M. V. Roloff, A. G. E. Wilson, W. E. Ribelin, W. P. Ridley, and F. A. Ruecker, eds. Taylor and Francis, Philadelphia.

Munro, I. C. 1989. A case study: The safety evaluation of artificial sweeteners. Pages 151-169 in: Food Toxicology: A Perspective on the Relative Risks. S. L. Taylor and F. A. Scanlon, eds. Marcel Dekker, New York.

National Academy of Sciences/National Research Council (NAS/NRC). 1981. The Health Effects of Nitrate, Nitrite, and N-Nitroso Compounds. Report of the Committee on Nitrite and Alternative Curing Agents in Food. National Academy Press, Washington, DC.

National Academy of Sciences/National Research Council (NAS/NRC). 1985. Evaluation of Cyclamate for Carcinogenicity. National Academy Press, Washington, DC.

National Academy of Sciences/National Research Council (NAS/NRC). 1989. Diet and Health: Implications for Reducing Chronic Disease Risk. National Academy Press, Washington, DC.

Nehrling, J. K., Kobe, P., McLane, M. P., Olson, R. E., Kamath, S., and Horwitz, D. L. 1985. Aspartame use by persons with diabetes. Diabetes Care 8(5):415-417.

Newberne, P. M., and Conner, M. W. 1986. Food additives and contaminants. Cancer 58:1851-1862.

Nichol, W. 1991. Food additive legislation within the European Community. Part A: The sweeteners directive. Food Saf. Notebook 2(2):19.

Nikonorow, N., and Karlowski, K. 1976. The effect of low-protein diet on the toxicity of butylated hydroxytoluene in rats. Roczniki Panstwowege Zakladu Higieny 27(6):605-609. (Nutr. Abstr. Rev. 47:7846)

Nitrite Safety Council. 1978. Nitrite. The Council, Washington, DC.

NutraSweet Co. 1987. NutraSweet: A Health Care Practitioners Guide. NutraSweet Co., Skokie, IL.

O'Donoghue, K. J. 1981. Developments with food additives. Food Technol. N.Z. 16(10):13 ff.

O'Sullivan, D. 1983. New sweeteners gain ground in Europe. Chem. Eng. News. 61(Jan 24):29-30.

Olsen, P., Billie, N., and Meyer, O. 1983. Hepatocellular neoplasms in rats induced by butylated hydroxytoluene (BHT). Acta Pharmacol. Toxicol. 53:433-434.

Österdahl, B. G., and Alriksson, E. 1990. Volatile nitrosamines in microwave-cooked bacon. Food Addit. Contam. 7(1):51-54.

Park, E.-H., Chang, H.-H., Cha, Y.-N. 1990. A comparative case-control study of colorectal cancer and adenoma. Jpn. J. Cancer Res. 81:1101-1108.

Pascal, G., Hitier, Y., and Terreine, T. 1979. Effects of intake of the antioxidant butylated hydroxytoluene on the metabolism of vitamin C and vitamin A in the rat. Int. J. Vitam. Nutr. Rev. 49(1):3-13.

Pim, L. 1981. The Invisible Additives. Doubleday, Garden City, NJ.

Potezny, J. N., McClure, J., Rofe, A. M., and Conyers, R. A. 986. The long term effect of dietary administration of refined sugars and sugar alcohols on plasma biochemistry, urine, and tissue histology in mice given a limited degree of dietary self-selection. Food Chem. Toxicol. 24:389-396.

Price, J. M., Biava, C. G., Oser, B. L., Vogin, E. E., Seinfeld, J., and Ley, H. L. 1970. Bladder tumors in rats fed cyclohexamine or high doses of a mixture of cyclamate and saccharin. Science 167:1131-1132.

Prochaska, H. J., De Long, M. J., and Thalay, P. 1985. On the mechanisms of induction of cancer protective enzymes: A unifying proposal. Proc. Natl. Acad. Sci. U.S.A. 82:8232-8236.

Reno, F. E., McConnell, R. G., Ferrell, J. F., Trutler, J. A., and Rao, K. S. 1975. A tumorigenic evaluation of Aspartame, a new sweetener, in the mouse. Toxicol. Appl. Pharmacol. 29:182-188.

Riggs, B. S., Harchelroad, F. P., and Poole, C. 1986. Allergic reaction to sulfiting agents. Ann. Emerg. Med. 15(1):77-79.

Robach, M. C. 1980. Use of preservatives to control micro-organisms in food. Food Technol. (Chicago) 34(10):81-84.

Roberts, H. J. 1990. Aspartame (NutraSweet): Is It Safe?. A Concerned Doctor's Views. The Charles Press, Philadelphia, PA.

Roberts, T. A., and McClure, P. J. 1990. Food preservatives and the microbiological consequences of their reduction or omission. Proc. Nutr. Soc. 49:1-12.

Rogers, P. J., and Blundell, J. E. 1989. Evaluation of the influence of intense sweeteners on the short-term control of appetite ad caloric intake: A phychobiological approach. Pages 267-290 in: Progress in Sweeteners. T. H. Grenby, ed. Elsevier Applied Science, New York.

Roquette Freres. 1990. Non-sucrose sweeteners (Alternatives to sugar). Tech. Bull. Roquette Corp., Gurnee, IL.

Rudali, G., Coezy, E., and Muranyi-Kovacs, I. 1969. Research on the carcinogenic effect of sodium cyclamate in the mouse. C.R. Hebd. Seances Acad. Sci. Ser. D. 269:1910-1912.

Rugg-Gunn, A. J. 1989. Lycasin and the prevention of dental caries. Pages 311-330 in: Progress in Sweeteners. T. H. Grenby, ed. Elsevier Applied Science, New York.

Salminen, S., and Hallikainen, A. 1990. Sweeteners. Pages 297-325 in: Food Additives. A. L. Branen, P. M. Davidson, and S. Salminen, eds. Marcel Dekker, New York.

Sela, M. N., and Steinberg, D. 1989. Glycyrrhizin: The basic facts plus medical and dental benefits. Pages 71-79 in: Progress in Sweeteners. T. H. Grenby, ed. Elsevier Applied Science, New York.

Sen, N. P. 1980. Nitrosamines. Pages 319-349 in: The Safety of Foods. H. D. Graham, ed. AVI, Westport, CT.

Sen, N. P. 1986. Formation and occurrence of nitrosamines in food. Pages 136-160 in: Diet, Nutrition and Cancer: A Critical Evaluation, Vol. II. B. S. Reddy and L. A. Cohen, eds. CRC Press, Boca Raton, FL.

Sims, R. J., and Fioriti, A. J. 1977. Antioxidants as stabilizers for fats, oils and lipid-containing foods. Pages 209-246 in: CRC Handbook of Food Additives, 2nd ed., Vol. II. T. E. Furia, ed. CRC Press, Cleveland, OH.

Siu, R. G. H., Borzelleca, J. F., Carr, C. J., Day, H. G., Fomon, S. J., Irving, G. W., La Du, B. N., McCoy, J. R., Miller, S. A., Plaa, G. L., Shimkin, M. B., and Wood,

J. L. 1977. Evaluation of health aspects of GRAS food ingredients: Lessons learned and questions answered. Fed. Proc. 36(11):2524-2562.

Skrypec, D. J., Gray, J. L., Mandagere, A. K., Booren, A. M., Pearson, A. M., and Cuppett, S. L. 1985. Effect of bacon composition and processing on N-nitrosamine formation. Food Technol. (Chicago) 39(1):74-76.

Slaga, T. J. 1981. Additives and contaminants as modifying factors in cancer induction. Pages 279-290 in: Nutrition and Cancer: Etiology and Treatment. G. R. Newell and N. M. Ellison, eds. Raven Press, New York.

Sloan, A. E., Powers, M. E., and Hom, B. 1986. Consumer attitudes toward additives. Cereal Foods World 31:523-532.

Smith, M. V. 1987. Food safety reform legislation: Dead or dormant? Food Technol. (Chicago) 41(6):119-123.

Smith, M. V., and Rulis, A. M. 1981. FDA's GRAS review and priority-based assessment of food additives. Pages 71-74 in: Overview: Food Ingredient Safety Review Programs. R. V. Lechowich and M. V. Smith, eds. Institute of Food Technologists, Chicago.

Spooner, T. F. 1990. The fate of potassium bromate. Baking Snack Systems 12(7):12.

Stamp, J. A. 1990. Sorting out the alternative sweeteners. Cereal Foods World 35:395.

Steffy, D. F. 1980. Properties and uses of sulfur compounds: A review. Pages 10-16 in: Update on Antimicrobial Agents. N.Y. State Agric. Exp. Stn. (Geneva).

Stegnick, L. D., and Filer, L. J. 1984. Aspartame: Physiology and Biochemistry. Marcel Dekker, New York.

Sturtevant, F. M. 1985. Use of aspartame in pregnancy. Int. J. Fertil. 30(1):85-87.

Sullivan, D. M., and Smith, R. L. 1985. Determination of sulfite in foods by ion chromatography. Food Technol. (Chicago) 39(7):45-46.

Sussmuth, R. 1982. Genetic effects of amino acids after chlorination. Mutat. Res. 105(1/2):23-28.

Tannenbaum, S. R. 1984. A Policy Perspective on Safety: Nitrite and Nitrate. Hoffmann-LaRoche Co., Nutley, NJ.

Tannenbaum, S. R., and Correa, P. 1986. Nitrates and gastric cancer risk. Nature 317(6039):675-676.

Taylor, S. L. 1986. Sulfites in food. Hazleton Food Sci. Newsl. 14(May/June):1-4.

Taylor, S. L., and Bush, R. K. 1986. Sulfites as food ingredients. Contemp. Nutr. 11(10):1-2.

Taylor, S. L., Higley, N. A., and Bush, R. K. 1986. Sulfites in food. Adv. Food Res. 30:1-76.

Thomson, A. W. 1980. Carrageenan toxicity. Lancet I(8176):1034.

Thomson, A. W. 1981. Carrageenan and the immune response. Lancet I(8221):671.

Tordoff, M. G., and Alleva, A. M. 1990. Effect of drinking soda sweetened with aspartame or high-fructose corn syrup on food intake and body weight. Am. J. Clin. Nutr. 50:963-969.

Tricker, A. R., and Preussmann, R. 1988. N-nitroso compounds and their precursors in the human environment. Pages 89-116 in: Nitrosamines: Toxicology and Micobiology. M. J. Hill, ed. Ellis Horwood, Chichester, England.

USDA. 1979. No-nitrite meats: Handle carefully. PA-1238. U.S. Government Printing Office, Washington, DC.

Van Pelt, D. 1987. Vitamin C relative is safer than sulfite. Insight, 3(Sept. 14):57.

Vecchio, A. J., Hotchkiss, J. H., and Bosogni, C. A. 1986. Ingestion of N-nitrosamines from fried bacon: A consumer survey. J. Food Sci. 51:754-756.

Vorhees, C. V., Butcher, R. E., Brunner, R. L., and Wootlen, V. 1984. Developmental toxicity and psychotoxicity of sodium nitrite in rats. Food Chem. Toxicol. 22(1):1-6.

Wattenberg, L. W. 1983. Inhibition of neoplasia by minor dietary constituents. Cancer Res. 43:2448s-2453s.

Weisburger, J. H., and Evarts, E. K. 1977. Inhibitory effect of butylated hydroxytoluene on intestinal carcinogenesis in rats. Food Cosmet. Toxicol. 15(2):139-141.

Weisburger, J. H., Barnes, W. S., and Czerniak, R. 1986. Mutagens and carcinogens in food. Pages 115-134 in: Diet, Nutrition, and Cancer: A Critical Evaluation. B. S. Reddy and L. A. Cohen, eds. CRC Press, Boca Raton, FL.

Wells, A. 1989. The use of intense sweeteners in soft drinks. Pages 169-214 in: Progress in Sweeteners. T. H. Grenby, ed. Elsevier Applied Science, New York.

Wightman, N. 1977. Saccharin—Are there alternatives? J. Nutr. Educ. 9(3):106-108.

Wilkens, W. F., and Gray, J. I. 1986. Reduce N-nitrosamines formation in bacon. Food Eng. 58(May):68-69.

Williams, G. M., and Weisburger, J. H. 1986. Chemical carcinogens. Pages 99-173 in: Casarett and Doull's Toxicology, 3rd ed. K. Klaassen, M. O. Amdur, and J. Doull, eds. Macmillan, New York.

Wursch, P., and Anantharaman, G. 1989. Aspects of the energy value assessment of the polyols. Pages 241-266 in: Progress in Sweeteners. T. H. Grenby, ed. Elsevier Applied Science, New York.

Wurzburg, O. B. 1986. Forty years of industrial starch research. Cereal Foods World 31:897-904.

Yagen, B., Hutchins, J. E., Cox, R. H., Hagler, W. M., and Hamilton, P. B. Aflatoxin B_1 with sodium bisulfite. J. Food Prot. 52:574-577.

Yamamoto, M., Iwata, R., Ishiwata, H., Yamada, T., and Tanimura, A. 1984. Determination of volatile nitrosamine levels in foods and estimation of their daily intake in Japan. Food Chem. Toxicol. 22:61-63.

Yang, W. H., and Purchase, E. C. R. 1985. Adverse reactions to sulfites. Can. Med. Assoc. J. 133:865-867 ff.

Food Colors and Flavors

CHAPTER 11

Enhancing the flavor and color and hence the acceptability of food has been a human goal throughout history. The desire for spices, salt, and other flavoring agents launched exploration and discovery that changed the course of human history. Unfortunately, not all flavors and colors have been used merely to enhance existing flavors or to make the food a more appealing color. At times, flavorings were used to disguise the flavor of inferior or less-than-fresh products. Colorants were used to disguise inferior products. Indeed, the sordid history of color use colors our current ideas about them.

FOOD COLORS IN HISTORY

The history of adulteration of food is filled with examples of unscrupulous use of food colors. Profit-hungry dealers at one time used poisonous copper sulfate to color pickles, alum to whiten bread, and poisonous plants such as *Cocculus indicus* to "naturally" color and flavor beer and molasses. Cheeses were colored with red lead, vermillion, and mercury sulfide. Candies were dyed with lead chromate, carbamate, red lead, and vermillion. Spent tea leaves were mixed with Prussian blue (copper arsenate) and turmeric and were sold as green tea. Lead chromate and indigos were also used to color tea (Francis, 1984; Marmion, 1984; McKone, 1990; NAS/NRC, 1971).

In England around 1900, milk was tinted yellow to prevent detection of skimming and watering. The practice was so widespread that when it was made illegal in 1925, people were afraid to buy untinted milk as they thought it was contaminated (NAS/NRC, 1971). It is easy to see that these practices were not only fraudulent but harmful, and potentially lethal in many cases. With this in the background, a skeptical acceptance of food colors is indeed understandable. Such a history emphasizes the need for regulation in the area of food colors.

In the United States, food color was first legalized by an act of Congress in 1886, authorizing the addition of coloring to butter. By

1900, coloring was used in jellies, syrups, flavoring extracts, butter, cheese, ice cream, sausage, pastries, noodles, confectionery, wines, liqueurs, and cordials. At that time, about 80 different food colors were used. However, no laws prevented the use in food of industrial colorants, principally textile dyes. Although consumer activists are still concerned about the few certified colors that remain, most would be forced to admit that this area of food is much better regulated now than in the past.

FOOD COLOR REGULATION IN THE UNITED STATES

Dr. Bernhard Hesse, working under Harvey Wiley and his "poison squad" in 1904, was concerned about the unregulated use of 695 available coal tar dyes. He elected to test further only those dyes for which there were no unfavorable reports from any source. Only 16 dyes met this criterion. From those, only seven were selected for use in food. The Food and Drug Act of 1906 subsequently established a voluntary program for certification of food color additives. Under the certification program, each color batch could be chemically tested by the Secretary of Agriculture. Ten additional colors were added between 1916 and 1929. The Food, Drug, and Cosmetic Act of 1938 instituted mandatory certification and extended federal control to drugs and cosmetics.

The Color Additives Amendment of 1960 was designed to replace the inconsistent and outmoded provisions of the 1938 act governing the use of colors in food. The amendment legally defined color additives as dyes, pigments, and other substances made from vegetable, animal, or mineral sources added or applied to food. To ensure the purity of the factory-made pigments used in food, the amendment continued the mandated certification of each color batch (see Box 11.1) but required premarketing approval of color additives, resulting in more toxicity tests on previously certified colors. Thus, all colors currently approved at the time were provisionally listed until testing and FDA evaluation could be completed. The amendment also included the Delaney clause prohibiting the approval of any additives (including color additives) known to cause cancer in any species.

The FDA instituted new requirements for the toxicity testing of color additives, including long-term testing in two rodent species. Tests for carcinogenicity and teratogenicity as well as multigenerational studies to establish that reproduction was unaffected were included in the requirements. Lakes, pigments produced by extending soluble colors on alumina hydrate, required independent listing for certification.

As such retesting took years to complete, many color additives were provisionally listed for more than 25 years. Consumer confidence about colors was partially eroded because the provisional status was misinterpreted by consumers to mean that evidence against the color additive placed it on the provisional list. However, color additives could remain on the provisional list only if they presented no known risk to public health. To change from provisional to permanent listing, a color additive had to undergo the same tests and procedures required for premarket clearance of other food additives. Eventually there were to be no provisional listings; colors were either to be permanently listed or deleted (Newsome, 1990).

Testing procedures not only require time, but are costly. According to the Certified Color Manufacturers Association (CCMA), the cost in 1985 dollars to test a color was over $6 million (Noonan, 1985).

Food Colors Permitted in the United States

The number of available colors has waxed and waned (NAS/NRC, 1971). Since 1906, 16 colors have been delisted. Only one color on the original list approved by Hesse (FD&C Blue No. 2) remains fully approved (Caldwell, 1990; Francis, 1984). Only seven certified FD&C colors are currently permitted in foods in the United States (Frick and Meggos, 1988). A list of these is given in Table 11.1.

Box 11.1

What Is a Certified Color?

Certification requires that the color manufacturer submit a representative sample from each color batch to the FDA for chemical analysis. If the sample complies with the specifications, a certificate is issued and that batch of color is released for use (Kassner, 1987). It is then labeled as, for example, FD&C Yellow No. 5. The certification process ensures that every new batch is chemically identical to the pigment used in the animal feeding studies upon which the approval of the color was based. Certified colors are available as either FD&C dyes or FD&C lakes. Dyes are water soluble and exhibit their color when dissolved in solvent. Lakes are insoluble pigments that color by dispersion. A lake is made by placing an aluminum or calcium salt of an FD&C dye on an alumina base. After a ruling in 1964, blends of certified colors no longer require further certification.

Only natural colorants and nature-identical colorants are exempt from certification (Table 11.3). Examples of nature-identical compounds include β-carotene, apo-carotenal, and canthaxanthin. The three carotenoid compounds are permanently listed by the FDA as uncertified color additives (Dziezak, 1987).

Even though the number of colors has declined over the years, it is still possible to obtain a desired color in a given food system. The colors in Table 11.1 can be mixed to obtain a desired hue. Also titanium dioxide, a white pigment permitted in foods, drugs, and cosmetics that is exempt from certification, may be used to contribute whiteness to dyes or opacity or to create pastels (Dziezak, 1987).

Safety of Certified and Other Food Colors

Safety questions about food colors arise from their history, from the association of their names with nefarious sounding coal-tar dyes, and from popular authors such as Dr. Ben Feingold, whose theory about hypersensitivity and color is discussed later in this chapter. What do we actually know about food colors now that the permanent listing of the colors is complete or nearly so? As usual, we know a great deal more about certified synthetic colors than we know about natural colors, which are exempt from certification, because the law mandated the study of the certified colors only. The paragraphs that follow constitute

Table 11.1. FD&C Colors

Color Additive	Common Name and Synonyms[a]	Foods in Which Color is Permitted[b]	Dye Type
FD&C Blue No. 1	Brilliant blue FCF	NR	Triphenylmethane
FD&C Blue No. 2	Indigotine, indigo carmine, CI[c] food blue 1 (73015)	NR	Indigoid
FD&C Green No. 3	Fast green FCF, CI food green 3 (42053)	NR	Triphenylmethane
D&C Red No. 3	Erythrosine, erythrosine bluish, CI flood red 14 (45430)	NR	Monoazo
FD&C Red No. 40	Allura red AC, CI food red 17 (16035)	NR	Monoazo
FD&C Yellow No. 5	Tartrazine, CI food yellow 4 (19140)	NR	Pyrazolone (azo)
FD&C Yellow No. 6	Sunset yellow FCF, CI food yellow 3 (15985)	NR	Monoazo
Citrus Red No. 2	CI solvent red 80 (12156)	Orange skin of mature green eating oranges; limit = 2 ppm	Monoazo

[a] Letters and numbers such as FCF that occur after the common name of the color were part of the name when the color was petitioned. Numbers in parentheses are the color index number.

[b] NR = no restrictions.

[c] CI is a notation used by the European Community.

a capsule summary of some of what we know and don't know about the safety of colors used in the food supply and some of the colorants that have been banned from use. It will be apparent that no general regulatory agreement exists among different countries as to what is and is not allowed.

Banned colorants

One of the first colors to be discontinued was the azo dye known as butter yellow. This colorant was used in margarines until 1940, when it was shown to induce hepatomas (liver cancer) in rats. Subsequent studies showed it to be carcinogenic in other species as well (Vettorazzi, 1980).

Another color banned in the United States is carbon black. This color is obtained by heating different types of organic material and thus may contain considerable ash and polycyclic aromatic hydrocarbons (PAHs) (Chapter 5). Concern over their PAH content caused them to be banned. Interestingly, this colorant may be used in foods for the European Community (EC).

FD&C Red No. 2—Amaranth. The most controversial of the color additives banned in the United States is amaranth (FD&C Red No. 2), which had been used in the food supply since 1908. Chemically it is an azo compound, which can be split at the azo linkage by bacteria in the gut. Any physiological effects of azo dyes are probably due to breakdown products rather than to the dye itself.

Amaranth was long regarded as the least toxic of the red food colors. Long-term feeding studies in both rats and dogs with color levels as high as 5% of the diet did not show any pathological effects or increase in the number of tumors (Concon, 1988; Radomski, 1974). Despite such data attesting to its safety, more political pressure was exerted on the FDA to remove this colorant than had ever been used before.

Armed with data from two Russian studies indicating that it might be carcinogenic or teratogenic, several consumer activist groups used

---

Box 11.2

What Is an Azo Dye?

The largest number of certified colors are of the azo type. They are characterized by one or more azo bonds (-N=N-). The certifiable azo colors can be subdivided into four groups: soluble unsulfonated pigments, soluble unsulfonated dyes, insoluble unsulfonated pigments, and insoluble unsulfonated dyes (Marmion, 1984).

this food color as a rallying point. In the emotional fever pitch, hundreds of studies showing its safety were overlooked. Even questions about the validity of the Russian data due to concerns about the purity of the material tested and the methods used did not diminish the consumer activist pressure. In an attempt to validate the Russian work, the FDA began a study. This ill-fated study only added to the debate, as there was evidence that control and experimental animals were placed in the wrong cages. Postmortems of the rats to assess tumors were not performed in sufficient time to prevent serious tissue breakdown or to enable accurate conclusions. Statistical analysis of the data revealed that, although the number of tumors did not increase, the ratio of malignant to benign tumors did increase in females fed amaranth as 3% of the diet. Despite questions about the scientific validity of the experiments, amaranth was removed from use in food in 1976, due partly to unresolved questions about its safety (Food, Drug and Cosmetic Law Reports, 1977) and partly to intensive lobbying efforts by consumer interest groups.

In 1980 a petition for permanent listing was denied by the FDA on

Table 11.2. Chronological History of Synthetic Food Colors in the United States[a]

Year Listed	Common Name	Food and Drug Administration Name	Year Delisted
1907	Ponceau 3R	FD&C Red No. 1	1961
	Amaranth	FD&C Red No. 2	1976
	Erythrosine	FD&C Red No. 3	In use
	Orange 1	FD&C Orange No. 1	1956
	Napthol yellow S	FD&C Yellow No. 1	1959
	Light green SF yellowish	FD&C Green No. 2	1966
	Indigotine	FD&C Blue No. 2	In use
1916	Tartrazine	FD&C Yellow No. 5	In use
1918	Sudan 1		1918
	Butter yellow		1918
	Yellow AB	FD&C Yellow No. 3	1959
	Yellow AB	FD&C Yellow No. 4	1959
1922	Guinea green B	FD&C Green No. 1	1966
1927	Fast green FCF	FD&C Green No. 3	In use
1929	Ponceau SX	FD&C Red No. 4	1976
	Sunset yellow FCF	FD&C Yellow No. 6	In use
	Brilliant blue FCF	FD&C Blue No. 1	In use
1939	Napthol yellow S potassium salt	FD&C Yellow No. 2	1959
	Orange SS	FD&C Orange No. 2	1956
	Oil red XO	FD&C Red No. 32	1956
1950	Benzyl violet 4B	FD&C Violet No. 1	1973
1959	Citrus red No. 2	Citrus Red No. 2	In use
1966	Orange B	Orange B	In use
1971	Allura red AC	FD&C Red No. 40	In use

[a] Adapted from Marmion (1984).

the grounds that the available data were insufficient to demonstrate safety, although it was acknowledged that they did *not* show amaranth to be a carcinogen.

Very few countries have taken the U.S. stance. The Canadian Health Protection Branch noted that the effects of Red No. 2 were not organ-specific, a property usually exhibited by a true carcinogen. Further, it noted that amaranth's chemical structure was not at all similar to that of other dyes found to cause cancer. The Canadians were also satisfied that amaranth is neither teratogenic (as their studies failed to show that it crossed the placenta) nor mutagenic, a property that accompanies teratogenicity (Vettorazzi, 1980). In addition to Canada, Japan and countries of the EC allow its use in foods. The international agency that evaluates carcinogens dismissed the Russian studies, as it felt the material used in these studies was only 65–75% pure (IARC, 1975).

Red No. 2 currently has an acceptable daily intake (ADI) level from the JECFA, the WHO/FAO expert committee, of 0.75 milligrams per kilogram (JEFCA, 1987). It is the most widely used red colorant in the world (Parkinson and Brown, 1981). The JECFA has required long-term studies on amaranth. Concern exists in the United Kingdom because the average intake of amaranth is nearly double the ADI (Drake, 1980) and because of possible allergenicity (Concon, 1988).

CERTIFIABLE FD&C COLORS

FD&C Red No. 40—Allura Red

Since the ban of amaranth, food manufacturers in the United States have been using FD&C Red No. 40, allura red. It was developed for use as a certifiable color and in 1970 became permanently listed, the first new color to be so listed in many years. It is the most used of any FD&C color in the United States (2.63 million pounds were certified in 1989).

Lifetime studies on rats and mice have indicated that allura red is not carcinogenic or teratogenic (Borzelleca, 1990; Borzelleca et al, 1989; Brown and Dietrich, 1983; Hazleton Laboratories, 1978; Parkinson and Brown, 1981; Vettorazzi, 1980). No consistent adverse effects have been shown except moderate growth depression in rats receiving the highest dose, over 5% of the diet (Borzelleca et al, 1989; Parkinson and Brown, 1981).

Red No. 40 has been reported to have some psychotoxicity, based on a controversial rat study (Vorhees et al, 1983). The rats were exposed to dye and then mated. The females were dosed during gestation, and the offspring were also fed this colorant. The offspring were then rated on a series of performance tests such as swimming. The test is controversial because doses as high as 10% of the diet were used. Although this test clearly indicates an adverse effect, opponents argue that these dose levels are particularly inappropriate for tests on psychotoxicity. Their rationale is that no person could ever eat anything approaching that amount of food color and that at lower levels no observable effects on performance could be seen.

Because of inadequate data, this particular color was listed by the Scientific Committee of the EC among those for which an ADI for humans could not be established and that are not toxicologically acceptable for use in food. However, an ADI was established by the JECFA in 1989 at 0-7 milligrams per kilogram, with a maximum anticipated daily intake calculated at 0-19 milligrams per kilogram (Borzelleca and Hallagan, 1991). Allura red was not allowed for food use in Canada until 1984 or in Japan until 1991. So for Red No. 2 and Red No. 40, contradictory regulations exist among the United States and many other Western countries.

Red No. 3—Erythrosine

This is a food colorant of rather unusual composition in that it is a xanthene dye that contains four iodine atoms. It is one of the more nontoxic food colors, since nearly all of the color is excreted unchanged in the feces. What little is absorbed is rapidly excreted through the bile.

It has been tested extensively in several species. In lifelong feeding studies, no direct carcinogenic or other pathological effects have been noted (Borzelleca and Hallagan 1987; Borzelleca et al, 1987; Brusick, 1989; Lin and Brusick, 1986). Its acute oral toxicity is low; the LD_{50} level in the rat is 7,400 milligrams per kilogram (Butterworth et al, 1976). The fact that it is poorly absorbed is one reason for its low toxicity (Daniel, 1962; Parkinson and Brown, 1981; Radomski, 1974; Webb et al, 1962).

FD&C Red No. 3 is partially deiodinated in the gut to lower-iodinated fluoresceins (Borzelleca and and Hallagan, 1991). Its high iodine content has made researchers question whether it would have any effects on the thyroid (JECFA, 1987). In one study, an elevation of protein-bound iodine was noted, although no effect on the thyroid or metabolism was observed. Another study showed that male rats fed Red No. 3 at 4% of the diet had an increase in thyroid tumors, but further investigation showed that the tumors were due to a secondary mechanism. The FDA has determined that the Delaney clause does not apply to substances that act secondarily or indirectly nor to those for which no-

Box 11.3

Abbreviations Deciphered

The Food, Drug, and Cosmetic Act of 1938 created three categories of coal-tar colors:

FD&C colors	Certifiable for use in food, drugs and cosmetics
D&C colors	Dyes and pigments considered safe in drugs and cosmetics when in contact with mucous membranes or when ingested
Ext. D&C colors	Considered safe when externally applied but not safe for ingestion

(Data from Marmion, 1984)

effect levels can reasonably be established (Anonymous, 1985; Malaspina, 1987.

In vitro studies indicate that erythrosine may inhibit neurotransmitters (Augustine and Levitan, 1980; Mailman et al, 1980). What these studies may mean in vivo has yet to be determined. Other adverse effects have been reported, including effects on blood and gene mutations in some strains of *E. coli*. In contrast, no mutations were seen using the Ames test (JECFA, 1987).

FD&C Red No. 3 itself is permanently listed in the United States and is approved for use by the EC. A temporary ADI was set at 0.05 milligrams per kilogram (JECFA, 1989). However, Red No. 3 Lake was never so listed and was banned by the FDA in 1990.

FD&C Yellow No. 5—Tartrazine

Tartrazine is the second most commonly used FD&C color. Most persons ingest tartrazine on a daily basis with a maximum per capita daily dose of around 16 milligrams. It was permanently listed for use in food in 1969 (Malaspina, 1987).

Chronic toxicity and carcinogenicity studies in rodents feeding at the 5% level failed to show any unfavorable effects of this colorant (Borzelleca and Hallagan, 1988a, 1988b). Mutagenic and other reproductive and developmental studies in rats and rabbits have not incriminated it (Burnett et al, 1974; Cordas, 1978; Pierce et al, 1974; Sobotka, 1977). However, it is associated with hypersensitivity reactions resulting in urticaria, eczema, migraine, and asthma (Millstone, 1985; Zlotlow and Settipane, 1977). Symptoms usually occur 1 1/2 hours after ingestion and last several hours. Reactions to tartrazine are attributable to elevations in histamine rather than to allergic (immunologic) reactions (Murdoch et al, 1987). It undergoes reduction in the gut, the major product of which is sulfanilic acid (Jones et al, 1964). This may have minor behavioral effects on young rats, but they cannot be extrapolated to humans (Goldenring et al, 1982).

In one study, 15% of people sensitive to aspirin were also sensitive to tartrazine (Tse, 1982). A more recent study using urticaria as an end point found a smaller number (Murdoch et al, 1987) than reported earlier. According to the EC Scientific Committee for Food, a very small number of people (0.03–0.15%) react adversely. Some reports also implicate tartrazine as a cause of problems in behavior and perception (Cordas, 1978). Others have failed to find any reactivity in double blind situations. More information is needed to either substantiate or refute these findings (IFT Expert Panel on Food Safety and Nutrition, 1986; Watts, 1984).

Because of the small percentage who react adversely, the FDA mandated many years ago that if FD&C Yellow No. 5 (tartrazine) is used in a packaged food, it must be specifically listed in the ingredient statement. The food could no longer be simply labeled as artificially colored. This requirement was extended to all FD&C certified colors in 1991.

FD&C Yellow No. 6—Sunset Yellow FCF

FD&C Yellow No. 6 is an azo food dye used in a wide variety of foods in the United States. Like other azo compounds, this one is split at the azo linkage by bacteria in the gut. Some of the products resulting from bacterial splitting are absorbed and excreted in the urine.

Long-term studies have been made in rats, mice, and dogs. No-effect levels have been determined to be 500–1,600 milligrams per kilogram per day in the rat and 2,600–11,400 in mice (CCMA, 1983). An ADI was established by the JECFA in 1982 at 0–2.5 milligrams per kilogram, with an anticipated maximum of 0.11 milligrams per kilogram.

Studies on its carcinogenicity were negative (CCMA, 1983; Malaspina, 1987), so it was permanently listed in the United States. Well before the requirement that all certified FD&C colors be listed in the ingredient statement, the FDA had proposed that FD&C Yellow No. 6 be so listed, because of reports of hypersensitivity to this color, but that specific rule was not adopted.

FD&C Yellow No. 6 is also used in the EC and approved for use in many countries throughout the world. The JECFA would like more information on its metabolism, as there is some question about its effect on the adrenals and the kidneys at very high doses. Like Yellow No. 5, it is poorly absorbed and undergoes azo reduction to sulfanilic acid (Goldenring et al, 1982).

FD&C Blue No. 1—Brilliant Blue FCF

FD&C Blue No. 1 is a triphenylmethane color. Like other colors of this chemical class, it is poorly absorbed (Brown et al, 1980). After administration, more than 90% is recovered in the feces unchanged. Feeding studies in rats for over 2 years with Brilliant Blue at concentrations as high as 5% of the diet showed no effect.

Lifetime dietary toxicity and carcinogenicity studies have shown no evidence of carcinogenicity (Borzelleca et al, 1990) and no adverse effects other than 15% less body weight and decreased survival of female rats receiving the highest dose (2% of body weight). No adverse reproductive or developmental effects have been exhibited (Burnett et al, 1974; Pierce et al, 1974). No-effect levels have been reported as 600–1,000 milligrams per kilogram per day in rats and 6,400–9,000 in mice.

As with other colors in this class, injection of the material leads to tumors at the site of injection. Much controversy exists over the significance of these findings, as data from feeding studies indicate no tumors or other adverse effects. Regulators in the United States decided that the feeding data were the data critical to decision making for food color safety.

Blue No. 1 is a permanently listed food color in the United States, although the FAO/WHO is critical of the adequacy of the research on it. It initially was not allowed as a food color in most EC countries or the United Kingdom, based on the tumors seen at the site of injection. The United Kingdom reinstated it after noting that many surface-active colors and substances caused sarcomas at the site of injection. Analysis of intakes in the United Kingdom showed that the average intake is at least 5,000 times less than the ADI (Drake, 1980). Its ADI according to JECFA is 0–12.5 milligrams per kilogram, with an anticipated maximum intake of 0.022 milligrams per kilogram per day.

Box 11.4

What Are Triphenylmethanes?

Triphenylmethane or triaryl dyes, such as FD&C Blue No. 1, consist of three aromatic rings attached to a central carbon atom. These dyes are all water soluble, making them readily usable in most foods.

FD&C Blue No. 2—Indigotine

Chronic toxicity studies even at the highest feeding levels in both rats, dogs, and pigs indicate that this substance is innocuous (Borzelleca and Hogan, 1985; Borzelleca et al, 1985; Parkinson and Brown, 1981; Radomski, 1974). It is poorly absorbed (less than 5%) from the gut in rats (Lethco and Webb, 1966). Like certain other food dyes, this one also produces tumors at the site of injection. In 1987 it was permanently listed in the United States. It is also allowed by the Scientific Committee of the EC, which states that the toxicological data are currently adequate. The ADI is 2.5 milligrams per kilogram. As a comparison, the ADI established by the JECFA in 1969 was 0–17 milligrams per kilogram, with a calculated maximum daily intake of 0.009 milligrams per kilogram. The average intake in the United Kingdom is 1,000 times less than the ADI (Drake, 1980).

FD&C Green No. 3—Fast Green FCF

This is a triphenylmethane color. It is used in foods in the United States and several other countries in relatively small quantities (only 4,000 pounds were certified in 1989, as contrasted with 1.6 million pounds each of FD&C Yellows No. 5 and 6).

Biochemical studies have shown that this color is poorly absorbed and almost completely excreted (Hess and Fitzhugh, 1955). It is not mutagenic by the Ames test (Parkinson and Brown, 1981). A two-year study feeding Fast Green FCF to rats, dogs, and mice at concentrations as high as 5% of the diet showed no carcinogenic or toxic effects (Radomski, 1974). No-effect levels have been reported as 1,500–4,000 milligrams per kilogram per day in rats and 8,800–11,800 in mice (Borzelleca, 1990). In 1986, JECFA established the ADI for humans at 0–25 milligrams per kilogram, with an anticipated maximum intake of 0.0003 milligrams per kilogram per day.

Repeated subcutaneous injections of Fast Green FCF have been shown to produce sarcomas in rats at the site of injection. The JECFA originally felt that this was cause for concern and that the toxicological data was inadequate to meet its requirements (Vettorazzi, 1980). Green No. 3 has been used in neither the United Kingdom nor the EC.

Other Colors Allowed for Specific Uses

Two other colors are allowed in the United States for specific uses in limited amounts. Orange B was restricted to use in the casings of sausages and hot dogs at not over 150 parts per million (ppm) in the finished food. However, it is no longer available or used. Citrus Red No. 2 is an oil-soluble red-orange dye that may be used only for the coloring of Florida orange skins. The amount must be less than 2 ppm by weight of the whole fruit.

Citrus Red No. 2 does not appear to have a clean bill of health. Chronic toxicity testing in rats, dogs, and mice have caused some bladder wall thickening and bladder tumors (Radomski, 1974), and some studies have also indicated its carcinogenicity (Vettorazzi, 1980). The JECFA has not approved this colorant for use, nor has the United Kingdom or the EC (Drake, 1980).

The industry argues that the amounts are low and that most people don't eat the skins of oranges (Concon, 1988). However, many do use orange zest as a flavoring and use rinds for marmalade. It is my contention that we need to reeducate the consumer to expect some green on orange skins rather than allow the use of this unnecessary food colorant.

NATURAL FOOD COLORS

Natural food colors are either obtained from direct use of approved pigmented animal, vegetable, or fruit products or concentrated extracts from these materials or are their synthetic equivalents (Table 11.3). The FDA has approved a number of such substances that are exempt from certification. Several inorganic colorants also do not require certification and thus are found in this group. They include ferrous gluconate, a highly available source of dietary iron, and titanium dioxide, a white nonabsorbable material. Although these do not require certification, the amounts used are limited (Freund, 1985). Certain vitamins, including riboflavin (vitamin B-2) and β-carotene (the vitamin A precursor), may also give a product some color. However, the use of riboflavin solely for color is not permitted.

Natural color extracts have neither the intensity nor the stability of certified colors; thus they often fade and change over time. They may be sold in many forms and strengths and have varying tinctorial power. Unlike the carefully controlled FD&C colors, these often lack precise chemical definition and composition.

Consumers express very little safety concern about natural colors that are exempt from certification, probably because of the popular belief that natural is either benign or beneficial. This may not always be the case, as was seen in Chapter 5. Unlike the certified colors, most of the natural colors have been subject to very limited toxicity testing. Regulatory scrutiny is just beginning for these natural color additives. The wide variation in the preparations causes problems for toxicity testing, as different preparations may actually contain different compounds.

Table 11.3. Natural Color Additives Permanently Listed for Human Food Use in the United States[a,b]

Coloring Material	Source	Color	Use Limitation
Algal meal, dried	Algae	Yellow	For use in chicken feed for enhanced color of chicken and eggs
Annatto extract	Annatto seed	Red/yellow	...
β-Apo-8'-carotenal	Fruits and vegetables	Yellow	GMP[c]
Beet powder	Beets	Red	Not to exceed 15 mg/lb of food
Canthaxanthine	Fruits and vegetables	Yellow	Not to exceed 30 mg/lb of food
Caramel	Heated sugar	Brown	GMP
β-Carotene	Fruits and vegetables	Yellow	GMP
Carrot oil	Carrots	Orange	GMP
Cochineal extract; carmine	Insects	Red	GMP
Corn endosperm oil	Corn	Yellow	For use in chicken feed for enhanced color of chicken and eggs
Cottonseed flour, partially defatted, cooked, toasted	Cottonseed	Beige, yellow-brown	GMP
Ferrous gluconate	Iron salt	Black	For black olives only
Fruit juices	Fruit	Various	GMP
Grape color extract	Grapes	Blue-red	For nonbeverage food only
Grape skin extract	Grapes	Blue-red	For beverages only
Paprika and its oleoresin	Spice and *Capsicum*	Red-orange	GMP
Riboflavin (vitamin B-2)	Plant and animal tissue	Yellow-green	GMP
Saffron	Crocus	Yellow	GMP
Tagetes meal and extract	Aztec marigold	Yellow	For use in chicken feed for enhanced color of chicken and eggs
Titanium dioxide	Minerals	White	Not to exceed 1% by weight of food
Turmeric and its oleoresin	Rhizome of *Curcuma longa*	Yellow	GMP
Vegetable juice	Vegetables	Various	GMP

[a] Data from Marmion (1984), NAS/NRC (1971), and Ockerman (1991).
[b] These colors are exempt from batch certification.
[c] According to Good Manufacturing Practice.

Annatto and Related Carotenoids

Annatto, a yellow carotenoid preparation obtained from the seeds of the plant *Bixa orellana* L., is a mixture of several different carotenoids related to the compound bixin. Its structure enables it to be made into a colorant that is both water-soluble and lipid-soluble.

Annatto has been used in foods—especially dairy products—for many years. The Scientific Committee for the EC has established the ADI as 1.25 milligrams per kilogram, based on no-effect levels in the rat (Vettorazzi, 1980). Data on the metabolism of annatto compounds are not available.

A similar carotenoid pigment, crocetin, is found in the spice saffron. Saffron is the hand-plucked stigma from the crocus (*Crocus savitus*). Its labor-intensiveness makes saffron one of the world's most expensive spices, selling for more than $300 per pound. Because of its high cost, the use of saffron to color food is obviously limited to gourmet items requiring the flavor characteristics imparted by this spice.

The same pigment found in saffron minus the flavor characteristics can be found in another plant, the Cape jasmine (*Gardenia jasminoides*). Extracts from gardenia fruits may offer promise for food coloring because the pigments appear to have some degree of stability (Francis, 1987), although their use was not permitted in the United States at the time this was written.

A red-orange carotenoid, capsanthin, is derived from paprika. It was shown to slow the heart beat and lower blood pressure when injected. Tests on feeding of this color must be undertaken. Paprika itself is a permitted colorant.

The safety of colors from turmeric (curcumin) and its oleoresins is uncertain. In vitro studies show that turmeric extracts or oleoresins cause chromosome damage. Short-term studies indicate that they may be safe, but long-term studies raise safety questions (Achaya, 1981). The JECFA (1990) asked for long-term and reproductive studies on this spice and its color principle. It established in 1986 a temporary ADI of 0.1 milligram per kilogram for curcumin and 0.3 milligram per kilogram for turmeric oleoresins.

Saffron also contains crocin, an oleoresin similar to curcumin. There are no data on the safety of this saffron colorant. It undoubtedly will not be used extensively in food because of the high price of the starting material.

Cochineal, Carmine, and Carminic Acid

Cochineal was used extensively as a red food color from the 1600s into the 1900s. It was produced before cheaper and more readily available sources of red food color became available. The color is an extract of the bodies of female coccid insects, collected just before egg-laying time. Several different insects from various parts of the world may be used to produce this extract. It takes nearly 100,000 insects to produce a kilogram of raw dried cochineal.

The pigment, when extracted as a simple hot water extract, is sold as extract of cochineal. Pigment treated with a protease for further purification is called carminic acid or carmine, the latter being the lake produced by extending carminic acid on alumina hydrate to produce a bright red color that is heat stable (Francis, 1987).

Chronic feeding studies of rats at up to 500 milligrams per kilogram of body weight daily for over two years showed no adverse effects on survival, growth, body function, blood chemistry, or reproduction. Likewise, tumor incidence was unaffected (Ford et al, 1987; Grant and Gaunt, 1987; Grant et al, 1987).

Natural Pigments from Commonly Used Foods

The following natural or synthetic potential colorants are considered safe because they are nutrients or derivatives of nutrients: β-carotene, β-apo-8-carotenal, caramel, ferrous gluconate, and riboflavin. Other natural products thought to be probably safe include: carrot oil, fruit juices, grape skin extract, toasted defatted cottonseed flour, and vegetable juices. In some instances safety data are nonexistent.

Carotenoids. These compounds are widely distributed in nature and impart color to a wide variety of species, including carrots, corn, tomatoes, egg yolks, salmon, shrimp, and the feathers of birds like canaries and flamingos. Fall's panoply of leaf colors is due to the display of carotenoids. Three nature-identical synthetic carotenoids (β-carotene, β-apo-8-carotenal, and canthaxanthin) are permanently listed and exempt from certification (FDA 1986a, 1986b, 1986c). These colorants are widely used to color margarine, cheese, baked goods, and other yellow items. Canthaxanthin imparts a red color and can be used in tomato products and many other foods (Dziezak, 1987). As a natural carotenoid in various crustaceans and salmon, it has been evaluated for safety by JECFA (1990). While not carcinogenic, it may cause changes in the liver and retina. Because of this, the temporary ADI was not extended.

These molecules may add nutritive value if they yield vitamin A in the body. Their use as colorants in food is not believed to contribute to vitamin A toxicity (Vettorazzi, 1980). In addition, there is some evidence that carotenes on their own are useful contributors to the diet as anticarcinogens (NAS/NRC, 1989). The ADI for this group of colorants has been set at 2.5 milligrams per kilogram.

Anthocyanins. Fruit juices and grape skin extract provide a wide spectrum of anthocyanins, which typically color foods blue, rose red, or violet. About 130 different anthocyanins exist, yielding colors that range from the purplish black of blackberries to the true blue of blueberries, to the blue red of raspberries, to the bright red of cranberries. This color source has some problems with stability and tinctorial strength.

Betacyanins. These are another group of red colors found in beets, chard, cactus fruits, and pokeberries. Pokeberries also contain saponins

---

Box 11.5

Per Capita Consumption of Food Colors and Other Items (pounds per capita)

Certified color additives, 0.024
Food, 1,393
Sugar and sweeteners, 130
Salt, 8.5
Alcohol, 17.1
Aspirin, 0.064

(Data from Marmion, 1984)

(see Chapter 5), so their use in coloring food may be limited. Pink lemonade can get its color from the addition of beet extract. Even though this colorant is natural, the lemonade is labeled as artificially colored because lemonade is not naturally pink.

Chlorophylls. These are the green colors in leaves, vegetables, and fruits. Chlorophylls have many stability problems that limit their use to achieve the desired color. Currently, they are not generally allowed as food color additives in the United States but are allowed in Canada (Francis, 1985). However, a copper complex of chlorophyll is allowed for specific applications in the United States.

Pigments from Carbohydrate Browning

Caramel color is the most widely used food color, accounting for 98% by weight of all food color used. It is used in the soft drink, brewing, and baking industries. The colorant results from the process of caramelizing (burning) sugar. Since the process may be stopped at any time and may occur with many different acids or bases present, an incredible array of different compounds is possible. Although their chemical compositions may vary, the amount of lead, arsenic, and mercury they may contain is limited by the Standards of Identity. In general, caramel colors contain 50% digestible carbohydrate, 25% nondigestible carbohydrate, and 25% PAHs—levels similar to those found in roasted coffee and on the crust of baked goods. The safety of PAHs was discussed in Chapter 5.

Over 20 years of studies with both new and old protocols in both rodent and nonrodent species indicate the safety of caramel colors. Questions that arose about their safety were resolved (Newberne and

Conner, 1986), although it is difficult to know that every preparation has a clean bill of health because each specific caramel contains its own array of chemicals. An ADI has been set at 100 milligrams per kilogram, based on rat studies, with a no-effect level at 10,000 milligrams per kilogram (Vettorazzi, 1980).

Miscellaneous Colors

Titanium dioxide, a white pigment, is used at levels up to 1%. It is believed to be safe because it is very insoluble, making it very poorly absorbed and rapidly and completely excreted.

Unusual sources of natural plant pigments have been sought. Microorganisms supply a red pigment that has been used in the Orient for hundreds of years. Other pigments have been sought and patents have been filed for extracts from algae, bacteria, and molds. In addition to these possible sources, plant tissue culture and biotechnology may become sources of natural pigments (Ilker, 1987). Currently, the FDA allows only the colors listed in Table 11.2. Petitions for colors from the unusual sources will require the same safety data required for any additive petition.

Petitions for food colors that are being used in other countries have been submitted to the FDA. These include D&C Yellow No. 10, quinoline, and carminosine red. The FDA may be repetitioned for the reinstatement of FD&C Red No. 2. Studies using adequate protocols were recently completed on all these colors. The CCMA hopes for more international cooperation in the testing and safety evaluation of color additives (Noonan, 1985).

ISSUES ABOUT COLOR USE

Labeling of Food Colors

When color is added to a food, the label must state "artificially colored" or "artificial color added." The term "natural" color may not be used even if the color is derived from nature. This labeling is to protect the consumer so that there will be no misunderstanding about whether this is the actual color of the item or a color that has been enhanced by addition of color of any kind. The alternate way to label the product is to declare on the label "colored with _____ " or "_____ color," where the blanks are filled in with specific names such as beet powder, β-carotene, or annatto. Declaration of the specific colorant FD&C Yellow No. 5 on the label has been required since 1986, as some people have a sensitivity to it. However, the Nutrition Labeling and Education Act passed by Congress in November 1990 required all certified colors to be shown on the label after November 1991.

Should Colors Be Added to Food?

Many consumers question the need for any use of food colors. This question deserves special consideration, as in many cases the addition of colors requires greater justification than the use of additives that preserve the food or enhance the nutritive value. The industry believes that there are reasons to add color. These apply 1) if the food has no color of its own or an unappealing color such as gray, 2) if the natural color is lost during processing or storage, or 3) if it creates more interest in the food.

Good data exist establishing that color affects food acceptability. In study after study, consumers unerringly choose food that is attractively colored over food that is not. In some cases, certain flavors are associated with certain colors (Christensen, 1983; IFT Expert Panel on Food Safety and Nutrition, 1986). One study showed that when lime-flavored sherbet was colored green, 75% of the taste panel could correctly identify the flavor; if colored purple, only 26% identified it correctly (Francis, 1984). Other studies testing the effects of color on flavor and aroma identification show similar results: color influences flavor perception.

Color has been shown both to alter the requirements for the amount of sugar and to affect consumer acceptability. Red drinks were perceived to be sweeter than identical drinks that were either colorless or of another color (Pangborn, 1960). Past experience shows that in the United States, people resist buying oranges with green skins even when they are ripe. One British manufacturer purposely decided to omit the color because consumers said they didn't want artificial colors; sales for that product dropped by 50% (Taylor, 1980).

Consumer reactions to food colors are mixed. Although few consumers in a Good Housekeeping Institute survey could name an FD&C color, 47% thought red colors were bad, 11% thought yellows were bad, and 5% thought all of them were bad (Sloan et al, 1986).

Several studies have shown that terminology for labeling color can be a powerful determinant of acceptability. "Natural color" or "no color added" was very acceptable to 80% of women polled, whereas "vegetable color" was very acceptable to only 44% of the women despite the fact that they described the same colorant (McNutt et al, 1986). Terms that connote approval such as "certified" and "FD&C" achieved moderate acceptability. "Coal-tar derivative" was completely unacceptable. In many cases, women surveyed were unaware of studies being done on the safety of food colors or of any pending FDA action. About 65% said the evidence should be studied to see whether food dyes really cause cancer. Nearly 20% said they should be banned just on the chance that they might cause cancer (McNutt et al, 1986).

Current and Future Color Use

Ten percent of the food in the United States contains added color (NAS/NRC, 1989). The average daily intake of food color in the United States, assessed on 12,000 diets by the National Academy of Sciences, is approximately 100 milligrams per day. The average daily intake is given in Table 11.4. There is some concern that the intake in children is higher than that in the general population, as they are heavier consumers of foods that contain more coloring. Unfortunately, data on the intake of food colors by various segments of the population are not available.

Several colors have been petitioned for food use in the United States. One is D&C No. 10. As its name implies, it is currently used in cosmetics and drugs. It is also widely used in South America and Europe. The CCMA has also petitioned for approval of carmiosine, which is also widely used in Central and South America and Europe.

Perhaps, rather than asking "Do we need coloring?" or "Are colors safe?" a more important question is "Do we need many of these items that contain coloring?" The data appear to indicate that most of the colors are safe when used in the proposed amounts. On the other hand, many foods with colors aren't filling a nutritional void but are in the category of fun foods. Selection of foods that provide needed nutrients and fiber often limits the intake of food colors to well within a safe level and yet allows for an occasional festively frosted birthday cake

Table 11.4. Maximum[a] Daily Intake of Certified Colorants[b]

Color Additive	Intake (mg/kg)
FD&C Blue No. 1	16
FD&C Blue No. 1 Lake	6.6
FD&C Blue No. 2	7.8
FD&C Blue No. 2 Lake	3.1
FD&C Green No. 3	4.3
FD&C Red No. 3	24
FD&C Red No. 3 Lake	15[c]
FD&C Red No. 40	100
FD&C Red No. 40 Lake	27
FD&C Yellow No. 5	43
FD&C Yellow No. 5 Lake	22
FD&C Yellow No. 6	37
FD&C Yellow No. 6 Lake	14

[a] Data represent the 99th percentile of persons over 2 years of age in the "eaters" group. An "eater" was defined as a person in the sample who ate one or more foods containing the additive in question. Thus 99% of the 12,000 people sampled were estimated to have intakes equal to or below the value shown.
[b] Data from NAS/NRC (1979).
[c] Currently not approved for use in the U.S.

or other artificially colored fun food. Excess colorant in the diet probably does not in and of itself create a problem, but it may flag a diet with a preponderance of foods that are low in micronutrients or high in salt, sugar, fat, or calories.

FOOD FLAVORS

Production of desirable flavor is crucial for food acceptance and enjoyment. During restaurant or home preparation of a food, natural flavors are developed through combinations of fully ripe, fresh ingredients; by processes such as browning in baked goods and meats; and by additions of spices, herbs, marinades, sauces, smokes, alcohol, and extracts. These embellishments are usually viewed positively.

Flavors used in processed foods, on the other hand, are viewed suspiciously. They make the consumer jittery because of a fear that flavors are added to replace what has been taken out or to make up for an ingredient not at its optimum. Further, there is worry about the safety of chemicals that serve as artificial flavor ingredients. This fear is heightened even though the actual intake of any one substance is low because a large number of substances is used. The remainder of this chapter, looks at concerns about flavor additives as a group and at a few controversial flavors or flavor enhancers.

Flavor Additives

Sixty-nine percent of the 1,547 additives and 1,175 GRAS substances recognized by the FDA in the United States food supply are food flavor ingredients. Although the number seems staggering, the quantities used are small (Table 11.5). The amount used in any one food must be extremely small or the taste of the food is ruined (Reineccius, 1989c). Statistics on pounds of flavor used substantiate this. Of the natural or synthetic flavorings used by the entire food industry worldwide,

Table 11.5. Levels of Certified Color Used in Various Food Types[a]

Category	Range (ppm)	Average (ppm)
Candies	10–400	100
Beverages	5–200	75
Dessert Powders	5–600	140
Cereals	200–500	350
Maraschino cherries	100–400	200
Bakery goods	10–400	50
Ice creams, ices, and sherbets	10–200	30
Snack foods	40–500	200

[a] Adapted from Marmion (1984).

78% were used in quantities of less than 1,000 pounds (0.0006 milligrams per day per capita); 61% in less than 100 pounds, and 38% in less than 10 pounds (Senti, 1983). So, although flavor additives are numerous, the quantities are usually insignificant. A prominent flavorist has calculated that flavor compounds already in food account for over 98% of ingested flavor, with added flavor compounds at less than 2% (Reineccius, 1989c).

Artificial Flavors

Most artificial flavor ingredients that are used as additives would be more accurately labeled "subtractitives." That is, artificial flavors are frequently characterized by the 8 or 10 predominant constituents. This is far less than the several hundred chemicals identified in miniscule amounts in a natural flavor.

No doubt those other flavor chemicals present in minute amounts give natural flavors the desirable balance that is not always achieved with artificial flavors. On the other hand, the artificial flavor may impart a stronger flavor that remains characteristic longer and has been thoroughly safety tested (Reineccius, 1989c).

Safety of Flavor Ingredients

Safety of flavors is a concern whether the flavor is natural or synthetic. Safety assessments have been attempted by many experts, including those in the United States, Britain, Germany, and Netherlands, as well as by the EC and the FAO/WHO's JECFA (Grundschober and Stofberg, 1986). There is general agreement that it is neither desirable nor possible to evaluate flavor ingredients in the same way as other additives because of their sheer number and the low concentrations used, with their very nature precluding overuse. Also, the vast majority of flavoring substances occur widely in traditional foodstuffs; the chemical structures of most of them are not related to substances of demonstrated toxicity (Grundschober and Stofberg, 1986); and very limited amounts of material are available for testing.

All the regulatory bodies involved agree that too many flavoring compounds exist to adequately test them all and that it would be an injudicious use of resources to assign equal importance to testing of materials with unequal risks. As a result, JECFA set up a ranking for testing based on the following:

1) The total amount of flavoring likely to be ingested;

2. Structural similarity to substances with known biochemical or toxicological properties; and

3) Whether the substance is nature-identical or totally synthetic (not

found in nature).

For example, flavoring ingredients that are totally synthetic and not nature-identical can be dealt with in the same way as any food additive. Flavorings derived from nature or that are nature-identical are subject to an evaluation based on their consumption ratio. This compares the quantity of natural flavoring substance to the quantity added as flavoring. If the ratio is greater than one, the amount of flavor found naturally in food predominates and the committee judges the flavoring to be food-predominant. If less than one, it is additive-predominant and toxicity testing is required.

To calculate the ratio, much information is needed. First, quantitative data on the natural occurrence of the flavoring in all foods and the per capita consumption of those foods are accumulated. This information must then be combined with data on the annual consumption of added flavorings.

Consumer Reactions to Flavors

Nearly 70% of the women in the Good Housekeeping Institute survey attributed better quality to natural flavorings, whereas 15% said there was no difference and 15% said they didn't know. Only one third said that they could not tell the difference between the tastes of the two types of flavorings. However, only 15% said that they insisted on natural flavorings at the store (Reineccius, 1989b). Almost 40% said that it didn't matter as long as the product tasted good. Over twice as many consumers said that artificial flavors were composed of chemicals but natural flavors weren't (McNutt et al, 1986). (It is interesting to speculate what these consumers thought natural flavors were composed of.)

This same flavor survey found that "naturalness" was more important for fruit, meat, chocolate, and tomato flavors than for strawberry, onion, vanilla, and butter. Natural flavors were least important in soft drinks and diet colas.

Artificial flavors were seen as beneficial for enhancing weak natural flavors; for allowing low-calorie soft drinks; for cutting calories, fat, or cholesterol in other foods (McNutt et al, 1986); or for increasing convenience in microwave products (Naude, 1989). Keeping the cost of food low was not seen by consumers as a good use for artificial flavors.

Flavor Labeling

In the United States, flavors derived from nature are labeled as natural flavors. Flavors that are nature-identical are labeled in precisely that manner in Europe, but in the United States they are labeled as artificial.

Further confusing the issue is the acronym WONF, which stands for "with other natural flavors." So natural strawberry flavor WONF contains flavor compounds derived from strawberries plus those derived from other natural products. For instance, it may contain other berry flavors plus perhaps extracts from more exotic plants, which in the food product taste like strawberries. Thus, one may well ask, "Are 'natural' flavors really natural?" (Reineccius, 1989a).

Understanding flavor labeling can be difficult for the manufacturer and totally confusing to the consumer. Let's consider five scenarios. First, consider strawberry ice cream that has no flavor added: all of its strawberry flavor comes from the presence of myriads of strawberries. It is labeled on the front panel as "strawberry ice cream."

Next, consider a strawberry ice cream prepared as described above but with fewer strawberries in the formulation. If this formulation

---

Box 11.6

Artificial Strawberry Flavor Formulation

	Formula (in Percent)	
Flavor Component	Nature Identical	Artificial
Amyl isovalerate	1.0	1.0
Ethyl acetate	2.0	2.0
Amyl butyrate	3.0	3.0
Ethyl butyrate	10.0	10.0
Vanillin	1.0	...
Maltol	2.5	2.5
Ethyl vanillin	...	0.3
Ethyl alcohol (solvent)	80.5	81.2

Ethyl vanillin, found in nature, is three times stronger than natural vanillin, hence the small difference in formulation. In the United States, these must both be labeled as artificial flavors. Natural flavor in the United States must be extracted from natural sources. In Europe there is a category of flavor called nature identical.

The true flavor of a strawberry as analyzed by gas chromatography has several hundred different compounds, not just six, as appears in either of these formulations.

(Adapted from Bauer, 1984)

still has enough strawberry flavor from the strawberries in the recipe so that strawberry is the dominant and recognizable flavor, additional natural strawberry flavor may be added to extend its flavor. This ice cream may have the same label on the front panel as the ice cream in the first case. On the side panel the ingredient statement must indicate that natural strawberry flavor was added.

In the third case, the ice cream contains insufficient strawberries to be the dominant flavor or no strawberries at all, but has natural flavor added; then it may be called "natural strawberry-flavored ice cream." The fourth case is like the third except that even with the addition of strawberry flavor, strawberry is not the characterizing flavor, so other natural flavors are added; then the principal display panel must read "strawberry-flavored ice cream with other natural flavors."

The fifth case is seemingly the most clear-cut but may actually leave the consumer with the least accurate information. If the ice cream contains *any* artificial flavors, then it must be labeled "artificially flavored" despite the fact that it may contain as many or more strawberries than the ice creams in all the previous four cases, or no strawberries at all. If the consumer reads the label on the front panel in combination with the ingredient statement, that may help some; however, it may not always tell the consumer which product has more actual strawberries (Bauer, 1984; Reineccius, 1989).

HEALTH CONCERNS ABOUT SPECIFIC FLAVORS AND GENERAL USES OF FLAVORS AND COLORS

Cinnamyl Anthranilate

Cinnamyl anthranilate was at one time widely used to impart either grape or cherry flavor to beverages, candy, puddings, chewing gums, and baked goods, generally in concentrations of 1,000 ppm or less. In a study at the National Cancer Institute, it was found to cause tumors in the kidney and the pancreas in mice when given at doses of 15,000 ppm or greater (Anonymous, 1981), and it was banned by FDA in 1985.

Caffeine

Caffeine is added to beverages because it imparts a bitter taste as well as for its stimulant properties. It is found naturally in more than 60 species of plants and foods from those plants, including coffee, tea, chocolate, and colas.

Caffeine received a category III rating from the GRAS review committee (IFT Expert Panel, 1987b). Such a rating indicates that there is no evidence available demonstrating that caffeine at current levels of use is a hazard, but uncertainties exist that suggest the need for additional studies. We will look at what is known about caffeine to try to understand what the uncertainties are.

Metabolism. Caffeine's rapid absorption is responsible for its quick stimulant response. It reaches a maximum concentration in the blood just 30 minutes after ingestion. It has easy access to all tissues and affects many systems (Leonard et al, 1987).

Physiological effect. People have widely different tolerance levels for this drug. Some appear to be unaffected by it, while others are very sensitive. Caffeine has a diuretic effect and can cause heartburn, upset stomach, and diarrhea. Its ability to stimulate is well known (Boecklin, 1988). It decreases drowsiness and increases the basal metabolic rate (Acheson et al, 1980). Up to 200 milligrams of caffeine increases alertness, helps problem-solving ability, and improves typing speed by making a more rapid, clearer flow of thought (Anonymous, 1984).

Caffeine levels of 200–500 milligrams can cause headache, tremors, nervousness, and irritability (Lecos, 1988). At higher levels, the caffeinism syndrome results (Leonard et al, 1987). Caffeinism can cause sensory disturbances. Depression, anxiety, and schizophrenia may be worse among high caffeine users. Children who are high caffeine consumers were more likely to be rated as hyperactive by their teachers than low caffeine consumers (Rapoport and Kruesi, 1983).

Caffeine is mildly addictive, causing a psychological craving. As with

Table 11.6. Caffeine Content of Some Beverages[a]

Beverage	Caffeine Amount	Range
Soft drinks, mg per 6 oz		
Cola	15–23	...
Citrus	0–32	...
Lemon-lime	0	...
Root beer	0	...
Decaffeinated cola	<1	...
Other beverages, mg per 5 oz		
Coffee		
Brewed, drip	115	60–180
Brewed, percolator	80	40–170
Instant	65	30–120
Decaffeinated, brewed or instant	2	1–5
Tea		
Brewed	50	20–110
Instant	30	25–50
Cocoa, hot	4	2–20

[a] Adapted from Lecos (1988).

other addictive chemicals, there is a tendency for the consumer to increase the dose and for adverse symptoms to occur when the substance is withdrawn. Its many effects on the cardiovascular system include increased cardiac output, heart rate, and regularity of the beat. The role of caffeine and coffee in coronary disease remains controversial (La Croix et al, 1986).

Mutagenicity and teratogenicity. Caffeine has been shown to be both mutagenic and teratogenic (NAS/NRC, 1989; Nolen, 1982). Mutations occurred in vitro with caffeine levels equivalent to over 25 cups of coffee per day (Lecos, 1980), but effects on higher organisms are uncertain (Joesoef et al, 1990; Linn et al, 1982; NAS/NRC, 1989). Pregnant and nursing women are advised to curtail their caffeine intake. Since caffeine crosses the mammary barrier and is found in breast milk, there are concerns about its possible effects on the infant (Martin and Brocken, 1982; Weatherbee and Lodge, 1979).

Carcinogenicity. Coffee and caffeine have both been suggested as contributors to various types of cancer. However, recent epidemiological findings revealed no link between the consumption of coffee and cancer at any site, including the pancreas (Gold et al, 1985; International Food Information Council, 1989; Jacobsen et al, 1986; MacMahon et al, 1981; Phelps and Phelps, 1988; Pozniak, 1985; Tarka and Shively, 1987).

Coffee has been the subject of numerous studies looking for a possible relationship to cholesterol and coronary heart disease (NAS/NRC, 1990). Some studies have found links; others have not. The effect of coffee drinking on coronary disease thus remains uncertain. Concerns about caffeine include how to accurately calculate the intake of caffeine from additives and natural sources. It is clear that too much caffeine should be avoided, no matter what its source.

Flavor Enhancer—Monosodium Glutamate

Monosodium glutamate (MSG) is one of those food additives that the consumer is sure is bad, but it is difficult to find nonanecdotal data to support what the public thinks it knows. Further, MSG is also one of the additives for which the amount in nature far exceeds its use as an additive.

MSG is the sodium salt of glutamic acid, one of the most common amino acids found in foods and in the body. In fact, glutamate makes up about one fifth of the body's protein. The range of consumption of free glutamate is 0.5-1.0 grams per day. Only one thousandth of the total body glutamate comes from free glutamate consumed as MSG. By contrast, an additional 20 grams of bound glutamate is ingested on average as a part of a day's normal food consumption (IFT Expert Panel, 1987a).

Flavor enhancement vs. basic taste. The use of MSG in foods dates back to ancient Chinese cooks, who used seaweed to make stock that was capable of enhancing the flavor of other foods cooked in it. Early in the 1900s, the Japanese discovered that the glutamate of seaweed was responsible for its flavor-enhancing properties. They then began producing MSG for use in food; it was not produced in the United States until several decades later.

For many years it was believed that MSG did not have a taste of its own but rather intensified the flavors already present in the food. However, recent sensory studies have shown that MSG is not simply a flavor enhancer but actually imparts a fifth basic taste that has been named *umani* (Yamaguchi, 1987). Studies with taste receptor physiology have substantiated the existence of a fifth basic taste (IFT Expert Panel, 1987a).

Metabolism. Free glutamate added to food as MSG is not preferentially absorbed over glutamate that is in food naturally. Plasma glutamate levels do not appear to change greatly after MSG use and are comparable to levels after protein feeding. For instance, plasma glutamate levels were the same after a meal of hamburger containing no MSG and a meal of sloppy joes containing MSG at 34 milligrams per kilogram. MSG in liquids such as water or tomato juice caused greater increases in plasma glutamate than MSG in nonliquid foods (Stegink et al, 1983).

MSG does not pass through the placenta. This was shown by experiments with extremely high plasma glutamate levels, obtained only by injection. MSG therefore cannot cause glutamate elevation in fetal blood. Although MSG was shown to pass through the mammary barrier, glutamate levels in milk were not elevated (Baker et al, 1979).

Neurological problems. Studies in 1969 using injected MSG produced neurological damage in young but not in mature rats (Olney, 1969). In contrast to the injection studies, chronic feeding studies did not produce any neurological lesions in young rodents, dogs, rabbits, or monkeys even with doses as high as 40% of the diet (Kenney, 1986). Human studies with MSG levels as high as 120 grams per day for extended periods showed only two minor effects, both of which might be considered neutral or beneficial. These were a fall in blood glucose concentration and a small drop in blood pressure (Bazzano et al, 1970).

Chinese restaurant syndrome. The personal description of one doctor's experience following a meal at a Chinese restaurant was published in the letter portion of the *New England Journal of Medicine* in 1968 (Kwok, 1968). The symptoms he reported were numbness that began in the back of the neck and radiated to the arms and back, weakness, and palpitations. He suggested that these might be related to the high salt content, MSG, or other components of the food.

Additional letters to the *Journal* continued to define the syndrome. Later, the group of symptoms he described were labeled the Chinese restaurant syndrome (CRS) (Schaumburg, 1968; Schaumburg et al, 1969). Symptoms were typically described as occurring 15–20 minutes after ingestion of MSG and disappearing after 2 hours. Other symptoms reported in the literature include flushing, paresthesia, chest pain, facial pressure, dizziness, sweating, headaches, nausea, and vomiting (Settipane, 1986).

Controlled challenges with MSG in persons who claimed to react to MSG gave very different results from those reported in the clinical letters. In the laboratory, individuals did respond to challenge doses of MSG, but puzzlingly, responses occurred only at levels of MSG much higher than would be used in food. Symptoms from high doses of MSG in the challenge studies included sensations of warmth, burning, tingling, numbness, and tightness. Headaches, nausea, gastric distress, and weakness occurred with equal frequency for placebo and MSG-containing foods. In double-blind challenge studies of individuals professing to have CRS, the response to the MSG challenge was identical to the response to the placebo (Kenney, 1986). One study showed that subjects were more likely to report CRS symptoms after ingesting coffee than after ingesting a 2% solution of MSG (Kenney, 1980). Another

---

Box 11.7

What Is a Double-Blind Placebo Controlled Study?

A double-blind placebo controlled study is considered to yield reliable results that are free of bias introduced either by the subject or by the researcher. In this type of study, neither the subject in the study nor the person evaluating the subject knows whether the test substance or a placebo has been administered. For the study to be valid, the placebo and the test substance must look and taste the same so that the "blindness" cannot be broken by the subject.

Often people who say that they react to various foods or food additives adversely fail to do this in double-blind tests. The blindness of the test eliminates what is known as the placebo effect. This effect is real and can be responsible for remission of some symptoms in some disorders. It has been measured as high as 50%. Stated another way, a strongly advocated curative recommended by a highly trusted physician or friends may, even though it has no active ingredients (e.g., contains only glucose), act as well in some people as items with proven therapeutic effects.

reported no flushing reaction when 24 supposedly CRS-susceptible subjects consumed as much as 3 grams of MSG at one time (Wilkin, 1986).

Thus, controlled studies testing the validity of CRS have failed to validate its existence. Part of the apparent discrepancy may be explained by the results of a survey conducted by the Market Research Corporation of America. It found that 43% of 3,000 individuals surveyed sometimes experienced some form of discomfort after eating anything (Kerr, 1979).

Kenney summarized the nearly 20 years of research on the adverse effects of MSG this way:

1. Symptoms of CRS occur in a small number of individuals when MSG is fed at concentrations greater than 3%. Even in these individuals, responses are seen only about one in every three times.

2. The symptoms experienced vary from day to day and are not correlated with plasma glutamate levels.

3. Reported symptoms such as warmth or tightness are not measurable by physical means such as altered tone or skin temperature.

Use of MSG in food. MSG is not a necessary nutrient and could therefore be omitted from foods. The only loss would be the loss of a unique flavor experience. However, it need be omitted only with justifiable reason. Some argue that it should be omitted because it contains sodium, as there is too much salt in the diet. Two factors run counter to that argument: not all sodium salts appear to affect blood pressure in the same way (Kurtz et al, 1987), and the amount of MSG that can be used in a food is limited, averaging less than one half teaspoon per six servings. Since the sodium content of MSG is lower than that of salt, use of MSG may actually lower the total sodium intake (IFT Expert Panel, 1987a). Furthermore, glutamates are also available as the calcium and potassium salts, and these would not contribute to the sodium intake at all.

Allergies and hypersensitivities to MSG do exist, as they do to many

Table 11.7. Comparison of Glutamate Ingested with Food with That Added to Food[a]

Food	Percent Total Glutamate Naturally Present	Recommended Percent Monosodium Glutamate (MSG) Added[b]	MSG as Percent of Total Glutamate
Beef	2.9	0.4	14.1
Chicken	3.3	0.3	9.1
Pork	2.3	0.4	17.2
Peas	5.6	0.2	3.6
Corn	1.8	0.2	11.3

[a] Adapted from data from the MSG Association, Atlanta, GA.
[b] Recommended level of addition figured as ½ teaspoon of MSG per pound of meat, or four to six servings.

other food components (Freed and Carter, 1982; Settipane, 1986). However, the portion (1-2%) of the population whose reaction to MSG can be actually measured in controlled studies (Kerr et al, 1979) remains unreconciled with the 25% suggested to exist by reports in the popular press. Just as food allergy appears to be overreported, CRS may also be overreported. Unlike the effects of allergy, those of CRS appear to be transient, usually lasting no more than an hour. There appears to be little factual basis for the likelihood that glutamate would cause more sensitivities than other food components, especially as natural glutamates are found in many foods, including tomatoes, mushrooms, and Parmesan cheese.

The JECFA, National Academy of Sciences, Federation of American Societies for Experimental Biology, Mayo Clinic, and FDA have all concluded that even when MSG is eaten in large amounts, it causes no harm to laboratory animals or humans (Schmitz, 1990; Tollefson, 1989). Gore and Salmon (1980) have pointed out that CRS may have been a hoax, perpetrated with no mal intent but still a hoax. The tragic thing about a hoax in the health area is that it takes years to stamp out a disease from which people love to suffer.

Regulatory agencies in a variety of countries deem MSG to be safe (JECFA, 1988). It is a GRAS additive and is used in some Standard of Identity foods as long as it is declared on the label. In Canada, MSG is considered a food ingredient and its use is dictated by Good Manufacturing Practices. In the state of Maine, a 1989 bill required restaurant consumers to be given notice if MSG was used (Schmitz, 1990). Such legislation might be proposed in other states. The MSG Association and others feel that such notice is neither desirable nor necessary since MSG has no health effect. However, some people also feel that consumers should be able to avoid components they choose to avoid, whether or not there is any rational reason to do so.

Possible Hyperactivity from Food Flavors and Colors

In 1973, Dr. Ben Feingold postulated a relationship between diet and hyperactivity. (The medical name for this condition is attention deficit disorder [ADD].) With the publication of the book *Why Your Child Is Hyperactive* (Feingold, 1975) and its promising claims, the postulate received widespread publicity. Talk show and media reports of rapid and dramatic improvement for 50% of hyperactive children given a diet free of artificial flavors and colors and natural salicylates furthered the public's belief in the efficacy of the diet. Anecdotal reports in the medical literature showed similar results.

One problem with analyzing the results of any of these reports was the very real problem of diagnosing ADD. Children are frequently labeled hyperactive without the difficult assessment required to make an actual diagnosis.

Systematic, controlled studies using carefully developed diagnostic criteria and crossover designs were done to test the Feingold hypothesis. Food for these studies was specially manufactured so that neither the child, parent, nor teacher could discern whether it contained additives. Dietary compliance was monitored. These studies failed to show the strong positive effects reported by Feingold (Conners, 1980; Conners et al, 1976; Harley and Mathews, 1979). If any improvement was noted, it was found among only a small subgroup of children and then only by the parents. Trained classroom observers with rating scales designed to assess hyperactivity could not find a difference in the behavior of the children placed on different diets. Further challenges were designed for those children who showed any beneficial effects of diets free of food dyes. Challenge cookies containing high doses of food colors were given to some children and a placebo to others. Both failed to cause deteriorating behavior even in the small subclass of children labeled as reactive in previous studies (Juhlin, 1983).

Three studies reported that children with sensitivity to salicylate

---

Box 11.8

What Is Hyperactivity?

In the medical community, hyperactivity is called attention deficit disorder or ADD. It is said to occur in 5-10% of young school-aged children. The child is usually very distractable and has difficulty staying on task. This can lead to academic problems in school and behavior problems in the home, school, and other environments. Diagnosis is difficult and requires a normed rating scale such as the Conner's Scale of Hyperactivity.

Anecdotal stories by parents often indicate that dietary modification made their child's disruptive behavior easier to manage. However, dietary shifts to eliminate additives and sugar have also improved the diet of the child, which may have been low in crucial vitamins and minerals. In addition, modifications of the diet have made the child with a behavior problem the center of attention by focusing the family's food preparation activity on his or her needs. Along with diets, families often make other changes in the routine that can have an impact on behavior. Thus, sometimes the elimination of additives and sugar from the diet is deduced as the cause of changed behavior but in fact all the changes in the family's behavior have had a positive impact on the child's behavior.

artificial flavors and colors had improved behavior after being on the diet (Cook and Woodhill, 1976). Several aspects of these studies have been criticized, including the nonblind aspects and the rating scales used to assess hyperactivity in the Cook and Woodhill study; the small number of positive responses, for instance, three of 19 alleged reactors (Wilson and Scott, 1989) or two of eight (Ward et al 1990); and the presence of other nutritional deficiencies (Ward et al, 1990).

The Kaiser-Permanente or Feingold diet excluded many common fruits and vegetables alleged to contain the natural salicylates that had also been condemned as causing hyperactivity. Analysis for salicylates by the highly sophisticated analytical technique of thin-layer chromatography failed to demonstrate their presence in a large percentage of the fruits and vegetables alleged to contain them (Ribon, 1982). Thus, if these fruits and vegetables do affect behavior as claimed, they must do so through a constituent other than salicylate.

The Nutrition Foundation's National Committee on Hyperkinesis reviewed data gathered from seven studies and found no consistent effect of food colors and salicylates on hyperkinesis (Nutrition Foundation, 1980). Further, they saw no need for a public policy change in either the school lunch program or the food labeling program.

In 1982, the National Institutes of Health held a Consensus Development Conference on Diet and Hyperkinesis and published the following conclusions (NIH, 1982): 1) The dramatic improvements reported by Feingold and in clinical reports were not seen in controlled studies; 2) diets may have been beneficial, but their effects could not be linked to food additives; and 3) some studies indicated that a very small number of children were less hyperactive on defined diets than on control diets (Swanson and Kinsbourne, 1980; Weiss et al, 1980). However, the decreases in hyperactivity were not consistently observed. Even children who reportedly benefitted from the diet failed to show deteriorating behavior when dietary challenges containing food colors were administered (Lewis and Mailman, 1984).

The challenge studies and the Consensus conclusion were criticized by Brenner (1986), who felt that the dose of food colors used in many of the studies was too low. The level chosen for many of the studies was the per capita daily consumption of 26 milligrams of food color. Brenner believes that some children may ingest three times that amount. One Canadian study with 20 hyperkinetic boys using much larger amounts (100–150 milligrams) of food color blend reported an adverse behavior effect that appeared after 30 minutes and lasted up to 3 1/2 hours. During this time, 17 out of the 20 boys made more errors on a learning test (Swanson and Kinsbourne, 1981).

Some criticism is also directed toward studies showing that certain food colors reduce neurotransmitter activity in vitro (Augustine and Levitan, 1980). These studies with rat neurons show that FD&C Red No. 3 blocks the uptake of material into the neuron (Weiss et al, 1980). Most discount the usefulness of this finding for two reasons: This effect did not occur in vivo in either normal or hyperkinetic rats, and Red No. 3 is not readily absorbed into the central nervous system and therefore could not be an important determinant of behavior (Freydberg and Gortner, 1982).

Additional criticism of the studies on hyperactivity stems from the fact that many studies focused on food colors and did not consider other additives. Both the NIH committee and critics of the report indicated that more research is needed on food colors and other additives to assess their effects on behavior as well as on other systems (Thorley, 1983). One study showed that challenges with food colors did change ratings on the Conner's scale (for hyperactivity) but that parents of the children tested were unable to perceive any behavioral difference (Pollack and Warner, 1990). A report from the EC said that we need to know more about what the EC scientific committee calls the "cocktail effect," i.e., the effect of many additives and chemicals together (Gormley et al, 1987). However, the Feingold explanation for hyperactivity is generally considered to have been disproven and discredited (Harper, 1987).

A workshop sponsored by the American Academy of Allergy and Immunology and the International Life Sciences Institute/Nutrition Foundation has laid out strict protocols for the assessment of adverse effects of food additives and other components of food on behavior as well as on other physiological systems (Metcalfe and Sampson, 1990).

SUMMARY

Many issues remain to be resolved. The bulk of the evidence indicates that colors are safe, but a few studies raise some warning flags. However, these are usually raised by studies testing amounts far greater than are found in the diet or using in vitro methods that may have little bearing in vivo. A bigger question is: When do we need colors? Is it a good use of food color if less sugar can be used to sweeten a product? Is it a good use of color and flavor if an item is more appealing, especially to groups that may need more calories such as the sick or the elderly? Can colors and flavors help make products palatable that we deem to have desirable nutritional attributes such as reduced fat, calories, and salt? These are issues that are not scientific but still must be decided, once it has been determined that levels used in the food supply do not create a hazard.

For flavors, a nagging fact is that we don't have enough information about them. We should, however, be somewhat reassured by the fact that the amount that can be used is self-limiting. We also need to gain perspective on the amount of flavor compounds already in the foods we eat and the amount added as artificial flavors. Perhaps more information on these aspects along with additional information on the flavor ingredients in question will be helpful. For instance, should caffeine addition be limited in foods that will be eaten by children? Does MSG cause any problem or does it allow the food to contain less sodium? What regulations should be set for the entire population because a relatively few individuals are hypersensitive to the color or flavor used? Do large amounts of color and flavor additives encourage consumption of diets low in basic food items such as fresh fruits and vegetables in exchange for items high in fat, sugar, and salt? It may be that our criteria for evaluating whether these should be used should include these questions: Does the food containing flavors and colors serve a useful purpose and make a contribution to the diet? Will addition of the flavors and colors make it more likely that certain necessary nutrients will be obtained? Finally, with these and all additives, we must gather data to see if any "cocktail effect" exists among the additives as well as within the mix of foods we commonly ingest.

REFERENCES

Achaya, K. T. 1981. Regulatory aspects of food additives. Indian Food Packer 35(3):11-14.

Acheson, K. J., Zahorska-Markiewicz, B., Pittet, P., Anantharaman, K., and Jéquier, E. 1980. Caffeine and coffee: Their influence on metabolic rate and substrate utilization in normal weight and obese individuals. Am. J. Clin. Nutr. 33:989-997.

Anonymous. 1981. Flavoring agent deemed carcinogenic. Chem. Eng. News 59(Jan. 5):21.

Anonymous. 1984. Making friends at coffee time. Food Chem. Toxicol. 22:917-918.

Anonymous 1985. The latest status of certified food colors. Red Seal Rep. (Warner-Jenkinson, St. Louis, MO) 47:1-4.

Augustine, G. J., and Levitan, H. 1980. Neurotransmitter release from a vertebrate neuromuscular synapse affected by a food dye. Science 207:1489-1490.

Baker, G. L., Filer, L. J., and Stegink, L. D. 1979. Factors influencing dicarboxylic acid content of human milk. Page 111 in: Glutamic Acid: Advances in Biochemistry and Physiology. L. J. Filer, ed. Raven Press, New York.

Bauer, K. J. 1984. Legal considerations in the development and application of flavors in the USA. Workshop held by the Am. Assoc. Cereal Chem., April 5. (Reprinted in Dragoco Report, Dragoco Inc., Totowa, NJ)

Bazzano, G., E'Elia, J. A., and Olson, R. E. 1970. Monosodium glutamate: Feeding large amounts in man and gerbils. Science 169:1208-1209.

Boecklin, G. E. 1988. Caffeine and health risk (letter). J. Am. Diet. Assoc. 88:366-372.

Borzelleca, J. F. 1990. Lifetime toxicity/carcinogenicity studies of FD&C Green No. 3 in rats. Food Chem. Toxicol. 28(12):813-819.

Borzelleca, J. F., and Hallagan, J. B. 1987. Lifetime toxicity/carcinogenicity study of

FD&C Red No. 3 (erythrosine) in mice. Food Chem. Toxicol. 25:735.

Borzelleca, J. F., and Hallagan, J. B. 1988a. A chronic toxicity/carcinogenicity study of FD&C Yellow No. 5 (tartrazine) in mice. Food Chem. Toxicol. 26:189.

Borzelleca, J. F., and Hallagan, J. B. 1988b. Chronic toxicity/carcinogenicity studies of FD&C Yellow No. 5 (tartrazine) in rats. Food Chem. Toxicol. 26:179.

Borzelleca, J. R., and Hallagan, J. B. 1991. The safety and regulatory status of Food, Drug, and Cosmetic color additives. Proc. Am. Chem. Soc. Conf. In press.

Borzelleca, J. F., and Hogan, G. K. 1985. Chronic toxicity/carcinogenicity study of FD&C Blue No. 2 in mice. Food Chem. Toxicol. 23:719.

Borzelleca, J. F., Hogan, G. K., and Koestner, A. 1985. Chronic toxicity/carcinogenicity study of FD&C Blue No. 2 in rats. Food Chem. Toxicol. 23:551.

Borzelleca, J. F., Capen, C. D., and Hallagan, J. B. 1987. Lifetime toxicity/carcinogenicity study of FD&C Red No. 3 (erythrosine) in rats. Food Chem. Toxicol. 25:75.

Borzelleca, J. F., Olson, J. W., and Rene, F. E. 1989. Lifetime toxicity/carcinogenicity study of FD&C Red No. 40 (allura red) in Sprague-Dawley rats. Food Chem. Toxicol. 27:701.

Borzelleca, J. F., Depukat, K., and Hallagan, J. B. 1990. Lifetime toxicity/carcinogenicity studies of FD&C Blue No. 1 (Brilliant blue FCF) in rats and mice. Food Chem. Toxicol. 28(4):221-234.

Brenner, A. 1986. Food additives and behavior. Md. Med. J. 35:344-345.

Brown, J. P., and Dietrich, P. S. 1983. Mutagenicity of selected sulfonated azo dyes in *Salmonella*/microsonic assay: Use of aerobic and anaerobic activation procedures. Mutat. Res. 116:305.

Brown, J. P., Dorsky, A., Enderline, F. E., Hale, R. L., Wright, V. A., and Parkinson, T. M. 1980. Synthesis of C-labeled FD&C Blue No. 1 (Brilliant Blue FCF) and its intestinal absorption and metabolic fate in rats. Food Cosmetol. Toxicol. 18:1.

Brusick, D. 1989. Addendum to review of genotoxicity of FD&C Red No. 2. Unpublished.

Burnett, C., Agersburg, T. P., Pierce, E., Kirshman, J., and Scala, R. 1974. Teratogenic studies with certified colors in rats and rabbits. J. Toxicol. Appl. Pharmacol. 29:121.

Butterworth, K., Gaunt, I., Grasso, P., and Gangolli, S. 1976. Acute term and short-term toxicity studies on erythrosine BS in rodents. Food Cosmetol. Toxicol. 14:525.

Caldwell, E. F. 1990. FD&C Red No. 3 Lakes banned. Cereal Foods World 35:423.

CCMA. 1983. Lifetime toxicity/carcinogenicity studies of FD&C Yellow No. 6 in rats and mice. Certified Color Manufacturers Association, Washington, DC.

Christensen, C. M. 1983. Effects of color on aroma, flavor and texture judgements of foods. J. Food Sci. 48:787-790.

Concon, J. M. 1988. Food Toxicology. Part B: Contaminants and Additives. Marcel Dekker, New York.

Conners, C. K. 1980. Food Additives and Hyperactive Children. Plenum Press, New York.

Conners, C. K., Goyette, C. H., Southwick, D. A., Lees, J. M., and Andrulonis, P. A. 1976. Food additives and hyperkinesis: A double-blind experiment. Pediatrics 58(2):154-166.

Cook, P. S., and Woodhill, J. M. 1976. The Feingold dietary treatment of the hyperkinetic syndrome. Med. J. Aust. 2:85-90.

Cordas, S. 1978. Experimental and clinical considerations with regard to tartrazine (FD&C Yellow No. 5). J. Am. Osteopath. Assoc. 77:696-707.

Daniel, J. 1962. The excretion and metabolism of edible food colors. Toxicol. Appl. Pharmacol. 4:572.

Drake, J. J.-P. 1980. Toxicological aspects. Pages 219-246 in: Developments in Food Colours, Vol. 1. J. Walford, ed. Applied Science, London.

Dziezak, J. D. 1987. Applications of food colorants. Food Technol. (Chicago) 41(4):78-89.

FDA. 1986a. Canthaxanthin. Code Fed. Reg. Title 21.82.51.

FDA. 1986b. Beta-apo-8'-carotenal. Code Fed. Reg. Title 21.73.90.

FDA. 1986c. Beta-carotene. Code Fed. Reg. Title 21.73.95.

Feingold, B. F. 1975. Why Your Child is Hyperactive. Random House, New York.

Food, Drug, and Cosmetic Law Reports. 1977. Notices of Petition Denied: FD&C Red No. 2. Commerce Clearing House, Chicago, IL.

Ford, G. P., Gopal, T., Grant, D., Gaunt, I. F., Evans, J. G., and Butler, W. H. 1987. Chronic toxicity/carcinogenicity study of carmine of cochineal in the rat. Food Chem. Toxicol. 25:897-902.

Francis, F. J. 1984. A food of a different color. Sci. Food Agric. 2(3):18-21.

Francis, F. J. 1985. Pigments and other contaminants. In: Food Chemistry. O. R. Fennema, ed. Marcel Dekker, New York.

Francis, F. J. 1987. Lesser known food colorants. Food Technol. (Chicago) 41(4):62-69.

Freed, D. L. J., and Carter, R. 1982. Neuropathy due to MSG intolerance. Ann. Allergy 48(2):96-97.

Freund, P. R. 1985. Natural colors in cereal-based products. Cereal Foods World 30:271-273.

Freydberg, N., and Gortner, W. A. 1982. The Food Additives Book. Bantam, New York.

Frick, D., and Meggos, H. N. 1988. FD&C colors—Characteristics and uses. Cereal Foods World 33:570.

Gold, E. B., Gordis, L., Diener, M. D., Seltser, R., Boitnott, J. K., Bynum, T. E., and Hutcheon, D. F. 1985. Diet and other risk factors and cancer of the pancreas. Cancer 55:460.

Goldenring, J. R., Batter, D. K., and Shaywitz, B. A. 1982. Sulfanilic acid: Behavioral changes related to azo food dyes in developing rats. Neurobehav. Toxicol. Teratol. 4:43.

Gore, M. E., and Salmon, P. R. 1980. Chinese restaurant syndrome, fact or fiction. Lancet 1:251-252.

Gormley, T. R., Downey, G., and O'Beirne, D. 1987. Food, Health and the Consumer. Elsevier, New York.

Grant, D., and Gaunt, I. F. 1987. Three-generation reproduction study on carmine of cochineal in the rat. Food Chem. Toxicol. 25:903-912.

Grant, D., Gaunt, I. F., and Carpanini, F. M. B. 1987. Teratogenicity and embryotoxicity study of carmine of cochineal in the rat. Food Chem. Toxicol. 25:913-917.

Grundschober, F., and Stofberg, J. 1986. Priority setting for the safety evaluation of flavoring substances. Pages 1-7 in: The Shelf Life of Foods and Beverages. G. Charalambous, ed. Elsevier, New York.

Harley, J. P., and Mathews, C. G. 1979. The Feingold hypothesis: Current studies. J. Med. Soc. N.J. 76(2):127-129.

Harper, A. E. 1987. Diet and behavior: An Assessment of Knowledge. Food Technol. Aust. 39(3, suppl.):ii-v, xi.

Hazleton Laboratories. 1978. Lifetime carcinogenic study in the ICR Swiss mouse: Allura red (FD&C Red No. 40). Submitted to Buffalo Color Corp., West Paterson, NJ.

Hess, S., and Fitzhugh, O. 1955. Absorption and excretion of certain triphenylmethane colors in rats and dogs. J. Pharmacol. Exp. Ther. 114:88.

IARC. 1975. IARC monographs on the evaluation of carcinogenic risk of chemicals to man. Vol 8. Int. Agency for Research on Cancer, Lyon, France.

IFT Expert Panel on Food Safety and Nutrition. 1986. Food colors. Food Technol. (Chicago) 40(7):49-56.

IFT Expert Panel on Food Safety and Nutrition. 1987a. Monosodium glutamate. Food Technol. (Chicago) 41(5):143-145.

IFT Expert Panel on Food Safety and Nutrition. 1987b. Evaluation of caffeine safety. Food Technol. (Chicago) 41(6):105-113.

Ilker, R. 1987. In-vitro pigment production: An alternative to color synthesis. Food

Technol. (Chicago) 41(4):70-73.

International Food Information Council. 1989. A decade of caffeine research produces a reassuring conclusion. Food Insight, Fall, p. 2-3.

Jacobsen, B. K., Bjelke, E., Kvale, G., and Heuch, I. 1986. Coffee drinking, mortality, and cancer incidence: Results from a Norwegian prospective study. J. Natl. Cancer Inst. 76:823.

Joint Expert Committee on Food Additives (JECFA). 1987. Food Colours. Pages 71-120 in: Toxicological Evaluation of Certain Food Additives and Contaminants. WHO Food Addit. Series 21. Cambridge University Press, New York.

Joint Expert Committee on Food Additives (JECFA). 1988. L-Glutamic acid and its ammonium, calcium monosodium and potassium salts. Pages 97-161 in Toxicological Evaluation of Certain Food Additives and Contaminants. WHO Series 22. Cambridge University Press, New York.

Joint Expert Committee on Food Additives (JECFA). 1989. Erythrosine. Pages 39-44 in: Toxicological Evaluation of Certain Food Additives and Contaminants. WHO Food Addit. Series 24. Cambridge University Press, New York.

Joint Expert Committee on Food Additives (JECFA). 1990. Evaluation of Certain Food Additives and Contaminants. WHO Tech. Rep. Series 789. World Health Organization, Geneva.

Joesoef, H. R., Beral, V., Rolfs, R. T., Aral, S. O., and Kramer, D. W. 1990. Are caffeinated beverages risk factors for delayed conception? Lancet 335:136-137.

Jones, R., Ryan, A. J., and Wright, S. E. 1964. The metabolism and excretion of tartrazine. Food Cosmet. Toxicol. 2:447.

Juhlin, L. 1983. Intolerance to food and drug additives. Handb. Exp. Pharmacol. 63:639-655.

Kassner, J. E. 1987. Modern technologies in the manufacture of certified food colors. Food Technol. (Chicago) 41(4):74-77.

Kenney, R. A. 1980. Chinese restaurant syndrome: Fact or fiction. Lancet 1:311-312.

Kenney, R. A. 1986. The Chinese restaurant syndrome: An anecdote revisited. Food Chem. Toxicol. 24:351-354.

Kerr, G. R., Wu-Lee, M., El-Lozy, M., McGandy, R., and Stare, F. J. 1979. Prevalence of the Chinese restaurant syndrome. J. Am. Diet. Assoc. 75:29-33.

Kurtz, T. W., Al-Bander, H. H., and Morris, R. C., Jr. 1987. "Salt-sensitive" essential hypertension in men: Is the sodium ion alone important? N. Engl. J. Med. 317:1043-1048.

Kwok, R. H. M. 1968. Chinese restaurant syndrome. New Engl. J. Med. 278:796.

La Croix, A. A., Mead, L. A., Liang, K. Y., Thomas, C. B., and Pearson, T. A. 1986. Coffee consumption and the incidence of coronary heart disease. New Engl. J. Med. 315:977.

Lecos, C. 1980. Caution light on caffeine. FDA Consum. 14(10):6-9.

Lecos, C. 1988. Caffeine jitters: Some safety questions remain. FDA Consum. 22(1):22-27.

Leonard, T. K., Watson, R. R., and Mohs, M. E. 1987. The effects of caffeine on various body systems: A review. J. Am. Diet. Assoc. 87:1048-1053.

Lethco, E. W. J., and Webb, J. M. 1966. The fate of FD&C Blue No. 2 in rats. J. Pharmacol. Exp. Ther. 154:384.

Lewis, M. H., and Mailman, R. B. 1984. Development disorders and defined diets. Cereal Foods World 29:152-154.

Lin, S., Schoenbaum, S. C., Monson, R. R., Rosner, B., Stubblefield, P. G., and Ryan, K. J. 1982. No association between coffee consumption and adverse outcomes of pregnancy. New Engl. J. Med. 306:141.

Linn, G., and Brusick, D. 1986. Mutagenicity studies on FD&C Red No. 3. Mutagenesis 1:253.

MacMahon, B., Yen, S., Trichopoulos, D., Warren, K., and Nardi, G. 1981. Coffee and cancer of the pancreas. New Engl. J. Med. 304:630-633.

Mailman, R. B., Ferris, R. M., Tang, F. L. M., Vogel, R. A., Kitts, C. D., Lipton, M. A., Smith, D. A., Mueller, R. A., and Beese, G. R. 1980. Erythrosine (Red No. 3) and its nonspecific biochemical actions: What relation to behavioral change? Science 207:535-537.

Malaspina, A. 1987. Regulatory aspects of food additives. Pages 17-58 in: Toxicological Aspects of Food. K. Miller, ed. Elsevier Applied Science, New York.

Marmion, D. M. 1984. Handbook of U.S. Colorants for Foods, Drugs, and Cosmetics. Wiley Interscience, New York.

Martin, T. R., and Bracken, M. B. 1982. The association between low birth weight and caffeine consumption during pregnancy. Am. J. Epidemiol. 126:813-821.

McKone, H. 1990. Copper in the candy. Today's Chem. 3(5):22-25.

McNutt, K. W., Powers, M. E., and Sloan, A. E. 1986. Food colors, flavors, and safety: A consumer viewpoint. Food Technol. (Chicago) 40(1):72-78.

Metcalfe, D. D., and Sampson, H. A. 1990. Workshop on experimental methodology for clinical studies of adverse reactions to foods and food additives. J. Allergy Clin. Immunol. 86(3, pt. 2):421-442.

Millstone, E. 1985. Food additives: The balance of risks and benefits. Chem. Ind. 21(Nov 4):730-733.

Murdoch, R. D., Pollock, I., and Naeem, S. 1987. Tartrazine induced histamine release in vivo in normal subjects. J. R. Coll. Phys. Lond. 21:257-261.

NAS/NRC. 1971. Food Colors. National Academy of Sciences, Washington, DC.

NAS/NRC. 1979. The 1977 Survey of the Industry on the Use of Food Additives: Estimates of Daily Intake, Vol. 3. National Academy Press, Washington, DC.

NAS/NRC. 1989. Pages 465-471 in: Diet and Health. National Academy Press, Washington DC.

Naude, A. 1989. Flavors challenged by convenience trend. Chem. Mark. Rep., June 26, pp. SR 16-18.

Newberne, P. M., and Conner, M. W. 1986. Food additives and contaminants. An update. Cancer 58(8 Suppl.):1851-1862.

Newsome, R. L. 1990. Natural and synthetic coloring agents. Pages 327-345 in: Food Additives. A. L. Branen, P. M. Davidson, and S. Salminen, eds. Marcel Dekker, New York.

NIH. 1982. Defined Diets and Childhood Hyperactivity. National Institutes of Health Consensus Development Conference Summary 4(3):1-8.

Nolen, G. A. 1982. A reproduction/teratology study of brewed and instant decaffeinated coffees. J. Toxicol. Environ. Health 10:769-783.

Noonan, J. E. 1985. An analysis of the factors affecting the future of FD&C colors. Cereal Foods World 30:265.

Nutrition Foundation. 1980. The National Advisory Committee on Hyperkinesis and Food Additives' final report to the Nutrition Foundation. October. The Foundation, New York.

Ockerman, H. W. 1991. Food Science Sourcebook, Part 2, 2nd ed. Van Nostrand Reinhold, New York.

Olney, J. W. 1969. Brain lesions, obesity, and other disturbances in mice treated with monosodium glutamate. Science 164:719-721.

Pangborn, R. M. 1960. Influence of color on the discrimination of sweetness. Am. J. Psychol. 73:229-238.

Parkinson, T. M., and Brown, J. P. 1981. Metabolic fate of food colorants. Annu. Rev. Nutr. 1:175-205.

Phelps, H. M., and Phelps, C. E. 1988. Caffeine ingestion and breast cancer: A negative correlation. Cancer 61:1051-1054.

Pierce, E., Agersberg, H., Borzelleca, J., Burnett, C., Eagle, E., Ebert, A., Kirschman, J., and Scala, R. 1974. Multigeneration reproduction studies with certified colors in rats. Toxicol. Appl. Pharmacol. 29:121.

Pollack, I., and Warner, J. O. 1990. Effect of certified food colors on childhood behavior. Arch. Dis. Child. 65(1):74-77.

Pozniak, P. D. 1985. The carcinogenicity of caffeine and coffee: A review. J. Am. Diet. Assoc. 85:1127-1133.

Radomski, J. L. 1974. Toxicology of food colors. Annu. Rev. Pharmacol. 14:127-137.

Rapoport, J. L., and Kruesi, M. J. P. 1983. Behavior and nutrition: A mini review. Contemp. Nutr. 8(10):1-2.

Reineccius, G. A. 1989a. "Natural" flavors. Cereal Foods World 34:292.

Reineccius, G. A. 1989b. Flavor labeling. Cereal Foods World 34:352.

Reineccius, G. A. 1989c. Flavor safety. Cereal Foods World 34:488.

Ribon, A. 1982. Is there any relationship between food additives and hyperkinesis? Ann. Allergy 48:275-278.

Rowe, K. S, 1988. Synthetic food colorings and "hyperactivity": A double-blind crossover study. Aust. Paediatr. J. 24(2):143-147.

Schaumburg, H. 1968. Chinese restaurant syndrome. New Engl. J. Med. 278:1122.

Schaumburg, H. H., Byck, R., Gerstl, R., and Mashman, J. H. 1969. Monosodium glutamate: Its pharmacology and role in the Chinese restaurant syndrome. Science 163:826-828.

Schmitz, A. 1990. MSG: A cause for alarmists. Hippocrates 4(6):24-25.

Senti, F. R. 1983. Insights on food safety evaluation. Regul. Toxicol. Pharmacol. 3:133-138.

Settipane, G. A. 1986. The restaurant syndromes. Arch. Intern. Med. 146:2129-2130.

Sloan, A. E., Power, M. E., and Hom, B. 1986. Consumer attitudes toward additives. Cereal Foods World 31:523-532.

Sobotka, T. J., Brodie, R. E., and Spaid, S. L. 1977. Tartrazine and the developing nervous system of rats. J. Toxicol. Environ. Health 2:1211.

Stegink, L. D., Baker, G. L., and Filer, L. J. 1983. Modulating effect of sustagen on plasma glutamate concentration in humans ingesting monosodium L-glutamate. Am. J. Clin. Nutr. 37:194-200.

Swanson, J. M., and Kinsbourne, M. 1980. Food dyes impair performance of hyperactive children on a laboratory learning test. Science 207:1487-1488.

Tarka, S. M., and Shively, C. A. 1987. Methylxanthines. Pages 373-425 in: Toxicological Aspects of Food. K. Miller, ed. Elsevier Applied Science, New York.

Taylor, R. J. 1980. Food Additives. John Wiley & Sons, New York.

Thorley, G. 1983. Childhood hyperactivity and food additives. Dev. Med. Child Neurol. 25:531-533.

Tollefson, L. 1989. MSG: FDA agency review of the literature. (Cited in Food Chem. News, Jan. 21, 1991, pp. 37-38)

Tse, T. C. S. 1982. Food products containing tartrazine. New Engl. J. Med. 306:11.

Vettorazzi, G. 1980. Handbook of International Food Regulatory Toxicology. Vol. I: Evaluations. SP Medical and Scientific Books, New York.

Vorhees, C. V., Butcher, R. E., Brunner, R. L., Wooten, V., and Sabotka, T. J. 1983. Developmental toxicity and psychotoxicity of FD&C Red dye No. 40 (Allura Red AC) in rats. Toxicology 28:207-217.

Ward, N. I., Soulsbury, K. A., Zettel, V. H., Colquhoun, I. D., Bunday, S., and Barnes, B. 1990. The influence of the chemical additive tartrazine on the zinc status of hyperactive children—A double blind placebo-controlled study. J. Nutr. Med. 1(1):51-57.

Watts, P. 1984. Tartrazine on trial. Food Chem. Toxicol. 22:1019-1026.

Weatherbee, P. S., and Lodge, J. R. 1979. Alcohol, caffeine, and nicotine as factors in pregnancy. Postgrad. Med. 66:165-171.

Webb, J., Finck, M., and Brouwer, E. 1962. Metabolism and excretion patterns of fluorescein and certain halogenated fluorescein dyes in rats. J. Pharmacol. Exp. Ther. 137:141.

Weiss, B., Williams, J. H., Margen, S., Abrams, B., Citron, L. J., Cox, C., McKibben, J., Ogar, D., and Schultz, S. 1980. Behavioral responses to artificial food colors. Science 207:1487-1489.

Wilkin, J. 1986. Does monosodium glutamate cause flushing (or merely "glutamania")? J. Am. Acad. Dermatol. 15:225-230.

Wilson, N., and Scott, A. 1989. A double-blind assessment of additive intolerance in children using a 12 day challenge period at home. Clin. Exp. Allergy 19(3):267-272.

Yamaguchi, S. 1987. Fundamental properties of umami in human taste sensation. Page 41 in: Umani: A Basic Taste. Y. Kawamura and M. R. Kare, eds. Marcel Dekker, New York.

Zlotlow, M. J., and Settipane, G. A. 1977. Allergic potential of food additives: A report of a case of tartrazine sensitivity without aspirin intolerance. Am. J. Clin. Nutr. 30:1023-1025.

Food Irradiation

CHAPTER 12

For many, food irradiation is a process that creates the same fear as its root word, *radiation*. Images are of devastation, debilitation, and danger to both life and the environment. They are formed from reports of nuclear incidents at Hiroshima (Japan), Chernobyl (USSR), and Three Mile Island (United States) and from discussion of nuclear waste disposal as well as of side effects of radiation (cobalt) treatments for cancer. Fear of food irradiation is built on a truly frightening image base, kindled by a few unknowns and by honest concerns on the part of the public and scientists alike, and it breeds on the resulting hysteria (Giddings, 1986; Pszczola, 1990).

Scientific view. Within the scientific and technical community, there is a vast range of opinion about the future of food irradiation. Some scientists believe it holds modest promise for use 1) as a food preservation method with low energy inputs and hence lower costs than freezing and refrigeration—an important consideration in the face of dwindling energy supplies, 2) as a "cold process," one that at low doses doesn't change the texture of the food as much as canning or freezing may, 3) as a viable alternative to fumigants, such as ethylene dibromide (EDB), which has been banned (Hatfield, 1985, Labuza, 1986), and 4) as a way to control salmonellosis, listeriosis, and other foodborne diseases (Pszczola, 1990; Sigurbjörnssen and Loaharanu, 1989; Urbain, 1989).

Others believe that irradiation holds tremendous promise—promise of reductions in postharvest losses so substantial that hunger will be eliminated in many parts of the world. One estimate suggested that 30% of all food storage losses could be prevented in this way. Food could be transported from regions where it can be produced most efficiently to where it is most critically needed without suffering substantial losses from infestations (Murray, 1990).

Business view. For still others in the scientific community and some in the business world, the promise of food irradiation is difficult to predict. It seems poised to become either a billion dollar industry or just another idea that quietly fizzles out due to technical problems, high start-up costs, and consumer resistance (Hall, 1989; Zurer, 1986).

Consumer activist view. To the consumer activist, irradiation is simply another rallying point (Lochhead, 1989). The Health and Energy Institute, a group based in Washington, DC, and the National Coalition to Stop Food Irradiation, a California-based group, are actively campaigning against food irradiation. An article in a health food magazine stated that many of the U.S. food giants such as General Foods and Campbell Soup—to list just a couple—have accepted irradiation as a method of extending shelf life (Kirkpatrick, 1986). While many companies may have accepted the method as a possibility, none is clamoring to use it because of high start-up costs and monumental consumer resistance, and some prominent ones have pledged never to use it. Currently very little food is irradiated on a commercial scale, and food treated with radiation must be labeled so consumers can't eat it unknowingly (Thomas, 1990; Tilyou, 1990).

WHAT IS FOOD IRRADIATION?

Most nonscientists don't realize that our food has been cooked by a form of radiation since humans began to use fire. Simple cooking involves the absorption of infrared radiation (heat) by the food. Investigations with shortwave radiation, which caused the inadvertent melting of a chocolate bar, led to the discovery that radio waves could be used to cook food. This in turn led to the invention of the microwave oven. In the microwave, radio waves (with shorter wave lengths and higher frequencies than those used for communication purposes) cause water and other polar molecules within the food to vibrate approximately 2.5 billion times per second, thus causing heat through molecular friction. Both infrared and microwave radiations are examples of nonionizing radiations used to heat food. Any biological effects, such as burning your tongue on hot coffee, result from the temperature increase in the food. The food does not and cannot become radioactive as a result of either of these radiation processes.

Food irradiation uses waves that have even shorter wavelengths and higher frequencies (Figure 12.1). Such high frequencies constitute ionizing radiation, which passes through the food without generating intense heat, as infrared and microwave radiation do, while disrupting the cellular processes associated with sprouting, ripening, or growth of microorganisms, parasites, and insects, by inhibiting or stopping them.

At the incredibly high levels of ionizing radiation emitted during a nuclear meltdown such as occurred at Chernobyl or that occur with nuclear weapons release, not only is growth of microorganisms dis-

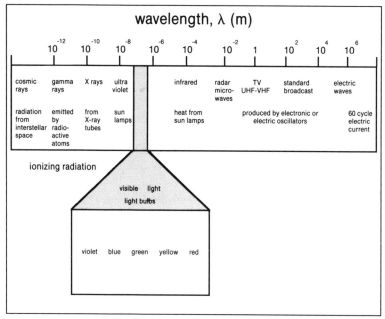

Figure 12.1. Common electromagnetic phenomena and their characteristic frequencies, wavelengths, and energy levels. (Reprinted, with permission, from Rogan and Glaros, 1988; ©The American Dietetic Association)

rupted, but the food itself can become radioactive. Thus, it is easy to see how consumers become fearful of a process that they erroneously associate with this potential. The real problem of radionuclides in food is discussed in Chapter 15.

Regulation

In many cases fear remains because the safeguards used for food irradiation are unknown. On the other hand, knowing some of the safeguards or something about other uses of radiation has not seemed to allay many people's fears. For instance, the Codex Alimentarius Commission of the Food and Agriculture Organization and World Health Organization (FAO/WHO), and also the FDA in its regulations, allows only forms of ionizing energy that are unable to cause the food to become radioactive (Kocol, 1987). These forms, X-rays and cobalt-60, are used in medicine. Even though many people are aware that X-rays and cobalt-60 treatments do not leave the patient emitting radioactivity, they still are fearful about food. However, ingestion of irradiated food does not increase one's cumulative exposure to radiation, as does occur with increasing numbers of X-rays.

Somehow consumers view food as different from medical devices, even though these may be used just as intimately. In fact, the processes proposed for the irradiation of food are the very same as those currently being used to sterilize more than 30% of disposable medical devices in the United States (Falk, 1985). Perhaps the difference has to do not so much with intimacy as with the almost spiritual quality that food has.

It ought to be obvious that officials of neither government nor industry would permit radioactive food to enter the food supply. Selfish concerns for their own survival as well as concerns for employees, consumers, the environment, and survival of the planet would preclude such profligate action.

Fate of ions produced. The real concern about irradiated foods should be directed toward the fact that ions are produced. These ions are highly reactive and can cause chemical changes within the food. Since most foods are high in water, the primary chemical reaction when food is irradiated is the splitting of water into its component ions. These may recombine to form water or hydrogen peroxide, or

---

Box 12.1

What Is Irradiation?

The Latin word *radius* means *ray* and is the root word for *radiant, radiate, radiation*, and *irradiation*. While the former two words have very positive connotations, the latter two are negative. Radiation to most consumers has nothing to do with rays of light, transmission of heat, or radio waves; even though these are all forms of radiation (Figure 12.1). To most it has only one connotation, that is the emission of harmful radioactivity.

Irradiation means exposure to radiation (rays) or radiant energy. Radiant energy or electromagnetic radiation is energy that moves at the speed of light but varies in intensity with both the electric and magnetic fields. It is measured by frequency in Hertz, i.e., the number of times per second the the wave completes its cycle in an electromagnetic field, or by the distance between corresponding points on successive waves in meters, i.e., wavelength. Figure 12.1 shows that, as the frequency increases, the amount of energy also increases. Gamma rays (from cobalt-60 or cesium-137), accelerated electrons (as produced by linear accelerators), and X-rays all have waves of high frequencies and high energies. These waves produce enough energy to strip an electron from an atom, leaving an ion (a particle with an electrical charge). These forms are therefore called ionizing radiation.

---

they may react with molecules in the food. Such reactions could lower nutritional value or produce undesirable flavors. However, improved processing techniques (such as exclusion of air from the radiation milieu, and/or holding foods that can withstand freezing at -20 to $-40°C$ during irradiation) can minimize nutritional losses and off-flavors (CAST, 1989). It has been theorized that ions might generate unique radiolytic products (URPs), but evidence for their existence is weak after 35 years of searching (Grootveld et al, 1990; Merritt, 1984; Nawar, 1986). Each of these issues is discussed later in this chapter.

Historical Perspective

Ionizing radiation was first recognized with the discovery of X-rays in 1895. Within a year of its discovery, ionizing radiation was suggested as a means to kill microorganisms in food (ACSH, 1989). This idea was patented in both the United States and France, one year before the passage of the first food and drug law in the United States. In 1920 a patent was granted for a method of eliminating trichinae from pork using ionizing radiation. Despite these patents, the technique stayed in the laboratory because of the scarcity and high cost of radiation sources before 1946. Research also showed that food taste and texture could be adversely affected by high doses of irradiation.

United States Army research. In the United States, most of the research was done by the Quartermaster Corps of the U.S. Army because of its intense desire to perfect shelf-stable field rations for troops. After many years of testing, the Army concluded that wholesome, economical, shelf-stable field rations could be provided through irradiation. It also noted that this process held promise for reducing both food handling costs and dependence on refrigeration.

Despite the positive conclusions by the Army, problems with the sensory quality of the food still existed. In fact, before 1960, irradiated meats were described in sensory tests as having a "wet dog aroma"! With sensory descriptions such as these, the U.S. food industry did not rush to adopt this process. Nor could it, for the United States legally defined irradiation in 1958 as an additive, not a process. The required FDA approval via the petition process was not immediately forthcoming. Commercial interest in the United States waned pending FDA approval of the irradiation of key food products, but it continued overseas, where irradiation was not subject to the crippling definition as a food additive.

Military research interest in the United States continued, with the building of a food irradiation facility at the U.S. Army's research laboratories in Natick, Massachusetts, in 1962. The Army continued its research on high-dose sterilization of meat products, sponsoring studies on the development of shelf-stable bacon, ham, pork, beef, hamburger, corned beef, pork sausage, codfish cakes, and shrimp (Skala et al, 1987). These studies led to the development of irradiated meats that could be held without refrigeration for several years (Durocher, 1985). In addition, the Army completed a 10-year safety testing program that culminated with the Surgeon General's pronouncement in 1965 that foods treated with irradiation were safe and wholesome.

This pronouncement was the go-ahead signal for the National Aeronautics and Space Administration (NASA), which soon became a user of irradiated foods. Foods developed for space flights combined heat and irradiation into a process known as "radappertization." As a result, the astronauts ate irradiated food as they orbited the earth and traveled to the moon (Josephson, 1983).

Neither the pronouncements regarding its safety nor its use by NASA encouraged any commercial interest in the United States. That was kindled in the 1960s, when studies with lower radiation dose levels showed promise. For instance, potato sprouting could be inhibited and insect infestation of stored grains could be controlled by low doses of irradiation. In the 1960s the FDA approved food irradiation for these purposes. Despite FDA approval for nearly three decades, no North American company chose to use this process commercially until Vindicator of Florida, Inc., did so in 1991. However, spices and condiments were at that time being processed by ionizing radiation for use in cured meats (Sigurbjörnssen and Loaharanu, 1989).

The Food Irradiation Process

There is one overriding requirement for an energy source to be employed in food irradiation. The energy levels must be below those that could possibly induce the food to become radioactive. After that requirement is met, sources are considered on the basis of their practical and economic feasibility. In addition, machine sources must produce radiation with relatively simple technology; isotopes must be sufficiently long lived and emit penetrating radiation.

Only four kinds of energy sources, two machine sources and two isotopes, meet the requirements. The machine sources are 5 million electron volt (MeV) X-ray beams and 10 MeV accelerated electrons from particle accelerators. Only two available isotopes, cobalt-60 (1.3 MeV) and cesium-137 (0.66 MeV), meet the requirements.

Each of the four systems has advantages and disadvantages. Machines offer the advantage of producing radiation only when desired (i.e., when turned on). However, they are more complex technically than isotope sources (Cessna and Rae, 1987; Doar, 1984). Furthermore, electron beams are limited in their applications since they do not penetrate very far into the food. For deeper penetration, the electrons are first directed at a heavy metal target, resulting in conversion to X-rays. For many applications, penetration to the center of a large bulky food item is necessary to ensure that insects and microorganisms are killed throughout the food as well as on the surface. Isotopes create a problem because they continuously emit radiation and cannot be turned off as machines can. Therefore, they must be shielded at all times. Also, their useful lives are limited because of radioactive decay. Regulatory controls of isotopes are necessarily very strict.

A typical gamma ray food irradiation facility. In a typical facility using cobalt-60 as an energy source, the radioisotopes are doubly encapsulated in 18-inch stainless steel tubes called pencils. When not being used, these are submerged in a 25-foot-deep pool of water located in a shielded chamber. The water itself also acts as shielding. Food to be irradiated is transferred on a conveyor into the shielded chamber automatically, so that no workers are anywhere near the radiation source. As the food passes through the chamber, the tubes are raised above the water so that everything within the chamber is exposed to gamma rays. The energy produced by the gamma rays slows or halts cell division so that insects or pathogenic and spoilage microorganisms can no longer proliferate in the food.

Figure 12.2 shows a facility in which the cobalt 60 source is not submerged but is located above the irradiation area. Food, in boxes,

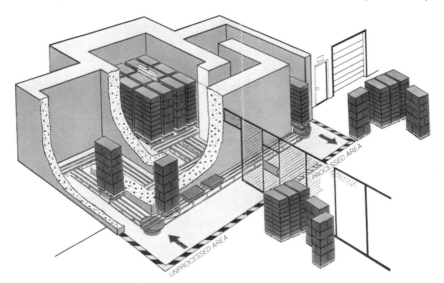

Figure 12.2. Radiation facility, showing shielded chamber, automatic movement of product into the chamber, and separation of irradiated and nonirradiated products. (Courtesy RTI Inc., Rockaway, NJ)

is loaded onto pallets, shrink wrapped to give the load stability, and moved automatically into the shielded chamber. A computer controls the movement of the pallets and the length of time they stay in the irradiation area. Irradiated products are kept physically separate from nonirradiated products.

Uses of Food Irradiation

Food irradiation has many uses in processing, depending on the dose level. There is a thousand-fold difference between the dose needed to inhibit sprouting and that needed to kill all microorganisms—i.e., for sterilization (Table 12.1).

Low-dose applications. These are the ones of interest to the consumer and the processor. A very low dose of irradiation (less than 0.1 kilogray [kGy]) inhibits sprouting not only of the potato, as established in the 1960s, but also of vegetables like onions and garlic (Bender, 1986; Elias, 1987).

Low doses (0.2–1 kGy) can kill insects infesting grain (Murray, 1990), thus appearing to offer an acceptable technical substitute for the use of pesticide fumigants for shipment and storage of grain products. Similar reductions of microbial and insect contamination occur during low-dose irradiation of fruits and vegetables, which are highly susceptible to infestation.

In fresh produce, low-dose irradiation may also be used as a fumigation alternative. Many imported fruits must be fumigated so that insects, such as the Medfly, are not imported along with the fruit. Produce growers are currently experiencing difficulty finding an adequate substitute for banned fumigants such as EDB. Irradiation offers such an alternative (Lochhead, 1989).

Deterioration of fresh fruits and vegetables can be delayed by irradia-

Table 12.1. Dose Levels for Food Irradiation[a]

Level	Function	Dose krad	Dose kGy
Low	Inhibit sprouts in potatoes garlic, and onions	5–15	0.05–0.15
	Delay ripening of produce or eliminate insect infestation	20–100	0.20–1.00
	Prevent trichinosis	30–100	0.30–1.00
Medium	Eliminate spoilage organisms	100–300	1.00–3.00
	Eliminate parasites and pathogens (except viruses and sporeformers)	300–800	3.00–8.00
High	Sterilization (reappertization)	2,500–5,000	25.0–50.0

[a]Adapted from ACSH (1988).

tion. Currently the normal process of decay ruins 25–30% of fresh produce. For instance, mold growth on fruits such as strawberries can be inhibited. Tropical fruits that are sensitive to chill injury, like mangoes and bananas, could be imported less expensively if ripening were controlled by treatment with low-dose irradiation (Kadar, 1986). Longer-lasting, lower-cost fruit for consumers and retailers may be among the benefits of this process.

Not all fresh produce is suitable for irradiation (Akamine and Moy, 1983). Mushrooms, potatoes, tomatoes, onions, mangoes, papayas, bananas, apricots, strawberries, and figs are good candidates for irradiation to extend shelf life. However, some foods actually degrade more easily or suffer losses in quality after irradiation. These include citrus, avocados, pears, cantaloupes, and plums. Cherries are interesting in that some varieties can have their shelf lives extended after irradiation, while other varieties appear to degrade more rapidly. Testing must be done on each product to see if it is a good candidate for irradiation.

Pasteurizing doses. Still higher doses of irradiation can kill or reduce populations of microorganisms that cause foodborne disease and food spoilage. Use of irradiation in this way is analogous to the process of pasteurization. For instance, *Salmonella* and *Campylobacter* can be eliminated from eggs, meat, and poultry and trichinae from pork in much the same way that pasteurization eliminates pathogens from milk (ACSH, 1988; Patterson, 1990; Shaffner et al, 1989). In 1964, Goresline et al recommended that food so treated to reduce spoilage organisms be called *radurized* (after *pasteurized*). This terminology has been used in the scientific literature (Elias, 1987).

Radurization coupled with refrigeration can delay food spoilage of highly perishable meats and fish. For instance, irradiated fish held

Control   2 Kilograys

Figure 12.3. Mold was inhibited on strawberries exposed to 2 kilograys of ionizing energy (right) and stored for 2 weeks at 3°C. (Reprinted, with permission, from CAST 1989).

under refrigeration can be kept fresh for 7–10 days (Diehl, 1981). It can also be coupled with modified atmosphere packaging (Patterson, 1990). Organisms such as parasitic liver flukes in fresh fish and shellfish can be destroyed, making some popular raw fish delicacies safer.

Cured meat products that are radurized require less nitrate to control the growth of botulinum. However, some people fear that radurization may also eliminate customary spoilage organisms that serve to indicate when the food is not fit to eat. Thus, a food may not appear "spoiled" but could still contain pathogens. Table 12.1 shows that lower radiation doses are required for some spoilage microorganisms than for some pathogens, especially sporeformers like botulinum. Hence there is a possibility that some spores will remain viable after other organisms are destroyed. To ensure that food spoils before it can become toxic, the radiation dose approved by FDA (e.g., 3 kGy maximum for fresh poultry) is low enough to leave sufficient surviving spoilage organisms in the food to bring about spoilage one week or more before it can become toxic.

Sterilizing doses. Goresline et al (1964) proposed the terms *radicidation*, for treatments that decrease numbers of nonsporing pathogenic organisms (other than viruses) to below detectable numbers, and *radappertization*, for treatments that sterilize food.

Such sterilization could potentially produce shelf-stable products closely resembling fresh food when served. For instance, food can be packed without liquid surrounding it, as is necessary in canning (Josephson, 1983). However, radappertization will most likely be used first in situations where the logistics of supply are difficult or storage facilities are limited, such as in military operations or camping foods, and in developing countries. Another limited but vital use of irradiation-sterilized food is in diets for immune-deficient patients, for example,

---

Box 12.2

How Is Ionizing Radiation Measured?

The most commonly reported unit in the past was the rad, which came from "radiation absorbed dose." The rad and various other units of measurement are listed in Table 12.2. In 1984, the International Committee on Radiological Protection adopted the Gray as the unit for the International System of Units (SI). The Gray (Gy) is equivalent to 100 rads or to 1 joule (100 ergs) in a kilogram. One kilogray (kGy) equals 100,000 rads or 100 Gy. It is the maximum amount that is allowed by the FDA for treating fruits and vegetables without further testing. One megarad equals 10 kGy or 1 million rads.

Table 12.2. Radioactivity Units and Their Meaning[a]

Unit	Meaning	Use
rad	Amount of radiation that would deposit 100 ergs (units of energy) in a gram.	
Curie (Ci)	Radiation given off by 1 gram of radium or 3.7×10^{10} disintegrations per second.	
Becquerel (Bq)	One disintegration per second (SI measure); 1 Bq $= -2.703 \times 10^{11}$ Ci	
Roentgen (R)	A unit of exposure related to the amount of ionization caused in air by gamma or X-radiation, measured in coulomb per kilogram. (The rad has more application and should be used whenever the absorbed dose is known.)	
rem	Stands for "radiation equivalent man." It expresses the dose equivalent and takes into consideration the absorbed dose as well as other factors such as the distribution of the dose within the target tissue.	
Sievert (Sv)	The SI unit for the rem; 1 Sv = 100 rem or 1 joule per kilogram.	

[a] Data from NCRP (1985).

patients with AIDS, on immunosuppresive drugs, or undergoing some types of cancer treatments. The high doses needed to sterilize the food are not yet approved for general use in the food supply (Pszczola, 1990).

QUALITY OF IRRADIATED FOODS

Sensory Quality

Undesirable changes in the taste and texture of irradiated food, particularly protein products, were common in early trials using the high doses of irradiation necessary to sterilize food (Murray, 1990). Studies indicate that off-tastes occur immediately in irradiated meat products. The following descriptors have been used for irradiated meat samples, primarily those prepared before 1964: burnt, metallic, bitter, cured meaty, cheesy, goaty, and wet-dog-like. Doses as low as 1.5–2.5 kGy can produce off-flavors, with the degree of off-taste increasing with increasing dose. Dairy products still are not candidates for irradiation because of off-odors (Kilcast, 1990; Skovgaard, 1989). In some cases, texture is also affected. Some panelists have described a prickly sensation to the tongue when they ate certain irradiated foods (Risvik, 1986). However, improved processing since the mid-1960s (low temperature during irradiation with exclusion of air and higher, better-controlled dose rates) is said to yield highly acceptable products, particularly meats

(Josephson, 1983).

In studies with fruits and vegetables, varying doses were found to preserve the sensory qualities of a variety of stone fruits, citrus fruits, and papaya. Low-dose treatment of some fruits and vegetables caused softening and other problems. For instance, irradiation of grapefruit and oranges caused perceptible changes in the aroma and texture, as well as an increase in the number of brown blemishes in the skins after 4–6 weeks (Murray, 1990).

Nutrient Quality

Irradiation causes chemical changes in food. Some of the changes involve destruction of nutrients. Concern has surfaced that widespread use of this process may lead to serious nutrient deficiencies, but in fact nutrient losses are in the same range as those caused by cooking, freezing, and even just storage. The U.K. Government Committee on Irradiated and Novel Foods (1986) felt that any nutritional losses were unlikely to have significant adverse effects when considered in the context of the diet (Bender, 1986).

Foods preserved with ionizing radiation may be nutritionally superior to those preserved by other means. Pork, for instance, has more thiamin remaining after radiation than after freezing. Many foods can be processed without the aid of any water, which eliminates obligatory losses of water-soluble nutrients into the cooking or blanching water.

Early studies on the effects of irradiation on nutrient retention showed that this process caused devastating losses. However, these studies were primarily done on model systems. Extrapolating these studies to what happens in an actual foodstuff may yield an incorrect result. For example, amino acids in dilute solution extensively decomposed when irradiated in the kilorad range but appear to be stable to megarad doses in many different protein foods. Even radappertization levels did not adversely affect some of the more process-sensitive amino acids such as methionine and tryptophan (Diehl, 1990).

Effects on macronutrients. Protein chains may split or aggregate when irradiated, but upon digestion these yield the same amino acids. Labile amino acids such as methionine may be affected, but amounts of available lysine in grains and legumes increase (Khattak and Klopfenstein, 1989b).

High-dose irradiation levels did not affect carbohydrate availability in representative fruits, vegetables, and cereals (Skala, 1987). Carbohydrate even protected certain labile amino acids such as cysteine that might be present (Diehl, 1981; Kocol, 1987).

The highly unstable double bonds of unsaturated fat may be attacked by the ions produced by irradiation. However, such loss of unsaturation is not of concern because the resulting off-flavor would cause the food to be rejected anyway (Sadler, 1990). Products formed by irradiation of fat have been shown to be the same as those formed during heating (Nawar, 1986). In actuality, more new compounds were produced by cooking fat than by treating it with irradiation.

Effects on micronutrients. Vitamins have differing sensitivities toward irradiation (Table 12.3). Water-soluble thiamin and ascorbic acid (vitamin C) and fat-soluble vitamins E and K are the most sensitive (Elias, 1987). Most reactions are considered similar to those in heated foods. Losses of thiamin occurred after some irradiation treatments. At radiation levels of 3–10 kGy (levels higher than allowed by the FDA), losses of thiamin measured from 0 to 94% depending on the food, its temperature during irradiation, the degree of exclusion of air, and the irradiation dose and dose rate. Irradiation at levels necessary to disinfest rice caused thiamin losses ranging from 0 to 22%. Niacin decreased only slightly at doses higher than 2.5 kGy in wheat and legumes (Khattak and Klopfenstein, 1989a).

Vitamin B-6 (pyridoxine), another somewhat labile B vitamin, showed losses in some foods. For instance, losses from fish were as high as 25%. Riboflavin, another B vitamin, was also lost under some conditions. Measured losses differ dramatically from trivial to significant according to the source of radiation, the food, irradiation dose, dose rate, and temperature at which the radiation is done (Bender, 1986).

Allegations that antimetabolites to thiamin and pyridoxine were formed during the radiation process have been disproven using enzymatic tests that are specific for the amount of vitamin and antivitamin activity (Skala et al, 1987). In general, losses of B vitamins due to irradiation were less than encountered with heat processing.

Vitamin C is the most easily destroyed of all the vitamins. It is labile to many food processing techniques, and irradiation is no exception. Losses may vary from 1 to 95% depending on the fruit or vegetable, specific cultivar, dose, exclusion of air, and temperature and duration

Table 12.3. Losses of Vitamins in Food During Irradiation[a,b]

Food	Percentage Losses of Vitamin						
	A	B-1	B-2	B-3	B-6	C	E
Milk	60–70	35–85	24–74	33	15–21	ND[c]	40–60
Beef	43–76	42–84	8–17	ND	21–25	ND	ND
Pork	18	96	2	15	10–45	ND	ND
Haddock	ND	70–90	4	ND	26	ND	ND
Potatoes	ND	ND	ND	ND	ND	28–56	ND

[a] Data from Brynjolfsson (1985).
[b] All losses in this table occurred at irradiation levels less than 10 kGy.
[c] No data was reported for this vitamin.

of storage. Low doses required for radurization and sprout inhibition leave losses of vitamin C in the range of 1–20%, but irradiated potatoes exhibited lesser losses upon storage than unirradiated controls (Quan et al, 1988). Where significant losses occur, as in orange juice, Bender (1986) suggested fortification as a possibility. Also care must be used in interpreting analytical results, as the oxidation product of ascorbic acid, dehydroascorbic acid, is also biologically active as vitamin C.

Radiation effects on fat-soluble vitamins have been studied less extensively than effects on water-soluble vitamins. One study found no effect of gamma irradiation on vitamin K in various vegetables. Studies of irradiation effects on vitamins A and E are contradictory. Vitamin A and E contents were unchanged in some studies that looked at the effects of ionizing radiation on nutrient retention. However, vitamin E losses during storage were greater in irradiated than in nonirradiated oats (Murray, 1990). Other studies showed decreases in the vitamins themselves or in the vitamin A precursor, carotene (Webb and Lang, 1987). As was demonstrated with the water-soluble vitamins, results may depend on the product and the conditions of irradiation.

The senior scientist for the Leatherhead Food Research Association in England summed up the information on nutritional losses as follows, "in contrast to opinions expressed by pressure groups and the media, the impact on nutrition will be insignificant. . . . At the low doses permitted, the losses will be small." (Sadler, 1990). David Maclean, U.K. Food Minister, was reported in that article as sharing that view.

SAFETY OF IRRADIATED FOODS

Although the nutrient quality of irradiated food was at one time of concern, nutrient losses have been shown to be no greater (and in some cases less) than from other forms of commercial processing. The safety of irradiated foods greatly overshadows any worries about its nutritional quality. There are three areas of concern with respect to safety. The first is that the ionizing process produces free radicals that could be damaging. Second, URPs might be produced. Third, food in which normal spoilage indicator organisms have been destroyed may become a source of inadvertently ingested pathogens. This last is the area of greatest concern to the FDA and other health regulatory bodies (Brynjolfsson, 1985; Hobbs and McClellan, 1986).

Free Radicals

These are formed when food is irradiated, but they are also formed by exposure to sunlight, frying, baking, grinding, and drying. More free radicals are created during the toasting of bread than through ionizing radiation. In wet foods, free radicals disappear within a fraction of a second; in dry foods, the free radicals are much more stable and don't dissipate as quickly (ACSH, 1988; Elias, 1987).

Multigenerational feeding studies using dry food treated with high doses showed no mutagenic or carcinogenic effects. In fact, these studies showed that irradiation brought about fewer adverse chemical changes than did traditional heat-processing of food.

Unique Radiolytic Products

Irradiated foods could contain either substances not present before irradiation—URPs—or increased concentrations of substances present before irradiation. Sixty-five radiolytic products were identified in a comprehensive search of the literature and of Army data by the FDA Irradiated Foods Committee, but identification of these products as unique is questionable. Twenty-three of the 65 could be found in higher quantities in foods that have been broiled, fried, baked, and microwaved than in foods that have been irradiated. Another 36 of the 65 are present naturally in other foods. Only six could not be verified as a constituent of some foodstuff, although they were chemically similar to natural food constituents (Pauli and Takeguchi, 1986). Since knowledge of the composition of natural foods is incomplete, there is a good chance that even these six products actually occur naturally in some foods (Lochhead, 1989).

Even if it is assumed that these six are unique, as in a worst-case scenario postulated by the FDA (Merritt, 1984), they are present in quantities much less than parts per billion (Kocol, 1987). In other words, 90% of the radiolytic products induced by food irradiation are already present in food in parts per million quantities without any apparent harmful effects (Josephson, 1990; Lecos, 1986). If a true URP could ever be identified, it might constitute a marker to determine whether a food had been irradiated, but an intensive search has not yet yielded such a marker.

To take the most conservative posture possible, the FDA Irradiated Foods Committee made two assumptions—that unique products are formed and that a 1-kGy dose would produce 30 milligrams of radiolytic product per kilogam of food, an amount much higher than actually measured. Thus, the maximum possible concentration of all such URPs combined would be 3 ppm. Further, the Committee noted that the compounds formed were very similar to other food components and that no single unique product could be formed in significant amounts at doses below 1 kGy. (This formed part of the rationale for acceptance of food irradiated with doses up to 1 kGy without animal testing. With these dosage limits, persons consuming irradiated food would consume a maximum of a few micrograms of URPs per day, if any [Pauli and Takeguchi, 1986; WHO/FAO, 1988].)

Most scientists feel that there will not be any surprises or cases of "shrinking zero" such as occurred with pesticides. From a chemical standpoint, the Army's search for unique products in irradiated food was very thorough. As a case in point, during their hunt for URPs they found traces of contaminants ultimately determined to be chemicals used to sterilize the hooks from which carcasses were hung in slaughterhouses (Brynjolfsson, 1985; Kocol, 1987).

Independent toxicologists and biochemists have also analyzed for radiolytic products and concluded that they would not be hazardous to consumers. The Federation of the American Societies for Experimental Biology (FASEB) and the United States Army Natick Laboratories stated that there were no grounds to suspect that radiolytic products would create a hazard (Brynjolfsson, 1985). A similar assessment was made by the British Ministry of Food (Maclean, 1990). However, they also cautioned that chemical studies alone were inadequate to establish safety; feeding studies should be undertaken.

Feeding Studies

Animal studies. Feeding studies to test the safety of irradiated meats were made as early as 1925. Over 1,300 studies on 300 different irradiated foods and feeds have been made. A comprehensive review of all 1,233 studies completed before 1979 was done by Barna (1979). Although some studies raise concerns about different adverse effects, many of these studies could be criticized for a variety of reasons. A common problem in some of the early ones was that such a large amount of an irradiated item was fed (in order to see the effect of the very small quantities of irradiated product) that a nutritional imbalance resulted in both the control and experimental groups. These experiments are often cited as evidence against the use of irradiated foods. However, the FDA in its deliberations does not consider studies in which the diet caused a problem for the control groups, and it ignores such studies when assessing the safety of food irradiation. In fact, one FDA official cited these studies as good examples of how not to do experiments.

Lifetime and generational studies have been done on rats, mice, dogs, and monkeys using irradiated foods as a large percentage of the diet. Some feeding studies have shown adverse effects, including lower birth weights, lower growth rates, and kidney damage. These studies have been counterbalanced by others failing to find an effect when all standard toxicological tests were performed. No metabolic, pathologic, genetic, teratogenic, or life-shortening effects were found in many different investigations. Food irradiated to an absorbed dose of 56 kGy was declared safe and nutritionally adequate after review of these numerous studies (Barna, 1979; CAST, 1989).

Chicken that had been been frozen, canned, and irradiation-sterilized at 47–71 kGy (58 kGy average) with cobalt-60 or with 10 MeV electrons was fed to mice, rats, rabbits, and dogs for eight years in a multigenerational study. No evidence of toxicity, mutagenicity, carcinogenicity, or teratogenicity was observed. Protein quality of the meat was not impaired. Irradiation had no effect on the response of the Ames and other tests for mutagens. The only effect seen was that the hatchability of fruit fly eggs reared on irradiated meat was less and followed a dose-response curve. The researchers did not know whether this had any biological significance for humans (Thayer et al, 1987).

Valid scientific evidence to support the apprehension that irradiated foods are carcinogenic is lacking. Multigenerational feeding studies in mammals showed no causal relationship between eating irradiated foods and cancer. Even studies showing differences in numbers of tumors could indicate only a trend, as the data were not statistically significant (Barna, 1979).

Human studies. One Chinese study actually used human volunteers. They consumed balanced diets in which 60–65% of the food had been treated with ionizing radiation. A broad spectrum of tests revealed no adverse effects after 15 weeks.

Figure 12.4. Exposing onions to 0.04 or 0.08 kilogray of ionizing energy inhibits sprouting during storage. (Reprinted, with permission, from CAST 1989).

Human volunteers for the United States Army ate diets comprised of 32–100% irradiated foods for seven 15-day periods within a year. Physical examinations and clinical tests upon completion of the experiments and 1 year later showed no adverse effects.

A study that is often cited by the critics of food irradiation was done with malnourished Indian children fed irradiated wheat. This study found a 1.8% incidence of polyploidy in malnourished children fed freshly irradiated wheat and 0% in those fed nonirradiated wheat (Bhaskaram and Sadasivan, 1975; Vijayalaxmi, 1978; Vijayalaxmi and Sadasivan, 1975). No polyploidy was observed in subjects fed irradiated wheat that had been stored for 12 weeks. Interestingly, 1.8% polyploidy is in the low normal range (usually the value is around 4% of the general population), and the finding of 0% is not only abnormal but improbable! Because of this, FDA and other bodies looking at the safety of food irradiation (including the Government of India) have discredited these studies (IFT Expert Panel on Food Safety and Nutrition, 1983; Maclean, 1990; WHO/FAO, 1988). However, irradiation critics state that supporters of irradiation such as the FDA discredit as inaccurate any work that raises suspicions about the safety of irradiation.

Box 12.3

How Much Radiation Are We Exposed To?

As a frame of reference, the average person in the United States is exposed to 31 mrem per year from cosmic radiation. Humans are exposed to gamma radiation from rocks and soils in the United States, with doses varying from 15 to 140 mrem, giving an average of 40 mrem per year. The absorbed dose can be easily doubled by changes in location or increased 50-fold through the use of diagnostic X-rays. Radioactive material is also found naturally in air, food, and water, with the dose from these sources averaging 24 mrem per year.

Using human data, the adverse effects of varying dose levels of radiation have been calculated. A dose of 100–200 rads produces nausea, vomiting, and diarrhea. A dose of 1,000–5,000 rads destroys the gastrointestinal tract and other body systems, with death resulting in a very few days. Survival at doses greater than 500 rads all over the body is not likely because bone marrow depression results in very low white counts, enabling many secondary infections (Lushbaugh, 1982). The most probable latent effects of ionizing radiation are cancer induction and gene mutation. There is no doubt that irradiation can be carcinogenic and mutagenic. However, there is still a great deal of controversy about the level of risk following low to very low doses of irradiation.

Regulation and Safety of Food Irradiation

In 1970 the International Food Irradiation Project was set up by 19 countries to sponsor research into the safety of the process. After over 10 years of study, the International Atomic Energy Agency and the Joint Expert Committee on the Wholesomeness of Foods and the Joint Expert Committee on Food Irradiation (JECFI) of FAO/WHO all concluded that food irradiation presented no health hazards or nutritional problems at doses up to 10 kGy. Further, they stated that irradiation is a process with uniform and predictable results.

The U.K. Government Advisory Committee on Irradiated and Novel Foods came to a similar conclusion and asked that the ban on irradiation be lifted. The committee reviewed a number of possible hazards and concluded that no special hazards were associated with food irradiation if it was properly controlled (U.K. Government Advisory Committee, 1986). However, they did not recommend sterilization doses because two large studies were not then complete, one in the United States on chicken and one in the Netherlands on ham.

The 1-kGy ceiling. In that same year an FDA committee in the United States took a much more conservative stance than those in other countries. Foods treated with irradiation doses less than 1 kGy (100 krads)—a dose 10 times less than that sanctioned by the JECFI— would be allowed in the United States without further testing. They based the ruling on comparative chemical analyses of foods that had been irradiated and those that had not. Below 1 kGy, the level of radiolytic products was so small (in parts per billion) that it was at the limit of chemical detection. Furthermore, they stated that after treatment with low doses, there was no identifiable difference between treated and untreated foods, as both contained miniscule amounts of radiolytic products. It was impossible to determine which of these were induced during irradiation and which were normal constituents resulting from metabolism within the food.

The maximum level of 1 kGy was also chosen to protect against the possibility that traditional spoilage organisms might be destroyed, eliminating warning signals that the food was unfit to eat. The FDA wanted to eliminate the risk that a person might consume a product that appeared fresh but actually harbored pathogens. At doses of 1 kGy or less, enough spoilage microorganisms survive to cause the food to spoil before becoming toxic from outgrowth of *Clostridium botulinum*, if present. Thus the consumer would not miss the clues that come from spoilage organisms.

Approval of higher doses. In 1983 the FDA approved the use of irradiation to control microorganisms and insects in spices, allowing doses of up to 30 kGy without animal testing. Note that this is a dose level 30 times greater than FDA's earlier ruling of 1 kGy and three times greater than the 10 kGy allowed by the JECFI. The rationale in this case is that spices are a particularly likely and troublesome source of infestation; they comprise under 0.1% of the diet; and the banning of effective fumigants leaves no other practical way to eliminate bacteria, parasites, and insects. Further, spices would not likely harbor botulism or other worrisome pathogens. A small but growing fraction of spices in the United States is currently being treated with irradiation.

Approval for specific foods. In 1985 FDA approved doses up to 1 kGy on raw pork to destroy trichinae. This enables export of U.S. pork to other countries where the organism is not found.

In 1986 the FDA permitted use of irradiation to inhibit growth and maturation of fresh fruit and to disinfest foods adulterated with insects. Along with this approval, the FDA issued rules regarding labeling of irradiated products (Bender, 1986; Kocol, 1987). All irradiated products must be labeled so that the consumer knows that the product has been irradiated and can chose whether or not to buy it. The label may say "treated with radiation" or "treated by irradiation" and must bear the logo (Figure 12.5). Fruit may be labeled in one of three ways: 1) a sticker on each fruit, 2) a shipping container with the label prominently displayed, or 3) a sign with the information placed in the display area of the treated produce. At the wholesale level, products must be labeled as above but they also must bear the statement "Do not irradiate again." In 1990 the FDA approved a 3-kGy maximum dose for poultry.

Worldwide, over 50 irradiated foods and food products are approved for irradiation in 36 countries. In each case, scientific deliberation by an expert body occurred before the process could be approved. Approval implied that the deliberating body found convincing evidence that irradiation of food posed no dangers to the consumer.

The FDA requires more testing before it will approve doses higher than 1 kGy; in contrast, the World Health Organization states that doses up to 10 kGy are safe for humans and do not necessitate additional

Figure 12.5. Logo required on irradiated foods.

testing. The Institute of Food Technologists, a professional society made up of scientists and technicians from government, academia, and industry, declared this technique to be safe in 1983; the American Medical Association followed in 1984 and the Massachusetts Medical Society in 1990.

Problems. One of the problems facing the regulatory agencies is that there is no commercially available way to determine whether a food has been irradiated and at what dose. Since irradiated products may retain a fresh appearance over an extended period of time, the consumer will be unable to determine which foods are fresh and which have been irradiated. Several groups are trying to develop ways to detect whether food has been irradiated. German scientists have been partially successful in developing a method to identify irradiated spices. Japanese scientists have introduced the use of ultraviolet light to identify irradiated fish. The U.S. National Bureau of Standards, working with the U.S. Food Safety and Inspection Service, has tried to use radio-chemistry to determine whether pork or poultry has been irradiated. Changes in DNA could be a possible method (Anonymous, 1990). The search for these methods is difficult because many of the chemicals induced by irradiation dissipate over time, and there are no URPs to serve as markers. Further, many of the changes are the same as those induced by other treatments (Morrison and Roberts, 1986).

The National Institute for Science and Technology (NIST) has had some success with a technique that can determine whether meats have been irradiated and at what dose by analyzing for a derivative of tyrosine that occurs in meat only if it has been irradiated. However, this derivative is found naturally in some other foods, so it could not be used as an indicator for all food products. To test compliance, FDA also feels that it is critical to be able to assess the dose level used. If a dose is too low, it may fail to produce the desired effect, such as eliminating a pathogen. If it is too high, it may exceed the allowed levels (Anonymous, 1986). The NIST technique, when perfected, might be able to determine the dose that meats received, but the USDA did not regard it as practical and dropped support of NIST's work on it. Other techniques must be developed for other foods in any event.

The ability to detect whether a food has been irradiated more than once—reirradiation—is also needed. Reirradiation has the potential of increasing radiolytic products but is also capable of degrading them. At approximately 10 kGy, any additional irradiation eliminates as many products as it produces (Kocol, 1987).

Food irradiation's effect on packaging material has also been added to the list of concerns (CAST, 1989). If high-dose irradiation is used to sterilize packaging, it may cause some components of plastic to undergo extensive degradation. Care must be exercised so that irradiated food-packaging materials do not create harmful substances that may migrate into the food and contaminate it (McGuinness, 1986).

Food Irradiation Fears Not Related to Food Safety and Quality

Other fears about this process have nothing to do with food safety. First, there is the fear of hazard to workers and the public from ionizing radiation. As explained in an earlier section, the process in the plant is carefully controlled so that the workers are separated from the radiation source. Safeguards actually make it impossible for a worker to enter the chamber when it is emitting radiation (Morrison and Roberts, 1986). A worker in this plant is no more likely to suffer from being irradiated than a person working in any new plant using electricity has a risk of being electrocuted. Nevertheless, the level of concern is probably important in assuring that safety standards are maintained at all times.

Controls for industries shipping and using isotopes are well developed. Plants using this technology in the United States are under the strict jurisdiction of the Department of Transportation, the Occupational Safety and Health Administration, and the Nuclear Regulatory Commission.

A San Francisco-based coalition to stop food irradiation states that radioactive sources used to treat the food will increase the risk of nuclear accidents and risks to workers. However, there are no effluents, noise, or thermal pollution to pose a risk to people living near a plant. Used cobalt-60 pencils with a half life of 5.3 years can be maintained in the facility for many years. Further, there is no possibility of a meltdown, as there is in a nuclear reactor, because the energies are much lower (ACSH, 1985).

Another fear is that this technology will add substantially to already high levels of nuclear waste. The small pencils of cobalt-60 to be disposed of would constitute only a minute portion of the 2.68 million cubic feet of low-level radioactive waste disposed of in 1984. In many cases, the decayed sources would have other industrial uses. Food irradiation would actually help utilize cobalt-60 that has insufficient intensity for medical uses. Thus, this industrial use reduces slightly the amount of waste (Morrison and Roberts, 1986). However, if the process gained wide acceptance and was used extensively, cobalt-60 might then be produced specifically for food uses. This would increase the environmental load of radioactive isotopes. However, if machine sources are used, there is no environmental waste.

Another idea being promulgated is that food irradiation is being pushed as a way to use radioactive wastes that have been produced as by-products of nuclear power and weapons production. This is clearly not the case: cobalt-60, the isotope of choice, is a created radioisotope used in medical centers and is not a by-product of nuclear power plants or of weaponry. While it is true that cesium-137 is a product of uranium fission and could be used for irradiation, it has not thus far been so used. In fact, the use of cesium-137 is fraught with problems, including its safe transport from the power plants to the food irradiation facilities (Radiation Technology, 1985). Furthermore, there is simply not enough of it available to replenish the supply if it were used as a way to process food (Morrison and Roberts, 1986).

Opponents also worry that possible mutated microorganisms will subsequently resist radiation. They point to colonies of mutant microbes that live inside the reactor at Three Mile Island in Pennsylvania. However, it has also been shown (CAST, 1989) that such radiation resistance is accompanied by a reduction in vigor of the organism and a decrease in competitive survivability.

Another fear is about the effect of this process on the environment and groundwater. With cobalt-60, groundwater contamination is not a problem because the isotope is not water-soluble. If cesium-137 were to be used, groundwater contamination could be a problem without special precautions in its use, transportation, and storage—another drawback to the use of cesium-137.

CONSUMER REACTIONS

Acceptance on a worldwide basis varies. Although the process has been studied actively in the United States for more than 40 years and is used in 32 countries, most American consumers are unfamiliar with and fearful of it. This is in contrast to South Africa, where 90% of consumers react positively and actually pay a premium for irradiated strawberries (Morrison, 1986). Russian consumers accept irradiated potatoes—perhaps because they believe the potatoes are wholesome and made more available, because they trust their government to protect them, because they have no choice, or because they has not been adequately informed. Japanese consumer groups oppose the use of irradi-

Table 12.4. Consumer Responses to the Statement: Food Can Safely be Exposed to Radiation to Make It Last Longer[a]

Year	Slightly Agree	Strongly Agree	Don't Know	Slightly Disagree	Strongly Disagree
1986	8	11	45	11	25
1988	8	13	36	10	33

[a] Data from a *Which* survey as reported by Thomas (1990).

ated potatoes because they question their safety and the safety of this process, but potatoes have nevertheless been irradiated on a commercial scale in Japan since 1973.

Box 12.4

Characterizing Consumers' Reactions to Food Irradiation

The Rejectors, 5–10% of consumers

> Beliefs: This group rejects any use of food irradiation because of opposition to the use of any nuclear power, environmental concerns, and the like.

> Educational approach: This group will probably not change, so an educational effort for this group is not likely to be worth the monumental effort required.

The Confused Consumer in the Middle, 55–65% of consumers

> Beliefs: This group is confused. They see food irradiation in terms of competing values. They would like meat and poultry without pathogens and may see irradiation as a means of achieving this end. They would like fruit and grain free of infestation without the use of pesticides. They are afraid of something new, technical, and not well understood.

> Educational approach: This group needs lots of credible information to address each of the concerns and to carefully outline the benefits. Any risks must be dealt with in an honest, forthright manner.

The Acceptors, 25–30% of consumers

> Beliefs: This group currently has a positive attitude but could be strongly influenced by groups stridently voicing concerns. Thus these are acceptors, but their acceptance is fragile.

> Education: This group needs the same types of messages as the group that is unsure—messages that give risks and benefits. Concerns must be listened to honestly, and data must be given that will support or refute the concerns.

(Adapted from Brand Group, 1986, and Marcotte, 1991).

Consumer resistance must be dealt with through an active education program. Accurate information presented in a credible manner must accompany the introduction of a high-quality product (Pauli and Takeguchi, 1986). In educating the consumer, it is critical that the process and its potential not be oversold. Further, the consumer must be convinced that the process has been done under conditions that are adequately controlled and regulated.

Consumer resistance can be diminished if the risks that are being eliminated are explained. For example, introducing the process to Canadian consumers as an alternative to pesticides and additives placed irradiation in a much more positive light than when it was simply introduced as another process. A successful program in the Netherlands took initial consumer perceptions that equated irradiated food with cancer and DNA damage and turned them into a preference for irradiated food over the use of chemicals. Most consumers welcome the knowledge that pesticide use can be decreased and that the irradiation process itself destroys pesticide residues that remain in food, with the degree of destruction dependent both on the dose applied and the food (Bachman and Gieszczyn'ska, 1982; Marcotte, 1991; Young, 1983).

Consumer resistance must also be dealt with by positively explaining the need for irradiation (Tilyou, 1990). Many Canadian consumers asked the question, "Why do we need it? We have safe, nutritious food now." This question points out that, like many consumers in the United States, they are unaware that herb teas and spices are infested with insects, larva, bacteria, and fungi and must be disinfested. Until it was explained in a program on network television, most American consumers didn't know that fish and chicken harbored microorganisms. Many assumed that the government inspection program certified the meat to be disease-free. This creates an interesting catch-22. In teaching about the benefits of irradiation, care must be exercised so that consumers don't become alarmed about the food they are currently eating.

Finally, some consumers will always resist it. Ecologically minded consumers are more wary than conventional consumers and may never accept it (Bruhn et al, 1986). For everyone, the law that products treated with irradiation must be so labeled should be an important point. No one should buy these products unknowingly.

SUMMARY

Nawar (1986) stated that the nutritional, toxicological, microbiological, and chemical data accumulated on the wholesomeness of irradiated food is unmatched by those on any other method of food processing. The following quote (Nawar, 1986) sums up most scientists' reactions to irradiation's safety as a food processing technique:

> The overwhelming weight of scientific data and evidence, gathered worldwide in an unprecedented quantity over decades, plus informed, professional judgement, attest to the safety and effectiveness of this physical process that has been in industrial use worldwide for well over a quarter of a century, including food treatments in various countries for over a decade. This is why knowledgeable and objective scientists and professional organizations including the American Medical Association and the World Health Organization and the International Union of Food Scientists and Technologists continue to endorse the process, recognizing that it is clearly much safer and more wholesome than treating foods with residue-leaving and environmentally-contaminating toxic chemicals like ethylene dibromide (EDB), ethylene oxide, and other chemical fumigants.

In contrast, critics of the process state that the effects of ionizing radiation are unknown. Nothing could be further from the truth. The effects are well documented, linearly related to dosage, and in most cases understood better than those for other toxic agents. So the common complaint that "no one knows what will happen to you if you experience irradiated food" is simply false.

For consumers to use irradiated food, they must believe that it is safe. Unless consumers believe this and are willing to purchase irradiated food, no grocer will be willing to stock it, and food processors will not invest in the technology. Safety is uppermost in the consumer's mind. The consumer must be aware of the benefits of irradiation. In low dose it can yield infestation-free grains, fruits, and spices without the use of pesticides; inhibit sprouting of root vegetables; and extend shelf life of fruits, vegetables, and fresh fish. Medium-dose irradiation may reduce spoilage microorganisms as well as yield poultry and beef that are free of *Salmonella* and *Campylobacter*. Other possible benefits include trichinae-free pork, tape worm-free beef, and liver-fluke-free fish. Foods such as steak tartare, carpaccio, ceviche, raw oysters, and sushi could be made safer because the "cold process" of irradiation kills microorganisms and parasites without necessarily giving a cooked taste and texture to the food. The U.S. Food Safety and Inspection Service believes that this process could sharply reduce occurrences and costs of foodborne illnesses (Roberts, 1985). The World Health Organization also recommends irradiating poultry as a way to deal directly with *Salmonella* and other pathogens. The director of its Food Safety Unit stated on this issue that "safety should come before prejudice" (Campbellplatt, 1990). Antibiotic-resistant organisms transmitted to humans through meats could be eliminated if killed by irradiation. Chemical additive use in food may also be eliminated or reduced in some cases.

Some additional advantages accrue. Irradiated wheat gives bread with larger volumes and irradiated beef is more tender. The gas-causing sugars in dried beans are reduced with irradiation.

Acceptance by food processors has been slow because of the negative initial experience with irradiated food and because of fears of consumer resistance. The processor must also consider whether irradiation is cost effective compared with alternative processes (Morrison, 1986; Roberts, 1986). The longer shelf life must offset the high cost of production. The Health Research Group, a group based in Washington, DC, stated that producers were not reacting enthusiastically because industry prefers cheaper techniques. The industry has a great deal of fear that a potentially major investment will reap no returns because of consumer resistance (Thomas 1990; Van Kooij, 1986).

The benefits of irradiated products must be considered against the risks that they pose. Spoilage mechanisms could be different in irradiated foods from those recognized in fresh or traditionally processed foods. Grains that have been irradiated to disinfest them of aflatoxin must be stored under conditions that won't allow reinfestation (Priyadarshini and Tupule, 1979). Consumer protection must be adequate so that scandals such as occurred in Europe (where a company irradiated prawns to conceal bacterial contamination and then sold them) are not repeated. Safeguards will be needed so that processors cannot use this process to pass off food that has been contaminated. Existing food law in the United States is strong enough so that this type of food would be seized as adulterated. Codes for Good Manufacturing Practices would apply to irradiated food facilities, equipment, raw materials, and the process.

An irradiation dose should be only that needed to achieve the desired effect and never more than the maximum allowed for the intended use (Pauli and Takeguchi, 1986). A low dose in combination with other preservation techniques may be one answer (Patterson, 1990). A successful way to monitor compliance must be in place (Kilcast, 1990).

The URP safety question must be laid to rest. Studies to date appear to indicate that URPs are no different from products produced by cooking or found in foods (Elias, 1989). Moy (1985) points out that we have always had unknown processing products: "If fire had just been discovered as a cooking method there would have to be FDA approval of unique cooking products (UCPs)" produced by this new processing method. Further study to answer any questions raised, even those from poorly controlled studies, will increase consumer confidence and decrease consumer fear.

REFERENCES

ACSH. 1985. Irradiated Foods. American Council on Science and Health, New York.

ACSH. 1988. Irradiated Foods. American Council on Science and Health, New York.

Akamine, E. K., and Moy, J. H. 1983. Delay in postharvest ripening and senescence of fruits. Pages 129-159 in: Preservation of Food by Ionizing Radiation, Vol. 3. E. S. Josephson and M. S. Peterson, eds. CRC Press, Boca Raton, FL.

Anonymous. 1986. Food irradiation detector. Technol. Forecasts Technol. Surveys 18(9):12-14.

Anonymous. 1990. Food irradiation—Challenges for the chemist. Nutr. Food Sci. 125:11.

Bachman, S., and Gieszczyn'ska, J. 1982. Effect of gamma irradiation on pesticide residues in food products. Pages 313-315 in: Agrochemicals: Fate in the Environment. Int. Atomic Energy Assoc., Vienna.

Barna, J. 1979. Compilation of bioassay data on the wholesomeness of irradiated food. Acta Aliment. 8:205-315.

Bender, A. E. 1986. Food irradiation. J. R. Soc. Health 3:80-81.

Bhaskaram, C., and Sadasivan, G. 1975. Effects of feeding irradiated wheat to malnourished children. Am. J. Clin. Nutr. 28:130-135.

Brand Group, Inc. 1986. Irradiated Seafood Products—Report of the National Marine Fisheries Service. No. NA84 AA H SK099. Natl. Marine Fisheries Serv., Washington, DC.

Bruhn, C. M., Schultz, H. G., and Sommer, R. 1986. Attitude change toward food irradiation and conventional and alternative consumers. Food Technol. 40(1):86-91.

Brynjolfsson, A. 1985. Wholesomeness of irradiated foods: A review. J. Food Saf. 7:107-126.

Campbellplatt, G. 1990. Food irradiation processes. Food Policy 15:447-448.

CAST. 1989. Ionizing energy in food processing and pest control: II. Applications Task Force Report 116. Council for Agricultural Science and Technology, Ames, IA.

Cessna, M. A., and Rae, C. R. 1987. Food irradiation technology: How safe is it? Home Econ. Forum 1(2):18-21.

Diehl, J. F. 1981. Irradiated foods—Are they safe? Pages 286-294 in: Impact of Toxicology on Food Processing. J. C. Ayres and J. Kirshman, eds. AVI, Westport, CT.

Diehl, J. F. 1990. Safety of Irradiated Foods. Marcel Dekker, New York.

Doar, L. H. 1984. Irradiation—Where are we? Food Eng. 56:56-57.

Durocher, J. F. 1985. The food irradiation controversy. Restaurant Bus. 84(10):158, 162.

Elias, P. S. 1987. Food irradiation. Pages 295-346 in: Toxicological Aspects of Food. K. Miller, ed. Elsevier Applied Science Publishers, New York.

Elias, P. S. 1989. New concepts for assessing the wholesomeness of irradiated foods. Food Technol. 43(7):81-84.

Falk, V. S. 1985. Irradiated foods. Wis. Med. J. 84(10):6-7.

Giddings, G. C. 1986. Food irradiation misconceptions. Chem. Eng. News 64(Oct 20):3.

Goresline, H. E., Ingram, M., Macuch, P., Mocquot, G., Mossell, D. A. A., Niven, C. F., and Thatcher, F. S. 1964. Tentative classification of food irradiation processes with microbiological objectives. Nature 204:237-238.

Grootveld, M., Jain, R., Claxson, A. W. D., Naughton, D., and Blake, D. R. 1990. The detection of irradiated foodstuffs. Trends Food Sci. Technol. 1:7-14.

Hall, R. L. 1989. Commercialization of the food irradiation process. Food Technol. 43(7):90-92.

Hatfield, D. 1985. Irradiation emerges as processing alternative. Environ. Nutr. Newsl. 8(12):1-2.

Hobbs, C. H., and McClellan, R. O. 1986. Toxic effects of radiation and radioactive materials. Pages 669-705 in: Casarett and Doull's Toxicology, 3rd ed. C. D. Klaassen, M. O. Amdur, and J. Doull, eds. Macmillan, New York.

IFT Expert Panel on Food Safety and Nutrition. 1983. Radiation preservation of foods. Food Technol. 37(2):55-60.

Josephson, E. S. 1983. Radappertization of meat, poultry, finfish, shellfish, and special diets. Pages 231-252 in: Preservation of Foods by Ionizing Radiation, Vol. 3. E. S. Josephson and M. S. Peterson, eds. CRC Press, Boca Raton, FL.

Josephson, E. S. 1990. Food preservation. Pages 148-150 in: McGraw-Hill Yearbook of Science and Technology. McGraw-Hill, New York.

Kadar, A. A. 1986. Potential applications of ionizing radiation in post-harvest handling of fresh fruits and vegetables. Food Technol. 40(6):117-121.

Khattak, A. B., and Klopfenstein, C. F. 1989a. Effects of gamma irradiation on the nutritional quality of grain and legume: 1. Stability of niacin, thiamin, and riboflavin. Cereal Chem. 66:170-171.

Khattak, A. B., and Klopfenstein, C. F. 1989b. Effects of gamma irradiation on the nutritional quality of grain and legume: 2. Changes in amino acid profiles and available lysine. Cereal Chem. 66:171-172.

Kilcast, D. 1990. Prospects for food irradiation. Chem. Ind. 5:128-130.

Kocol, H. 1987. Food irradiation. Assoc. Food Drug Officials Q. Bull. 51(1):13-17.

Labuza, T. P. 1986. Future food products: What will technology be serving up to consumers in the years ahead? Assoc. Food Drug Officials Q. Bull. 50(2):67-84.

Lecos, C. W. 1986. The growing use of irradiation to preserve food. FDA Consumer 20(6):12-15.

Lochhead, C. 1989. The high-tech food process foes find hard to swallow. Food Technol. 43(8):56-59.

Lushbaugh, C. C. 1982. The impact of estimates of human radiation tolerance upon radiation emergency management. In: The Control of Exposure of the Public to Ionizing Radiation in the Event of Accident or Attack. Natl. Council on Radiation Protection and Measurements, Bethesda, MD.

Maclean, D. 1990. Rays fiction. Nutr. Food Sci. 125:9-10.

Marcotte, M. 1991. Consumer Acceptance of Food Irradiation. Norden International, Inc., Kanata, Ontario.

McGuinness, J. D. 1986. Migration from packaging materials into foodstuffs. Food Addit. Contam. 3(2):95-102.

Merritt, C., Jr. 1984. Radiolysis compounds in bacon and chicken. Eastern Regional Res. Center, Agric. Res. Serv. Document 83. Available from Natl. Tech. Info. Serv., Springfield, VA, PB 84-187905.

Morrison, R. M. 1986. Irradiation's potential for preserving food. Natl. Food Rev. 33:1-6.

Morrison, R. M., and Roberts, T. 1986. Food irradiation policy issues. Natl. Food Rev. 33:11-13.

Moy, J. H., ed. 1985. Radiation Disinfestation of Food and Agricultural Products. University of Hawaii Press, Honolulu.

Murray, D. R. 1990. Biology of Food Irradiation. Research Studies Press, Somerset, U.K.

Nawar, W. W. 1986. Volatiles from food irradiation. Food Rev. Int. 2(1):45-78.

NCRP. 1985. SI Units in Radiation Protection and Management. Rep. 82. Natl. Council on Radiation Protection and Measurement, Washington, DC.

Patterson, M. F. 1990. The potential for food irradiation. Lett. Appl. Microbiol. 11:55-61.

Pauli, G. H., and Takeguchi, C. A. 1986. Irradiation of foods: An FDA perspective. Food Rev. Int. 2(1):79-107.

Priyadarshini, E., and Tulpule, P. G. 1979. Effect of graded doses of γ-irradiation on aflatoxin production by *Aspergillus parasiticus* in wheat. Food Cosmet. Toxicol. 17:505-507.

Pszczola, D. E. 1990. Food irradiation: Countering the tactics and claims of opponents.

Food Technol. 44(6):92-94.

Quan, V. H., Oularbi, S., Langerak, D. I., Wolters, T. C., and Tayeb, Y. 1988. Effect of wound healing period and temperature, irradiation, and post-irradiation storage. Rep. 68. Int. Facility for Food Irradiation Technology, Wageningen, Netherlands.

Radiation Technology. 1985. Food irradiation: Facts and fantasy. Radiation Technology, Inc., Rockaway, NJ.

Risvik, E. 1986. Sensory evaluation of irradiated beef and bacon. J. Sensory Stud. 1(2):109-122.

Roberts, T. 1985. Microbial pathogens in raw pork, chicken and beef: Benefit estimates for control using irradiation. Am. J. Agric. Econ. 67:957-965.

Roberts, T. 1986. Irradiation's promise: Fewer foodborne illnesses? Natl. Food Rev. 33:7-10.

Rogan, A., and Glaros, G. 1988. Food irradiation: The process and implications for dietitians. J. Am. Diet. Assoc. 88:833-838.

Sadler, M. 1990. The greening of consumer demand. Food Policy 15:448-450.

Schaffner, D. F., Hamdy, M. K., Toledo, R. T., and Tift, M. L. 1989. *Salmonella* inactivation in liquid whole egg by thermoradiation. J. Food Sci. 54:902-905.

Sigurbjörnssen, B., and Loaharanu, P. 1989. Irradiation and food processing. Pages 13-30 in: Nutritional Effects of Food Processing. J. C. Somogyi and P. Loaharanu, eds. Karger, New York.

Skala, J. H., McGown, E. L., and Waring, P. P. 1987. Wholesomeness of irradiated foods. J. Food Protect. 50(2):150-160.

Skovgaard, N. 1989. Irradiation of food. Maelkeritidende 102:380-384. (Cited by CAB Abstracts D 365828).

Thayer, D. W., Christopher, J. P., Campbell, L. A., Ronning, D. C., Dahlgren, R. R., Thomson, G. M., and Wierbicki, E. 1987. Toxicology studies of irradiation-sterilized chicken. J. Food Protect. 50:278-288.

Thomas, P. A. 1990. Food irradiation and the consumer. Radiat. Phys. Chem. (Part C) 35:342-344.

Tilyou, S. M. 1990. Food irradiation buoyed by regulatory and scientific acceptance but held back by perception problems, J. Nucl. Med. 31(9):A11.

U.K. Government Advisory Committee on Irradiated and Novel Foods. 1986. Letter to the editor. Int. J. Radiat. Biol. 49:1039-1040.

Urbain, W. M. 1989. Food irradiation: The past fifty years as a prologue to tomorrow. Food Technol. 43(7):76.

Van Kooij, J. G. 1986. International trends in and uses of food irradiation. Food Rev. Int. 2(1):1-18.

Vijayalaxmi, C. 1978. Cytogenetic studies in monkeys fed irradiated wheat. Toxicology 9:181-184.

Vijayalaxmi, C., and Sadasivan, G. 1975. Chromosomal aberrations in rats fed irradiated wheat. Int. J. Radiat. Biol. 27(2):135-142.

Webb, T., and Lang, T. 1987. Food Irradiation—The Facts. Thorsons Publishing Group, Rochester, VT.

WHO/FAO. 1988. Food irradiation: A technique for preserving and improving the safety of foods. World Health Org., Geneva.

Young, M. 1983. Consumer acceptance of irradiated food. Tech. Paper GPS 334. Consumers' Assoc. of Canada, Ottawa.

Zurer, P. S. 1986. Food irradiation: A technology at a turning point. Chem. Eng. News 64(May 5):46-56.

Pesticides

CHAPTER 13

PESTICIDE, HERBICIDE, FUNGICIDE, RODENTICIDE: WHAT KIND OF -CIDE IS COMMITTED WHEN WE EAT?

Pesticide. The very word sends chills down the spine, comprised as it is of two root words, *pest* and *cide*, both of which resurrect images that we would rather leave buried. The root word *pest* conjures up images of pestilence and famine dating back to biblical times. The biblical account of the plague of locusts is one of several examples (Exodus 10:14,15) in which the available "pesticides" against insects were flails or walls of burning brush. As one can well envision, and the various accounts bear out, these methods were far from successful, and famine was the frequent result. Mosaic law permitted the eating of locusts,[1] presumably because they were the only thing left after the plagues.

Pests. To the U.S. farmer, the word *pest* evokes a vague recollection of the over 10,000 species of insects that at one time or another cause significant agricultural damage and creates real fears about the 200 or so species that are serious pests annually (Mandava, 1985). Worldwide, farmers must cope with over 100,000 species of pests. Their destructive power generates big news annually. For instance, recent locust swarms in Ethiopia caused the destruction of 167,000 tons of grain in only one month—enough grain to feed a million people for a year!

To the American consumer, visions of pests still create a ghastly specter, although pests are not associated with real fears of food shortages and famine. Visions include seemingly invisible worms that lurk on the broccoli, waiting to float to the top of the cooking water when the boiling begins; rodent hairs and other unmentionable rodent

---

[1]Another interpretation is that "locust" referred to the locust bean.

traces in the carton of Freakies; and flour that has so many moving parts that it is not only self-rising but self-mixing! Despite the fact that everyone consumes roughly 1 pound of insects annually in a variety of common products such as jams, peanut butter, and tomato paste, most consumers are revolted by the mere mention of insects in food.

To various government bodies, pests are something to regulate. The number of rodent hairs and insect parts that are allowed on certain foods are part of government regulations, as specified by defect action levels. If the specified levels of contamination are exceeded, the product is deemed adulterated and is seized. Most consumers do not think allowing even a small number of rodent hairs is protecting the food supply. They don't want any filth in their product and are irate that the government allows any at all!

To the citizens of the world, the news accounts of famine caused by drought and pests create haunting visions of dying animals, malnourished children, and human suffering.

Cides. Visions created by the root word *cide* are more similar for the various constituencies and are directed by other words with the root *cide*, such as *genocide, suicide, homocide,* or *matricide.* All these words, of course, mean some form of killing. Pesticides occupy the unusual position that they are deliberately added for the purpose of injuring some life form. In addition, over 100 deaths in the United States and 20,000 deaths worldwide occur annually due to careless use, misuse, or mishandling of pesticides (Pimentel et al, 1990). In some developing countries, pesticide chemicals are chosen as the agents of suicide. It is hardly surprising that people are scared when they hear that -cide X (no matter how small the quantity) is part of their entree!

The consumer desires the ideal—food that contains neither pests nor pesticides. In fact, however, both the quality and quantity of food on the dinner table are determined by the balance struck between pests and -cides. The total absence of one of these may well mean the presence of the other. Humans and pests vie for the same food supply: pests must therefore be controlled to assure an adequate human food supply. A senior chemist for the Environmental Protection Agency's (EPA's) Office of Pesticides and Toxic Substances, N. B. Mandava (1985), assessed the situation in the United States in the following words, "The U.S. has been blessed with high quality dependable supplies of low cost food and fiber, but few people are aware of the never-ending battle that makes this possible."

In addition to thoughts about pests and -cides in food, other issues about pesticides are highly charged because they affect many facets of daily life, including contemporary agriculture and the environment, causing concern for birds, wildlife, air, waterways, and soil. Knowledge that some pesticides are persistent, not only in the environment but also in the fat pads of the human body, creates nagging questions. Further disquiet is generated by research showing that pesticides may be carcinogenic, neurotoxic, and/or genotoxic in experimental animals. Charges by migrant workers that the pesticide sprays are responsible for a high incidence of cancer and other diseases are also alarming.

Four different visions. Thus, four very different visions are created by pests and pesticides, and these very different visions each suggest a different course of action. The consumer wants either to be oblivious of these things in food or wishes their complete elimination. Consumer activists are aware of the consumer position and have used this issue to garner support for a variety of issues concerning the food supply and the environment. Governments see the need to regulate these materials, as their elimination would create as many problems as it would solve. The world agricultural community, concerned about feeding the planet, sees the need for judicious use of pesticides.

This chapter explores the pesticide controversy from its beginning to the present. It looks at which issues fuel the debate and attempts to find some common ground among alternatives of being forced to consume locusts, learning to appreciate the joys of mealy bugs and rodent parts, and involuntarily ingesting foods sprayed with substances that seem akin to hemlock. Pesticides are discussed with a focus on food production and food safety, but the discussion touches on issues of environmental safety and disease control.

HISTORY OF PESTS AND PESTICIDES

Neither pests, pesticides, nor pestilence are new. All are described in the Bible, by Greek and Roman philosophers, and in ancient Chinese writings. With regard to pesticides, Homer wrote of the fumigant value

---

Box 13.1

What Are Pesticides?

Pesticides are chemicals or mixtures of chemicals used for the prevention, elimination, or control of unwanted insects, plants, and animals. Also included are plant growth regulators, defoliants, and desiccants.

Pesticides are usually organic chemicals but may also be inorganic compounds. They can be produced in the laboratory, and some are produced naturally by the plant itself.

of burning sulfur and Pliny wrote about the insecticidal use of arsenic. Historical Chinese writings showed that they also used arsenic as well as tobacco extracts as pesticides.

With regard to pests, there are accounts of pestilence throughout history. The ancient Egyptians suffered plagues of locusts, as did the Mormons during the settling of the American West. The Irish potato famine in the latter half of the 19th century had a major impact on the settlement of North America. The blight on the potato necessitated the emigration of one third of the population of Ireland—approximately 6 million people. Starvation claimed another million (Merliss, 1975).

About 1850 German farmers began using sodium chloride and lime to combat weeds. Near the turn of the century, a few pesticide chemicals were emerging in the United States and Western Europe. French grape growers began using an arsenic compound called Paris green, a somewhat more romantic name than is used for pesticides today. Also iron sulfate, sulfuric acid, and copper nitrate were used to control broadleaf weeds in cereal crops. In the United States, copper arsenate was used against the potato beetle and sodium arsenate was used to control weeds. Seed stock in Germany began to be treated with mercury to prevent sprouting before planting. By 1933 growers in the northern United States were controlling broadleaf weeds in cereal crops with dinitrophenol compounds and other weed killers. All weed killers available up to the discovery of 2,4-dichlorophenoxyacetic acid (2,4-D) had the disadvantage of being nonselective, expensive, and frequently damaging to the crop itself as well as to the weeds.

Pesticide Euphoria—The 1940s

Pesticide euphoria began in 1939 with Swiss chemist Paul Meuller's realization that dichlorodiphenyltrichloroethane (DDT) functioned as a pesticide (Concon, 1988). It continued with the commercial production of pesticides and herbicides in the United States and Britain. The worldwide reduction of cases of malaria and typhus following use of DDT and the fungicide carbamate made these scientific discoveries among the most humanitarian of acts. Because of the unheard of degree of pest control on farms, these new substances were viewed as miracle chemicals.

The first U.S. applications of DDT caused stunning increases in yields. Potatoes grown in Wisconsin increased yields by 68%; the potato crop in the United States rose by an average of 100 bushels per acre. Tomato yields rose from 5 to 10 tons per acre between 1945 and 1950. Apple damage in Illinois orchards declined from 69 to 2.2% between 1956 and 1968. In some cases, crop yields increased more than fourfold. Farmers who did not adopt these revolutionary new chemicals were at risk for their livelihood (Hassall, 1982). Meuller was awarded the Nobel prize for his discovery.

Pesticide Disillusionment—The 1950s

The euphoria was soon replaced by the first round of postpesticide blues when the Senate Select Committee on Chemicals in Food and Cosmetics stated that there was insufficient evidence that pesticides were safe and that more regulation was needed. These proceedings resulted in the passage of the Pesticide Chemical Act of 1954.

Information about the adverse effects of these chemicals in the soil and water generated feelings of alarm, especially since they were found to be persistent in the environment. Further disillusionment about pesticides was generated by reports that new crop infestations were being caused by pests that in the past were not considered problematic— so-called secondary infestations. These resulted from the killing of the natural insect predators that customarily controlled populations of insects capable of causing infestation. Beneficial insects such as honeybees were also unintentional victims of nonselective pesticides.

Total disillusionment in the United States came during the Thanksgiving season of 1957, with the aminotriazole scare in cranberries. Rat studies indicated that aminotriazole at 100 ppm induced thyroid cancer in four out of 26 rats. At that time we didn't know that the rutabagas that might also be served at Thanksgiving dinner naturally contained 100 times more thyroid cancer-inducing activity than badly contaminated cranberries. In any event, cranberries were reluctantly banished from this traditional feast, and with this banishment came the loss of kindly thoughts about pesticides.

Pesticide Damnation—The 1960s and 1970s

The specter of a birdless environment painted by biologist Rachel Carson's 1962 book *Silent Spring* created dire concern bordering on damnation. This book shocked the public, generated legitimate fears for the environment, and caused a flurry of government and scientific activity that resulted in environmental protection laws. Legislation was easily passed amidst reports that 5 million fish were poisoned in the lower Mississippi River due to the pesticide dieldrin.

Environmental concerns about pesticides were not limited to their effects on birds. Other concerns had to do with overspraying and drift. Although drift was not initially a cause for concern, it has been subsequently learned that once pesticides are airborne either from drift after spraying or from wind erosion of sandy soils, they can be distributed worldwide. Drift has carried pesticide residues to pristine areas such as glacial lakes and polar ice caps (Kearney and Helling, 1982;

Woodwell, 1981).

Reports about drift and other adverse effects of pesticides made them an emotionally charged issue. In several instances, people reported adverse effects of pesticide spraying even when no spraying had occurred. In Germany the forest service was deluged by calls about dead fish after one group learned that spraying was scheduled for their area. The calls came before the application of the pesticide, as spraying had been postponed because of wind conditions. A similar incident was reported in Jamaica, where a large number of gardeners claimed they had garden damage due to aerial spraying. In fact, the claims were made before any pesticide spraying—the first flight was made using water just to calibrate the spray equipment (Barrons, 1983).

The uncontrolled overuse of pesticides caused other problems to surface. Insects began to become resistant or immune to the effects of pesticides. Many operators responded with more and heavier pesticide applications in an attempt to keep the insect populations down. Cross resistance, in which insects become resistant not only to one pesticide but also to many chemically similar and some not-so-similar ones, began to occur and to create even greater problems (CAST, 1983).

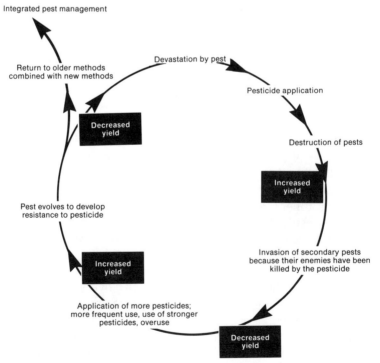

Figure 13.1. The pesticide-yield cycle, leading to Integrated Pest Management.

In addition to pesticide resistance, secondary pests were also a problem. The story of pest control in cotton dramatically illustrates this. Initially, the boll weevil was cotton's major pest. The boll weevil was nearly eliminated by early use of pesticides, and cotton yields increased dramatically. Subsequently, a secondary pest, the bollworm, began to flourish since its natural predator, the boll weevil, was no longer able to keep the worm population in check. The bollworm then evolved to be pesticide-resistant and wreaked havoc on the crop despite 15–18 pesticide applications.

Fears about human and animal health also rose with the knowledge that DDT could degrade into dichlorodiphenyldichloroethane (DDD) and dichlorodiphenyldichloroethylene (DDE). The fact that these metabolites were stored in the fat pads in perpetuity was certainly not comforting. Raging controversies about the safety and use of the herbicide Agent Orange in the war in Vietnam escalated the uncertainties in the 1970s. Finally, damage to the eggshells of eagles, grebes, and other fish-eating birds led to the U.S. banning of DDT just 30 years after its introduction.

Pesticide Balance Sought—The 1980s and 1990s

Attempts to maximize benefits and minimize risks to public health, agriculture, and the environment sparked a current ecologically sound system for pest control. It is called Integrated Pest Management (IPM). IPM reactivates many of the techniques that farmers and manufacturers used before the advent of so-called miracle chemicals, but these are used together with judiciously applied pesticides (Gianessi, 1991). In some cases, pesticide levels are reduced by up to 50% without decreasing crop yields or increasing storage and other postharvest losses. Many new and alternative pest control techniques are also being sought

---

Box 13.2

Methods Used to Control Pests
Before Chemical Pesticides Were Developed

Crop rotations
Intensive cultivation
Destruction of crop refuse
Trap crops
Pruning and defoliation
Varying planting dates
Hand removal of weeds and insects

(El-Zik and Frisbie, 1985).

IPM is just one of many tools being used to develop sustainable agriculture. The goal of sustainable agriculture is to produce adequate amounts of high quality food while protecting resources. Thus, its techniques must be both environmentally safe and profitable. In addition to IPM, sustainable agriculture seeks to use pest-resistant varieties of plants and other practices that maintain yields while reducing pesticide use (IFT's Office of Scientific Public Affairs, 1990; Reganold et al, 1990).

The pesticide scares of the late 1980s have left the consumer afraid. A cover of *Time* magazine (March 28, 1989) that asked rhetorically whether anything was safe accurately transcribed what was on consumers' minds. For the consumer in the 1990s, several things are needed. Scars from pesticide abuses of the past need to heal. Scaremongers need to be disarmed by clear and understandable translations of the risks and benefits of pesticide use. Further research is needed to answer unresolved safety questions.

EFFECTS ON YIELD, FOOD COST, QUALITY, AND AVAILABILITY

Even with the current annual use of nearly 1 billion pounds of pesticides, crop losses occur. The latest available figures show the annual loss of crops in the United States to be 37%, with 13% loss due to insects, 12% due to pathogens, and 12% due to weeds. The loss due to birds and mammals averages less than 1%, but it can be as high as 50% depending on the crop and conditions (Pimentel, 1990; Pimentel et al, 1990).

Losses in some countries can be significantly greater than those reported for the United States. Annual worldwide losses of cereal grains alone are calculated at a million metric tons. Pimentel and Levitan (1986) figured that 45% of the total food production was lost to pests, with preharvest losses at 30% and postharvest loss at 10–15%. Percent losses due to various causes are given in Table 13.1.

Pesticides, along with many other modern agricultural techniques, increased yields 230% in the period between 1947 and 1986 (Szmedra, 1990), thus enabling a single farmer in North America or Europe to feed 60 people. This is in contrast to what was the norm at the turn of the century or what currently occurs in developing countries, where a single farmer feeds only four to 10 people (Food Watch Facts, 1991; Hayes and Laws, 1991). These figures have dramatic implications for the amount of income that consumers must devote to food purchases. In the Third World, 50–70% of income goes toward food, whereas in

North America and Europe the amount is 15-30% (Food Watch Facts 1991; Hassall, 1982). Without pesticides, the amount spent on food would double in North America and Europe. Thus, the consumer in Western countries has directly benefited from modern agricultural methods, including pesticides, which provide a low-cost food supply of dependable quality (Bruhn, 1991).

The Risk-Benefit Equation

According to 1990 figures, the economic benefits of pesticide use are estimated to be about $10,900 million per year. Balanced against this are the direct costs of $2,800 million, which include the indirect costs of resistance, loss of natural enemies, and the like (Pimentel et al, 1990). This is a threefold return to the producers. Other calculations indicate $3-$5 in direct return to the farmer for every dollar spent (Pimentel et al, 1990). Use of nonchemical controls and IPM may make the returns for producers even higher (Pimentel, 1987a). However, the actual effect of pesticides on profitability and yield depends on the crop in question. In a study of yields for 10 different crops, raised with and without pesticides, crop losses were reduced with chemical treatment in all cases. Reductions ranged from 17% for tomatoes to 78% for apples and pears. The increase in yield for strawberries was 1.3-12 times greater than for apples (Melnikov, 1971). Thus, the degree of benefit derived from pesticide application appears to be crop dependent.

The risk benefit equation changes slightly if reviewed in the light of data gathered by using the techniques of sustainable agriculture. With these methods, wheat yields were 8% lower than those seen on some conventional farms, but these yields were not lower than the average yield for the region (Reganold et al, 1990). In this case, profits were about the same as those of a conventional farm, as the input costs were less. Thus, the combination of techniques appears to effect the actual benefit.

Table 13.1. Losses from Pests in the World's Major Crops[a]

Crop	Percent Losses			
	Insects	Diseases	Weeds	Total
Rice	27.6	8.9	10.8	47.3
Wheat	5.0	9.1	9.8	23.9
Corn	12.4	9.4	13.0	34.8
Sorghum/millet	9.6	10.6	17.8	38.0
Cassava	7.7	16.6	9.2	33.5
Sweet potatoes	8.9	5.0	11.7	25.6
Potatoes	6.5	21.8	4.0	32.3
Soybeans	4.5	11.1	13.5	29.1

[a] Adapted from Bifani (1987).

The risk-benefit equation is also affected by climate and region. Differences in yield in two areas with different climate types are shown in Table 13.2. Pesticide effects on decreases in crop losses and increases in crop yields were more dramatic in some developing countries than in the United States. In India, fourfold increases in wheat yields were reported by Murkerjee (1982). Lever (1982) calculated the benefit of pesticide use in monetary and yield terms for several countries. Without pesticides, Tanzania would lose 30% of the stored grain, with a product value of £100,000 sterling (almost $200,000). Treatment of the stored crop would cost only £4,500. In Pakistan, a 30% increase in sugar cane yield resulted from a $77,000 expenditure on insecticides. Thus a $77,000 expenditure produced a $7.2 million additional sugar yield. In Nigeria, 50% of stored cowpeas were attacked by a species of beetle, and 50% of the dried fish catch from Lake Chad was destroyed; reduction of losses of these kinds would have a tremendous impact both on availability of food and on profitability. Murkerjee calculated that avoidable losses in Indian agriculture due to pests equaled an annual loss of 60 billion rupees. The rupee value of this loss is staggering, but even more staggering is how far the destroyed agricultural product could have gone toward feeding the people of India.

The Yield Issue

Table 13.3 compares pesticide use and yields in Thailand and the Philippines. It is clear from the table that pesticide use correlates well with increased yield. Other countries show the same pattern with respect to yields (Hayes and Laws, 1991).

Efficiency of land use. Another way to look at the yield benefit ascribed to pesticides is to calculate how much additional land would need to be cultivated to compensate for expected crop losses. The financial paper *Barron's* calculated the amount of land required to make up the yield deficit for 16 common U.S. crops raised without chemical pest control. It would be a land mass equivalent to the combined areas of Minnesota, Iowa, Missouri, Arkansas, South Dakota, Nebraska,

Table 13.2. Average Yields (kg/hectare) in Temperate and Tropical Regions[a]

		Tropical	
Crop	Temperate	Amount	Percent of Temperate
Rice	4,109	1,958	48
Wheat	2,984	1,363	46
Corn	3,993	1,351	34
Sorghum	2,270	1,249	55
Soybean	1,620	1,249	55

[a] Adapted from Bifani (1987).

Oklahoma, and Louisiana (Klingman and Nalewaja, 1984). This calculation obviously does not include any additional costs or environmental effects, such as the increased fossil fuel needed to plant and harvest these additional acres. These examples of increased yield and profitability are impressive and rarely disputed.

Effect of curtailing pesticide use. Disputes do exist about the actual amount of food that would be available if pesticide use were curtailed. Nobel prize winner Norman Borlaug (1972) calculated that crop losses would increase by 50% and food prices would increase four- to fivefold. In contrast, Cornell University's Pimentel calculated that crop losses would be on average only 7% greater. He did point out that losses might be much greater for some crops than for others (Metcalf, 1980).

The lower values for Pimentel's predictions than for Borlaug's reflect three factors. First, pesticides are not currently used on all crops. According to 1985 figures, of the 890 billion acres in cultivation in the United States, only 24% was treated with herbicide, 9% with insecticide, and 1.5% with fungicide. Second, increased insect resistance has diminished the predicted effectiveness of pesticides. Third, in Borlaug's calculation, losses due to insect infestation were calculated first and then added to those attributed to weeds. The same plant might be calculated as suffering loss due to both weeds and insects, making the loss greater than actually would occur (Pimentel, 1985, 1990).

Crop losses. Predicted crop losses can also vary widely depending on the basic assumptions chosen. Different crops can show radically different percentages of loss. For example, the Food and Agriculture Organization calculated worldwide average losses of wheat at 35%, potatoes at 40%, sugar beets at 24%, strawberries at 51%, and apples at 30% (Consumers Association, 1980). These average percentages of loss also vary markedly depending on the soil and climatic conditions of the region, the specific variety chosen, and the cultivation techniques

Table 13.3. Effect of Pesticide Use on Yield in the Philippines and Thailand[a]

		Yield per Hectare			
		Philippines		Thailand	
Fertilizer Level	Weed Control Level	Tons	U.S. Dollars[b]	Tons	U.S. Dollars[b]
Low	Low	3.0	128	2.4	78
Low	High	2.9	121	2.5	76
High	Low	3.4	137	3.0	87
High	High	4.2	171	3.9	127

[a] Adapted from Lever (1982).
[b] Return above variable costs.

used. For instance, stressed crops such as those grown in a drought are more susceptible to infestation. In a normal year, only 1% of the soybean crop in the midwestern United States is treated with insecticides, but during the drought of 1988, up to 40% was treated (Pike 1989). Certain pests thrive during droughts. Also, small grain crops such as wheat, oats, and barley receive substantially less pesticide per acre in the United States than in Europe.

New higher-yielding varieties may be more susceptible to disease. In fact, early plant-breeding programs chose plants for maximum responsiveness to fertilizer and pesticides. Thus, loss calculations would be much greater if they were based on the planting of all high-yielding varieties. Breeding programs now seek resistance to insects and other pests as well as response to fertilizer.

Crop losses are also dependent on cultivation techniques. The trend today toward monocropping (i.e., planting large fields of a single crop such as corn) has increased the need for pesticides (Pimentel, 1985). The reason is obvious. Any self-respecting pest will choose a field with an immediate, seemingly everlasting food supply, rather than a field where there are only a few rows of a favorite entree and where the pest is forced to use all the energy just obtained to move on to the next field.

Decreased crop rotation may increase plant loss unless pesticides are used to kill the organisms that grow in the material left in the field from the past year. Similar problems may be seen when farmers use the environmentally sound practice of conservation tillage. This laudable practice, which reduces fossil fuel consumption and decreases soil erosion, is being used on 31% of the cropland. This tillage method has the potential for decreasing yields unless pesticide levels are increased because the leftover crop material has great potential to breed infestations and weeds (Klingman and Nalewaja, 1984).

Cultivation of certain crops (such as potatoes) in warm climates necessitates a substantially greater use of pesticides than when the same crop is grown in cooler regions. Predicted yield losses for crops grown without pesticide should be determined by which region was being considered and what the cultivation techniques were.

Herbicides

The pesticides used to control competing plants and plant pathogens are called herbicides. Of known crop pests, an estimated 160 species are bacteria, 250 are viruses, 8,000 are fungi, and 30,000 are weeds! Every major United States field crop may be infested with 10–50 species of weeds. These are destructive because they compete with crops for minerals, water, light, space to grow, and essential gases from the air. Weeds interfere with harvest because they clog equipment and are difficult to separate from the grain. Thus, seeds and stems from weeds find their way into the grain bin. Their high moisture content interferes with grain drying, which increases the potential for grain spoilage and may result in off-flavors, lower crop values, or worse still, toxic factors such as aflatoxin.

Losses due to weeds. The USDA estimates that 12% of the U.S. crop, amounting to $12 billion a year, is lost because of weeds (Pimentel, 1990; Sanders, 1981). An additional $6.2 billion is spent to control weeds, which brings the annual cost of weeds to $18 billion (Sanders, 1981). Without herbicides, other costs of production increase dramatically because more weeding, fertilizer, and irrigation are necessary. A single herbicide, 2,4-D, has been estimated to increase the production of bread grains by at least a billion bushels per year (Sanders, 1981). This greater output of grain is enough to to make a dozen 1-pound loaves of bread for each person on our planet.

Yields of corn and soybeans are greatly affected by the use of herbicide. Corn and soybeans account for half of the herbicide use in the United States (Gianessi, 1991). Over 90% of the acreages of corn, soybeans, peanuts, and rice are treated with herbicides (Szmedra, 1990). However, the actual poundage of chemicals is increasing (EPA, 1988).

In the tropics, crop yields might be reduced to zero without herbicides (Lever, 1982). As is the case for other pesticides, crop losses due to weeds vary by crop and by region. Estimated crop losses due to weeds are 8-24% in the United States, 6-50% in the United Kingdom, 20-40%

Box 13.3

Reasons for Crop Losses Despite Intensified Insecticide Use

Planting of crop varieties that are more susceptible to insect pests
Increase in pests resistant to pesticides
Reduction in crop rotations
Increase in monoculture and decrease in crop diversity
Lowering of defect action levels for insects and insect parts in food
More stringent cosmetic standards by fruit and vegetable processors
    and retailers
Reduction in tillage, which leaves more crop residue
Growing of crops in climates where they are subject to insect attack
Use of herbicides that increase vulnerability to pests

(Adapted from Pimentel, 1991)

in the USSR, and 30–100% in India (Lever, 1982).

Herbicides vs. other weed control. Cost comparisons of plots that have been hand weeded with those treated with herbicide also show striking results for the economic benefit of herbicides both to consumers and to producers. Weed control costs for 1 acre of cotton were figured to be $4.00 per acre with herbicides. To be price competitive, the hand weeder with a hoe would need to be paid $0.40 per day! Similar calculations for strawberries showed that herbicides decrease costs of production by $100 per acre. Some crops may need as much as 100 hours of hoeing per acre and add costs of over $300 per acre for a season (Klingman and Nalewaja, 1984).

Since herbicides reduce the amount of mechanical cultivation that is required, they also reduce fuel use. This obviously reduces costs and reduces the burning of fossil fuel—another important environmental concern.

Risks of Pesticide Use on Yields and Crop Quality

Although increased yields from the use of pesticides are often glowingly reported, yield increases do not always occur. Data from Finland collected from 1973 to 1981 showed that while yields of chemically treated barley increased, carrot yields decreased 20–60% in chemically treated plots from the level in hand-weeded plots (Heinonen-Tanski, 1986, 1985). Pimentel (1987a) found similar data with a few varieties of corn. Herbicide-treated corn produced more protein, which actually made it more attractive to insects.

The increased yield seen with pesticides has diminished somewhat in recent years due to the development of pesticide resistance. Because of resistance, many pesticides are impotent. Recent estimates showed that 447 species have become resistant to insecticides, 55 to herbicides, and five to rodenticides since their use was begun just 40 years ago. Major pests such as the Colorado potato beetle and the diamondback moth are proving impervious to every chemical pesticide (Anonymous, 1986a). Malaria-carrying mosquitoes are resistant to DDT. In some cases, the problem is compounded because pesticides have wiped out the natural predators of the current crop of insect pests. As a result, the National Research Council of the National Academy of Sciences (1986) recommends that farmers limit their pesticide use to every other year and that they use old-fashioned alternatives such as importing natural predators into the fields. One of the members of the Council, Dr. Bruce Levin from Amherst, Massachusetts, was quoted as saying that we are reaching an end to the era of chemical control of pests (Anonymous, 1986a). Obviously, further insect resistance and cross-resistance will have a detrimental effect on yields and on protection against vectorborne disease.

Benefits for Preventing Disease

Early successes in disease prevention were just as dramatic as those in increased yields. For example, in Sri Lanka (Ceylon) in the mid-1950s, 2 million cases of malaria occurred; in 1963 after a DDT spraying program, 17 cases were reported. Then spraying was banned because of environmental concerns. In 1967, there were 3,000 cases, and by 1969 the number returned to 2 million (Merliss, 1975). Other diseases caused by insect vectors—including malaria, louse-borne typhus, bubonic plague, sleeping sickness, and yellow fever—have been significantly reduced through the use of pesticides. In areas where these and other vectorborne diseases are a problem, the very real risk of contracting them may far outweigh any toxicity problems that pesticide chemicals might cause. However it must be pointed out that pesticide use for any purpose does add to the pesticide load present in the groundwater and the environment.

Marini-Bettòlo (1987) stated it this way:

Pesticide use is indispensible in the struggle against hunger and disease. Emphasis should therefore be given to judicious and safe use rather than on banning or restriction of pesticides. Nevertheless, in view of the adverse effects resulting from pollution of the environment by these chemicals . . . there is greater need to monitor pesticide usage.

Consumer Demands for Pesticide Use

While some consumers feel that the only beneficiary of pesticide use is the agricultural producer, many enjoy low-cost, readily available food as a benefit of pesticide use. However, for some consumers, even lower cost is not an acceptable reason for pesticide use.

In some cases, consumer choices in the marketplace necessitate pesticide use. Pesticides are often applied to fruits and vegetables because consumers desire an unblemished product. Nearly 75% of fruits and vegetables are pesticide treated to meet stringent defect action levels that are set for cosmetic appeal of the food and have no role in increasing available food (Eilrich, 1991; GAO, 1986; Thonney and Bisogni, 1989). These high pesticide levels are mandated because many consumers reject any produce that may have some truly organic, yet unwanted, components. For instance, broccoli contains 100 times more pesticide than is required for production of the product (Jennings, 1991). Orange trees are sprayed to prevent a mite that causes brown spots on the skin of the fruit but absolutely no change in the yield or quality on the inside of the orange. Such pesticides are used because often the same person who argues against pesticides selects perfect-looking oranges.

Consumers want a perfect, unwormy apple without pesticides, yet the apple varieties that are the most popular—Red Delicious, Golden Delicious, and McIntosh—are extremely susceptible to apple scab and require seven to ten sprayings per season. Simply by choosing another kind of apple, such as the Baldwin, which is not susceptible to scab, consumers could reduce levels of pesticide in apples (and in the environment). As yet they neither select nor demand Baldwin apples. Furthermore, a perfect-looking peach and a badly blemished one may have had equivalent levels of pesticide treatment if the blemished fruit is damaged because it was poorly handled between the orchard and the supermarket. Consumer judgments about pesticide level cannot be made simply by looking at the produce (Jennings, 1991).

Thus, we may have a somewhat schizophrenic idea about what we want in our food. On the one hand, we want food without pesticides and, on the other hand, food without pests. Worse still, it looks as if we don't make consistent choices at the supermarket even when they are offered.

PESTICIDE RESIDUES IN FOOD

Monitoring Residues

In 1954, the Miller Amendment (PL518) to the Food, Drug, and Cosmetic Act was passed, stemming from concerns that these chemicals in the food supply need to be regulated and controlled. This bill gave the FDA power to set allowable levels of residues.

At the time the law was passed, no residues of organic pesticides were allowed in foodstuffs, although tolerances were in place for residues of inorganic pesticides such as lead arsenate used on apples. However, analytical capability for detection of organic residues advanced in quantum leaps. What was formerly defined as zero residue became measurable by sophisticated methods in parts per billion or parts per trillion.

The FDA had initially allowed no residues of organic pesticides in food but then had to reverse itself. This caused many consumers to lose faith in the government's ability to protect the consumer and to be a credible source of information. Subsequently, this was called the "case of the shrinking zero."

Setting tolerance levels. New tolerances were set for organic pesticides. Levels that could not even be detected by earlier methods became the maximum residue levels allowed when measured by the the higher-resolution methods. These maximum residue levels (MRLs) are far below the acceptable daily intake (ADI). The Code of Federal Regulations contains over 7,000 tolerances. Such a large number is needed because tolerance levels are set for each pesticide and agricultural commodity. Just because one tolerance is set for a grain does not mean that the same level, or for that matter, the same pesticide is allowed on fruit (Carnevale, 1987).

Levels are kept as low as possible (Turnball, 1984). If the pesticide is used in more than one commodity, the worst case scenario is considered in in setting the MRL: it is calculated as if an individual ingested a daily maximum of all treated food items. To make certain that the utmost conservatism is used in setting the MRL, the calculation is made assuming 1) that the residue at the time of consumption is maximum, 2) that all of the crop has been treated with the product resulting in the residue, and 3) that the person consumes daily the maximum per capita amount.

Estimating maximum intake. Such underlying assumptions grossly and purposely overestimate the likely maximum intake (Benbrook, 1991). Overestimates are the rule because usually only 10–30% of a crop is treated. Levels at the time of consumption may be only 20–40% of what they are at the time of harvest, as processing, cooking, frying, and baking reduce residue levels. Values allowed refer to the whole commodity, but the pesticide might be in a portion not usually eaten, such as the peel. An underlying principle in all these deliberations is that since food is the universal route of exposure, tolerances must err on the side of being too strict (FAO, 1984). Foods with residues exceeding the MRL would not necessarily be toxic, any more than driving over the speed limit means that there is certainty of having an accident. With the margin of safety employed in setting these allowed levels, pesticides at the tolerance levels in crops, milk, and meat do not constitute hazards (Turnball, 1984).

It is unfortunate that these "worst case" risk scenarios designed to maximize public safety are often interpreted literally, leading to the perception that the government expects some individuals to be adversely affected (Petersen and Chaisson, 1988).

Data from Italy, the United Kingdom, Canada, and the FDA's own Market Basket Surveys of purchased foods from 20 cities in the United States (FDA, 1990) verified that pesticide levels are far below a level that should give any concern (Campanini et al, 1980; Consumers Association, 1980; Gartrell et al, 1985b).

The actual analysis of the Market Basket Surveys (based on intakes of 19-year-old males, the theoretical maximum food consumers) showed that the fats and oils and the meat and poultry groups were most likely to contain pesticide residues. None of the intakes approached the ADI; the highest measured intake was for dieldrin at 16% of the ADI (Gartrell et al, 1985b). This value decreased with each passing year, and by 1988 was about one twentieth of what it had been 20 years earlier. For all the organophosphates, the mean and median intakes were less than about 1% of the ADI. Even consumers who were in the 90th percentile for consumption consumed only 2.16% of

---

Box 13.4

Residue Reductions During Marketing and Processing

The following examples are approximate reductions. The actual values may vary dramatically, depending on the crop and also on the amount of residue per se.

	Percent Reduction
Shipping to Supermarket	
Peppers	14
Celery	86
Cabbage	86
Lettuce	87
Cucumbers	100
Tomatoes	100
Washing	
Apples	14
Grapes	36
Peaches	73
Tomatoes	83
Trimming	
Lettuce	89
Cabbage	93
Heating	
Potatoes	
Baked	61
Boiled	68
Chipped	86
Beans	
Canned	72
Frozen	92

(Data from Eilrich, 1991)

the ADI (Gorchev and Jelinek, 1985). In summary, daily pesticide intakes averaged less than 1% of the ADI (Pennington and Gunderson, 1987).

In a similar survey of foods that an infant would conceivably ingest, no pesticide levels approached the ADI. Dieldrin was highest for this population group also. It was 48% of the ADI for infants and 36% for toddlers. For these young consumers, the foods most likely to contribute to residues would be fats, oils, and whole milk (Gartrell et al, 1985a). In infants and invalid diets monitored in a more recent study (Georgii, 1989), average intakes were no more than 0.33% ADI and intakes of persistent pesticides continued to decrease. In addition, these data confirm that the levels of residues in food have not changed or have decreased since the previous data were gathered (Black, 1988; Gartrell et al, 1985a, 1985b; Georgii et al, 1989; McLoed et al, 1980). All data show that levels have not increased, as some writers imply. Similar data for private laboratories and from the Grocery Manufacturers of America show declines in pesticide levels (Brown, 1986).

The good news is that a variety of commonly used cooking and processing procedures reduce pesticide levels. Washing, peeling, cooking, canning, baking, drying, freezing, deodorizing, juicing, and irradiating reduce or eliminate residues (Chin, 1991). Loss of residues in meat drippings also significantly reduces the residue value below that found in the raw commodity. Reported values are, therefore, very likely higher than consumed values (Tomerlin and Engler, 1991).

Areas of special concern. One area that causes concern is imported foods. Many countries allow pesticides that have been banned in the United States. Although the FDA samples nearly 50,000 shipments from foreign countries, this is only 1% of the arriving shipments (Anonymous, 1987a). According to a General Accounting Office (GAO, 1986) report, as much as a quarter of the incoming fruit tested had higher residue levels than the allowed tolerance levels.

Because of the need for increased monitoring, the FDA has increased the total number of imported and domestic foods it monitors. In 1989, over 18,000 samples of food were analyzed; 57% of these were imported foods. No detectable residues were found in 61% of the samples. Less than 4% of the samples were in violation. Over 3% of the violations were caused not because the residues were present above allowed levels but because specific pesticide residues were found in foods for which the pesticide was not registered (Shank et al, 1989). Even though the total number of foods analyzed increased by 25% and a larger portion of the foods was imported, the violation rate was similar to that found in earlier years.

Another area of extraordinary concern is pesticide residues in human milk. Unfortunately, human milk contains from five to ten times more organochlorines than does cow's milk. The reason for this is as follows: humans have only short lactation periods to excrete many years of pesticide accumulation, while cows undergo intense lactation for a lifetime (Blanc, 1981). The long-term effects of these pesticide levels in breast milk need to be established. The many benefits of breast-feeding should not be discarded because of concerns about residues in human milk, but it is crucial that more be known of the effects of these pesticides on human infants (Jensen, 1983) and that their concentrations in breast milk be monitored (Black, 1988; Gallenberg and Vodicnik, 1989).

Government Bodies That Regulate and Monitor Pesticides

In addition to the FDA's role in pesticide monitoring, all pesticides must be registered with the USDA under the Federal Insecticide, Fungicide, and Rodenticide Act. Registration requires that the manufacturer give the accurate chemical composition of the pesticide, its proposed use and documented effectiveness, and acute and chronic toxicity data. An environmental impact statement giving evidence that it does not create an undue hazard to fish or wildlife is also needed. Analytical procedures for measuring residues in crops and commodities are also required (Hassall, 1982; Upholt, 1985).

The public also has a role in determining which pesticides are allowed. The Rebuttable Presumption Against Registration is a procedure that allows the public to enter into risk-benefit discussions before a pesticide is registered or cancelled (Upholt, 1985).

The National Academy of Sciences generates data on the safe use of the pesticides; the Occupational Safety and Health Administration regulates the safety of workers applying or manufacturing pesticides; and the Federal Aviation Administration regulates pesticide spraying. The EPA regulates environmental hazards posed by pesticide use, and ensures that pesticide residues in foods do not pose an unreasonable risk to human health (Tomerlin and Engler, 1991). The Food Safety and Inspection Service of the USDA has, as part of its National Residue Program, a program for pesticide monitoring in meat and poultry products.

The General Accounting Office has suggested that Congress revise the pesticide law. Congress, in turn, has mandated more data on many diverse aspects, including inherent toxicity, residue levels, and ecological and environmental effects of the 600 active ingredients and 900 inert ingredients in currently registered pesticides. With current funding levels, the Congressional data mandate will not be completed until sometime in the 21st century (GAO, 1986). Since recent evaluations of federal pesticide monitoring programs have highlighted the gap between the number of pesticides that might be present in food and the number that can be routinely measured, the Office of Technology Assessment (OTA) was asked to look at currently existing technologies and to make suggestions for improving analytical capabilities of the FDA and other agencies. The OTA summarized these suggestions in a report to Congress in 1988 entitled *Pesticide Residues in Food: Technologies for Detection* (Parham et al, 1988).

Monitoring systems at the state and federal level for raw agricultural commodities show that a wide variety of commodities do not exceed tolerances (Clower, 1991). CONTAM, a multistate monitoring system, looked at many foods in all categories (13,041 samples in all) and found over 10,147 samples with no residues detectable by current methods. Of the 2,894 samples with detectable residues, only 2.1% had violative levels. The largest single numbers of foods in violation were in the "other" category, which included spices, alcoholic and nonalcoholic beverages, and bottled water (Minyard and Roberts, 1991). California state laboratories perform more analyses than any other state (Moye, 1991; Wells and Fong, 1991) and in 1989 found illegal residues in only 0.71% of the products tested (Carpenter, 1991).

Monitoring also occurs on an international level. In the early 1960s the WHO Expert Committee on Pesticides and the FAO Panel of Experts on the Use of Pesticide Residues in Agriculture recommended that studies be conducted to evaluate hazards to humans arising from the occurrence of pesticide residues in foods (Vettorazzi, 1980). Since that time, conferences have been held at 2-year intervals to summarize current data available on pesticides and establish or reset ADIs. Proceedings of the conferences are published by the FAO. The FDA participates in these conferences and relies on their findings, together with other data, to establish or modify regulations in the United States.

MEET THE PESTICIDES

More than 600 pesticide chemical active ingredients are registered for use on food crops in the United States; about 350 of these account for over 98% of the total poundage of pesticides applied (Benbrook, 1991). Space simply doesn't allow the introduction here of all the pesticide players in the United States, much less in the world. Anyone interested in a specific pesticide would find one of several dictionary-like volumes very helpful—*The Pesticide Manual: a World Compendium* (Worthing and Hance, 1991; Worthing and Walker, 1983), *Crop Protection Chemicals Reference* (Anonymous, 1985), and *Handbook of Pesticide Toxicology* (Hayes and Laws, 1991). All give common and trade names, companies, chemical names, structures, properties, uses, toxicological data, and methods for analysis.

The Organochlorines

Several major chemical groupings need to be introduced. The first and perhaps most notorious is the organochlorines. Within this group, there are several famous players including DDT and its metabolite DDE, dieldrin, chlordane, benzene hexachloride (also known as HCH), aldrin, endrin, heptachlor, and lindane. Of these, the most well-known is certainly DDT. Few compounds have been the subject of so much emotional criticism or learned controversy.

Dichlorodiphenyltrichloroethane (DDT). At its birth, DDT was praised as the perfect pesticide because it was strikingly effective against numerous pests, was initially thought to be safe to humans and warm blooded animals, and controlled vectorborne disease. No chemical made by humans—not even penicillin, streptomycin, or the sulphonamides—was credited with saving as many lives as DDT. Billions of people have been exposed to DDT, yet it is necessary to search long and hard to find any serious human side effects except for accidental poisonings and suicides (Hassall, 1982).

While mouse and rat studies indicate that DDT is a liver carcinogen (Williams, 1981), data on dogs and monkeys are inconclusive, and in humans and hamsters no cancers have been related to food sources of DDT (Anonymous, 1990a; Austin et al, 1989; Concon, 1988). The fact that DDT is metabolized in rodents by a pathway different from that in other mammals and that DDT's cancer-promoting ability varies with species (Mehendale, 1989) may be important in understanding the seemingly different effects. Despite this discrepancy, the EPA has tried to estimate human cancer risk from rodent data. Toxicological rather than epidemiological data are used by regulatory agencies like the EPA because human data are often insufficient on chemicals for which predictions are needed (Doull, 1989). The EPA's extrapolation attributed 153,000 cases of human liver cancer from exposure to three liver carcinogens—DDT, dieldrin, and aflatoxin. The estimation clearly overpredicts, as there are fewer than 8,000 cases of liver cancer per year in the United States from all sources of exposure.

Persistence. Another problem with organochlorines is that these organic, lipid-soluble compounds are highly stable! Put another way, these molecules are persistent. In the soil they can persist months or even years, especially when applied at high doses to clay or rich organic soils. In plants, they may accumulate in edible parts, especially in root crops. In animals and humans, they persist in the fat pads (Concon, 1988).

Through agricultural runoff, these compounds can make their way into rivers, lakes, and streams (Marini-Bettòlo, 1987). Although highly water-insoluble, they can associate with floating particulate matter or sediment (Coffin and McKinley, 1980). Figure 13.2 shows how pesticides get into the water. Groundwaters in areas where runoff is a problem exceed recommended concentrations for pesticides in water supplies (CAST, 1986). Despite these high concentrations, no adverse health effects have been seen in these regions—perhaps due to the 100-fold safety factor used in setting the recommended concentrations. CAST calculated that a child would need to drink 26 gallons of water containing residues above such concentrations to attain a level comparable to that which caused an adverse effect in experimental animals.

Concentration. In addition to being persistent, organochlorines are concentrated in the food chain. For instance, fish eat plankton, which

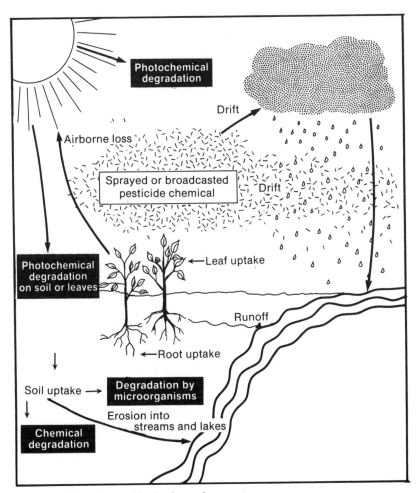

Figure 13.2. Fate of pesticides in the environment.

might contain a residue level of 0.00003 parts per million (ppm), and aquatic invertebrates, which might contain 0.001 ppm, and eventually attain a fish tissue concentration of 0.5 ppm. The fish are eaten by fish-eating birds. Birds are very susceptible to damage by organochlorines, since these pesticides interfere with shell production. Inadequate shell production disrupts the reproductive cycle of birds such as eagles, grebes, and herons.

Although the effects on mammals and humans are less than on fish-eating birds, an analogous example of concentration in the food chain occurs in human foodstuffs. Plants eaten by grazing animals can have organochlorine residue levels of 0.02 ppm. These animals store organochlorines in the fat, giving meat residue levels of 0.2 ppm. If a great deal of meat is consumed, human adipose tissue can have pesticide levels of 6 ppm—levels so high that the current laws governing residues would deem human flesh unfit and make it subject to FDA seizure in the United States (Hassall, 1982). Once deposited in the fat of mammals including humans, organochlorines remain there indefinitely.

Adverse effects on humans. Adverse effects on humans related to DDT's use as a pesticide or as a residue in food do not seem to surface despite its widespread use (Durham, 1987; Murphy, 1986). Human data come from persons accidentally or occupationally exposed. Even then, almost all studies of factory workers involved in the manufacture of DDT and DDT sprayers showed no apparent adverse effects from this pesticide. Only one recent report indicates any association between DDT (and other pesticides) and cancer of the pancreas due to industrial exposure (Anonymous, 1990).

Organochlorines, unlike organophosphates, are poorly absorbed through the skin. This fact may account for DDT's rather good safety record despite its numerous uses and sometimes careless handling by applicators. In countries where DDT was used regularly for control of diseases such as malaria, the intake levels were documented to be below the ADI.

Ninety volunteers taking large doses of DDT of 35 milligrams per day (200 times the normal exposure, but one fifth the dose that caused mild transient sickness in animals) for over 2 years showed no adverse effects (Durham, 1987).

DDT at toxic levels that occur through accidental ingestion causes irritability, dizziness, disturbed equilibrium, tremor, and even convulsions. Symptoms occur several hours after exposure.

Consumers find little reassurance in the lack of data on adverse effects or in the current belief that DDT stays trapped in the fat pads and does not enter working body tissue. There is a gnawing fear that this will be another case of the shrinking zero, where science and the FDA reverse their findings. To add to the unrest, safety unknowns exist for DDT. For example, even though organochlorine residues may stay in the fat pads, what happens if a person reduces the fat pads? During weight loss or fat loss from lactation, would stored organochlorines be released into the active metabolic pool? The possible release of organochlorines during lactation is serious, as it could contribute to high levels in breast milk (Monheit and Luke, 1990). High levels of organochlorines have been measured in breast milk, but little is known about the impact of these on the human infant (Jensen, 1983). Studies of both overall and cancer mortality showed no relation to serum DDT levels (Austin et al, 1989).

Other organochlorines. Other organochlorines—dieldrin, endrin, and aldrin—have higher acute toxicity than DDT. In fact, there are reports of accidental contamination of flour resulting in convulsions and death of some individuals using the flour. These organochlorines have been suspected of being carcinogenic. Because of this, their manufacture and use has been drastically curtailed. Most uses have been banned in many industrialized countries, but they may be found in illegally imported food from Third World countries, where their application is still permitted.

Other organochlorines, including chlordane and the closely related heptachlor, have also been banned from use in the United States. These were used here only for the control of termites and not on food crops. Their acute toxicity involves the central nervous system and the liver. Their chronic toxicity has caused problems with hematopoiesis (red blood cell synthesis) in a subpopulation of susceptible individuals. Since chlordane is a suspected human carcinogen (Mehendale, 1989), it was banned in the summer of 1987 despite protests that it is the most effective pesticide for the control of termites.

Lindane, the most toxic isomer in the hexachlorocyclohexane group of chlorinated hydrocarbons, is not registered for use in the United States but is used in the Third World. It is a suspected carcinogen (Mehendale, 1989).

Mirex and the structurally related Kepone are also banned organochlorines. Mirex was used primarily for control of fire ants and as a flame retardant. Although an incident of milk being contaminated with Mirex occurred in 1974, the concern now is that Great Lakes fish may be polluted with it. These pesticides appear to be estrogenic (Concon, 1988).

Mechanism of toxicity. Data on the chlorine-containing herbicides 2,4-D (2,4-dichlorophenoxyacetic acid) and 2,4,5-T (2,4,5-trichlorophenoxyacetic acid) indicate greater human toxicity than in DDT, probably due to the contaminant 2,3,7,8-tetrachlorodibenzo-*p*-dioxin (known as 2,3,7,8-T, dioxin, or simply TCDD). Epidemiologists have concluded that 2,4-D causes non-Hodgkins lymphoma in exposed farmers, even though animal studies failed to show any relationship or even to indicate mutagenicity. Thus, questions about the relationship remain (Doull, 1989).

Animal feeding studies by Dow Chemical Company showed that these herbicides decreased total weight gained, increased kidney weight, and caused hypertrophy of liver cells (Murphy, 1986). The last two changes indicate that the body is attempting to detoxify something. Other indicators of detoxifying mechanisms were also found, including the stimulation of a variety of drug-metabolizing enzymes and the inhibition of energy-producing enzymes (Alvares, 1984). In fact, acutely toxic doses appear to make the animal look as if it were starving to death.

TCDD is one of the most toxic substances, with an LD_{50} of 0.6 microgram per kilogram of body weight for the guinea pig, the most sensitive laboratory animal. However, acutely toxic doses vary as much as 1,000-fold among species. Extremely small doses of TCDD can cause cancer, birth defects, depression of the immune system, and death to experimental animals (Concon, 1988).

Use of the TCDD-containing defoliant Agent Orange in Vietnam has raised many questions for Vietnam veterans, especially as TCDD levels in blood and fat were 10 times higher in veterans than in controls. However, several studies showed no greater cancer incidence among the troops who sprayed Agent Orange, the so-called Ranch Hands. This group will continue to be monitored until 2002. Studies showed that Marines who served in Vietnam had an unexpectedly high cancer death rate, whereas Army personnel who served there had fewer cancer deaths than those who served elsewhere (Anonymous, 1987b; Hanson, 1987). The Centers for Disease Control published the results of a study showing more stress, alcoholism, and depression among veterans than others but no physical differences that could be attributed to Agent Orange. It is not known whether the contamination of herbicides with TCDD gives sufficient concentration to exert a carcinogenic effect.

Accidental human exposure to TCDD resulted from an industry mishap in Seveso, Italy, in 1976; 37,000 people of all ages were exposed to considerable amounts. No human deaths were attributed to the incident, although some chloracne and reversible liver toxicity occurred (Tschirley, 1986). The dose level, if extrapolated from the dose toxic to guinea pigs, should have killed every human being within miles of the spill. It appears from this incident that humans are much less susceptible than laboratory animals.

Corroborating data comes from workers who have been exposed to

TCDD in industrial accidents. Their death rates were lower than expected, and no increase in either cancer or cardiovascular deaths occurred despite the fact that TCDD does increase the numbers of tumors found in laboratory animals.

Frightening data show that key enzymes of DNA synthesis and nerve transmission are inhibited by some herbicide components. TCDD can cause psychiatric disturbances as well as neurotoxic effects that decrease levels of brain serotonin and exert long-term effects on behavior and learning in newborn rats (Fagan and Pollack, 1984). The neurotoxicity seen when these herbicides were administered to newborn rats may be due to a long half-life in neonates (for example, a half life of 97 hours in a neonate versus 3.5 hours in the adult). Nerve cell replication is thought to be impaired under conditions where the herbicide remains in the metabolic pool for a long time. If these data in rats are indicative of what may occur in humans, young infants and children would be more adversely affected by these herbicides than adults would be.

Organophosphates

Organophosphates, particularly malathion, received much press when the spraying to control the Medfly was done in California. These widely used insecticides have an array of structures, each with different properties and agricultural uses. They can protect crops against aphids and soft-bodied insects from early planting until near harvest.

Box 13.5

Possible Costs to Include in the Cost of Using Pesticides in the United States

	Estimated Cost in Millions of Dollars
Human poisonings	184
Animal poisonings or contaminated feed	12
Reduction in natural enemies or pesticide resistance	287
Honey bee poisonings	135
Losses to other vegetation and wildlife	91
Costs of pollution control programs	140

(Data from Pimentel et al, 1990)

Organophosphate fumigants are used to combat weevils and thus enable the shipping and storage of grain crops.

Unlike the organochlorines, organophosphates are not persistent, so they are not concentrated in the food chain. That's good news for the environment but bad news for pest control and pesticide applicators, because more frequent applications are required to get the same degree of pest control throughout the crop cycle. As with some other pesticides, various pests are evolving to be resistant to these chemicals.

Toxicity of organophosphates. As a group, organophosphates are more acutely toxic than organochlorines. In fact, when the big switch was made from organochlorines to organophosphates, an increase occurred in human and animal poisonings—both accidental and intentional. The toxicity of the compounds in this group varies as widely as do their structures and functions. The most toxic is parathion; it also is the most persistent and is classified by WHO as extremely hazardous (Ekström and Åkerblom, 1990). Toxic doses of parathion can be absorbed through the skin. Malathion is toxic to insects but much less toxic to humans than parathion.

The primary cause of toxicity is inhibition of the enzyme acetylcholinesterase at the nerve junctions. Severe poisoning results in muscle twitching, convulsions, and failure of the diaphragm muscles, which results in death (Taylor, 1984). It is also cardiotoxic and causes abnormalities in an electrocardiogram. The effects on the heart appear to be dose dependent (Marosi et al, 1985).

Mild poisoning leads to dermatitis, depression, immunosuppression, insomnia, irritability, and impaired ability to concentrate (Murphy, 1986; Vettorazzi, 1980). Subtle neurological impairment has been measured in studies of persons occupationally exposed to high levels of organophosphates. Subjects scored lower in reading comprehension and verbal fluency and performed less well on motor skill tests and problem solving tasks than matched controls not occupationally exposed. Also the exposed group suffered a greater frequency of accidental trauma-related deaths.

Organophosphates have been shown to be teratogenic (especially to birds, less so to mice) and mutagenic in tissue culture of human chromosomes; thus the question of whether these compounds are also carcinogenic needs to be asked (Anonymous, 1986b). Tests by the National Cancer Institute have indicated that methyl parathion, parathion, trichlorfon, and malathion are not carcinogenic (Young, 1987). Only tetrachlorvinphos showed evidence of carcinogenicity when fed at 16,000 ppm—a dose that exceeded the maximum tolerated dose. While malathion may not be carcinogenic itself, some studies indicate that it inhibits other carcinogen-inactivating factors and thus helps to increase tumor incidence and decrease tumor latency (Mehendale, 1989).

Much controversy about the safety of organophosphates exists. Part of it comes from an allegation by the United Farm Workers (UFW) that the death of a 43-year-old grape sprayer was due to the parathion, paraquat, and other pesticides that he sprayed. The attending physician said that death was caused by acute hepatitis due to alcohol, but the family disputes the death certificate. Because of incidents such as this, the UFW made a mass appeal urging consumers to boycott grapes. The grape growers say that the allegations of the UFW have no basis in fact. As of mid-1991, no legal judgment had been made in the case, although evidence has accumulated that toxicity in spraying paraquat is unlikely (Smith and Elcombe, 1989).

Thus, many questions remain about the safe use of organophosphates. These questions must be answered, as these pesticides are currently being chosen because they have fewer long-term adverse environmental effects.

Carbamates

Carbamates are often used when insects have become resistant to organophosphates. So far, pests appear to be less well able to evolve resistance to carbamates; however, the these pesticides cost twice as much as the organophosphates.

Carbaryl (Sevin) is a popular carbamate. Large oral doses are required before any toxic effects are seen in humans. Its LD_{50} is 630 milligrams per kilogram. It can therefore be used near harvest. At near toxic doses in dogs, but not in other experimental animals, carbaryl was teratogenic (Durham, 1987). It has not been associated with carcinogenicity (Mehendale, 1989) but may affect cellular resistance.

The most toxic of the currently used carbamates is aldicarb. Its high toxicity is flagged by its low maximum residue tolerance (0–1 ppm) and the limited number of crops on which it can be used. It has the lowest LD_{50} of any registered pesticide—i.e., it is highly toxic—and, along with six other pesticides, is classified by WHO as extremely hazardous (Ekström and Åkerblom, 1990). However, it is not a suspected carcinogen (CDC, 1986). Aldicarb poisoning occurred on watermelons in California in 1985 because growers illegally applied this pesticide to a crop that was not one of the designated crops on the registration certificate. Other poisonings may have occurred (Goldman et al, 1990). Aldicarb is a type of pesticide that is systemic and therefore cannot be washed off. However, heating reduces the residue level (Mott and Snyder, 1988).

Severe carbamate poisoning causes constriction of the pupils, muscle weakness, and spasms, lowered blood pressure, respiratory failure, convulsions, and cardiac arrest. Like organophosphates, carbamates inhibit cholinesterase (Murphy, 1986). Less severe incidents result in headache, abdominal pain, vomiting, diarrhea, confusion, twitching, blurred vision, excessive salivation, and decreased ability to learn. Impairment in the learning ability was accentuated if the person was on a low protein diet. Most carbamates have not been shown to be carcinogenic (Durham, 1987), although some isomers may be (Woo and Arcos, 1991) or may cause formation of nitrosamines (Concon, 1988).

Fungicides

Soil and seed fungi are frequently controlled by chemicals containing mercury, copper, and tin. Seed that is to be planted should, of course, never be eaten. There have been several incidents of accidental poisonings when humans and livestock have ingested treated seed grain. In one case, the seed grain was sold under the ruse that it was more "natural" because it was the actual material that the farmer would plant. Since by law the fumigant-treated seed grain must be colored, it was easy for authorities to recognize that this was not wheat for human consumption and to seize it.

Many other fungicides are used for a wide variety of purposes. Fungus such as apple or pear scab is controlled by compounds such as captan, which is also used on strawberries, almonds, and some seeds before planting. Captan is found as a residue in grapes. It has low acute and chronic toxicity in mammals, although it appears to be more toxic to fish. Its toxicity increases when diets are low in protein. Captan residues can be removed by washing and destroyed by heat during cooking and canning.

In 1985 the EPA severely restricted the use of this pesticide because of data indicating that it may be immunotoxic, mutagenic, teratogenic, and carcinogenic (Murphy, 1986). However, Paul Lapsley of the EPA's pesticide review program said that if a person consumed an average diet contaminated to the maximum possible extent for a 70-year lifespan, the chances of developing cancer from exposure would range between one in 1,000 and one in 100,000 (Anonymous, 1991).

The fungicide ethylenebisdithiocarbamate has low relative toxicity but does have antithyroid activity. Adverse effects are made worse by the simultaneous ingestion of alcohol. The EPA has initiated a review of this fungicide because of possible risks to farmers and workers who mix or apply it. Although a test will be made to assess the carcinogenic potential of residues of this fungicide in food, no increased health risk from it is expected to be found (United Fresh Fruit and Vegetable Association, 1990)

Fumigants

Ethylene dibromide (EDB) was a widely used fumigant before its banning in 1984. It was used to fumigate soil, grain in shipment and storage, milling machinery, and fruit that came from outside the United States and might contain insects that could wreak havoc if allowed to enter the country.

EDB was banned because several long-term studies showed that it did increase cancer rates in all experimental animals tested. It was also mutagenic by the Ames test. EDB is a carcinogen of medium potency, i.e., much greater carcinogenicity than saccharin and much less than aflatoxin. The amount ingested was estimated to be 5-10 micrograms per person—a very small amount indeed. The American Council on Science and Health (ACSH, 1985) calculated that a person would need to eat 400 tons of food per day to reach the dose fed to the experimental animals. So although we are talking about a carcinogen of medium potency, the amount ingested is extremely small. Dr. Bruce Ames et al (1987) calculated the hazard index from ingesting this substance in grain products and found it to be nearly 10,000 times less than drinking 12 ounces of beer.

Other fumigants have also been banned or severely curtailed. Farmers and grain elevator operators are seeking alternates to keep infestation levels low.

Plant Growth Regulators

Daminozide (Alar) is a plant growth regulator that was front page news in 1989 (Smith, 1990). It is used to reduce shoot formation, control flowering, and increase plant resistance to disease. In the United States, it is primarily used on apples and cherries but sometimes on other fruits and tomatoes.

Animal feeding studies indicated that daminozide causes a variety of cancers, as does one of its metabolites (a hydrazine) that is produced during storage and processing. The EPA stated its concern that this compound was occurring in significant levels in the marketplace. Of particular concern was dietary exposure of young children who consume treated fruit and peanut products.

This product is systemic, so careful washing does not render the fruit safe. A consumer group and a group of pediatricians asked EPA to ban it. The petition was turned down, as the existing data do not show risk from low-level exposures. Doses causing cancer in experimental animals would be equivalent to eating 50,000 pounds of apples per day for life! However, despite the data, some one-sided reporting of the facts on alar by lobby groups and news magazine programs caused all U.S. apple growers to cease use of this product.

Box 13.6

A Special Look at Alar

Alar was registered as a pesticide for use on apples and peanuts in 1963. It was used on the apple crop because the apples stayed on the trees longer, retained their crispness during marketing, and attained more perfect shape and color. In 1985, after reviewing feeding studies, the EPA proposed to cancel Alar. However the agency's Scientific Advisory Panel created by Congress reported that the studies had some flaws and were inconclusive and that the chemical did not pose a hazard. The panel recommended further feeding studies that involved both Alar and its breakdown product unsymmetrical dimethylhydrazine (UDMH). These were begun by the manufacturer, Uniroyal.

Early data from these high-dose mouse-feeding studies indicated that Alar itself was not carcinogenic. However, UDMH caused vascular tumors when fed at overtly toxic levels. Still, data from these studies did not make clear whether the tumors were from UDMH itself or from a general toxicity to the excessively high levels of UDMH, so more studies were scheduled. Further, the vascular tumors that were seen are notably rare in humans. Epidemiological studies did not indicate any increase in the incidence of this tumor type, even among workers exposed to rocket fuel, which would markedly increase UDMH exposure.

The real question then became: what was the risk if the dose was many times lower, as was the case with Alar. Market Basket data indicated that the exposure to UDMH was 0.000023 milligrams per kilogram of body weight per day.

Originally the EPA had planned to decide about cancellation only after all the data were evaluated. However, in October of 1988, an environmental action group, the Natural Resources Defense Council (NRDC), hired a public relations firm to publish its forthcoming report "Intolerable Risk: Pesticides in Our Children's Food." The firm subsequently arranged to break the story to CBS's "60 Minutes" and lined up placement of articles and interviews in a wide array of popular media. As a result, most of the population became aware of the report, and sales of applesauce dropped by 25% and apple juice by 31%. In two and a half months, apple growers lost $100 million. Some stores posted signs over their apples stating, "The fresh apples sold in our stores are not treated with Daminozide (Alar)."

The data presented in the NRDC's case are in themselves subject to question because the study assumed that all apples are treated with Alar at the maximum allowed level. In actuality, the amount found was one-twentieth of that amount. The highest level measured in juice by the Consumer's Union, using an unapproved method, was 1.8 ppm. For every 1 ppm Alar there is a maximum of 1 ppb UDMH; thus the maximum in the juice was 18 ppb. The NRDC also calculated cancer risk using a linear extrapolation model, which can overestimate the risk.

What the EPA Did

Because of the upcoming media attention, on February 1, 1989, the EPA announced its intention to initiate cancellation of Alar, using wording that called Alar a carcinogenic threat to human health. Ironically, the EPA's announcement probably unintentionally led American consumers to accept the NRDC's conclusion that Alar posed a threat to children and further undermined consumer confidence in the government's ability to protect the safety of the food supply (Weinstein, 1991).

The pronouncement by the EPA was frought with controversy in the scientific community in that many well-respected scientists and bodies stated that the level of Alar was safe and that any residues found in apples were toxicologically insignificant. The British Advisory Committee on Pesticides stated that no risk to health existed, even from excessive consumption by infants and children of Alar-treated produce (Weinstein, 1991).

Views of the Public and of Consumer Activists

The public viewed the risk of Alar differently. Even though the upper lifetime limit of incremental cancer risk is 45 cancer cases per million persons, the public does not view these numbers as an insignificant or very slight risk. People believe it likely that they or their children will get cancer from eating treated apples. This belief is heightened because the various groups disagree as to the actual risk and because most consumers are simply unwilling to accept any cancer risk that they know about, no matter how small. Hypothetically, a child faces a greater health risk playing for an hour in a chlorinated swimming pool (Wilkinson, 1990). Chester Groth, Associate Technical Director of the Consumer's Union stated the consumer's position this way: "We should ban carcinogens, however small the risk, because people think they are dangerous."

In their letter "Pesticides, Risk, and Applesauce," Dr. Bruce Ames in collaboration with Lois Swirsky Gold (1989) summed up the Alar controversy in this way: "Regulation of low-dose exposures to chemicals, based on animal cancer tests, may not result in significant risk reduction of human cancer, because we are exposed to millions of different chemicals— almost all natural—and it is not feasible to test them all."

Casting Out One Demon

The withdrawal of Alar may increase the use of other pesticides. Alar applied to apples in July would prevent premature fruit drop. A variety of pests that cause premature fruit drop could be tolerated at much higher levels if Alar were used on the apple crop (Gianessi, 1991). Has the removal of this pesticide made the situation safer, or have we cast out one demon and left a place for seven to rush in?

Herbicides

Herbicides are important weapons against weeds, especially in areas where rainfall is heavy. Approximately 40% of the crop protection dollar goes toward weed control (Klingman and Nalewaja, 1984). Herbicides are designed to disrupt specific metabolic processes that occur in plants. Some are selective because they disrupt the metabolism of a specific plant type. With some exceptions, herbicides are not as toxic to animals as pesticides are. Their high LD_{50} levels reflect their lower toxicity. Commonly used herbicides such as atrazine and simazine are nontoxic to a variety of animal species.

Dinitrophenol (dinoseb), which was among the first organic herbicides, has been shown to cause lassitude, headache, increased perspiration, thirst, dyspnea, fever, and profound weight loss in humans (Leftwich et al, 1982). The EPA issued a warning to women of childbearing age to avoid exposure to dinoseb, as the chemicals cause birth defects in experimental animals. The EPA indicated that this was a warning to the agriculture community, not to consumers. The report stated that residues were far below levels needed to exert an effect. Reports from the UFW indicate that this pesticide causes loss of vision and yellow staining of the tissue. The UFW has put it on its "Deadly Dozen" list and wants it removed from agricultural uses.

RISKS

Cancer and Pesticides

Many methods have been used to assess whether pesticide residue exposure results in increased cancer risk in humans. The delayed effects of low-dose pesticide exposure have been difficult to determine and are perhaps below the power of epidemiological tests to detect. Epidemiological results leave us with two possible answers. The first and hoped for answer is that there are very few effects at all. The second, more skeptical, answer is that the test may be unable to detect the effect.

Epidemiological studies involving occupationally exposed persons are difficult to interpret because of conflicting results and confounding factors in the study such as cigarette smoking or higher accident frequency (Sharp and Eskenazi, 1986). For example, herbicide workers in Sweden showed increased cancer risk with increased phenoxy herbicide exposure. Studies using the same methods in New Zealand showed no significant association between exposure to the herbicide and cancer rate, while studies in the United States gave mixed results.

Farmers on the whole have lower rates of cancer than the U.S. population as a whole. However, several studies have indicated an increased risk of leukemia and other cancers among farmers and their families (Donham and Mutel, 1982). These increased rates have been attributed to increased exposure to agricultural chemicals, including pesticides (Blair, 1982; Burmeister et al, 1982; Moses, 1989; Reif et al, 1989). On the other hand, a recent study of New York state farmers failed to show any excess cancer at any site and, in fact, found the cancer rates lower than those for the comparison population, comprised of others living in the same state but not in New York City (Stark et al, 1990).

Dioxin. Dioxin exposure in Vietnam is just now reaching the minimum cancer latency period of 20 years or more (Sharp and Eskenazi, 1986). The next 10 years will yield data on "occupational" exposure to these chemicals, which were widely employed as defoliants during the late 1960s and early 1970s. A study by the National Institute for Occupational Safety and Health of a possible link between 2,3,7,8-T and cancer suggests that dioxin causes cancer in workers exposed at high levels but may not pose a substantial risk at lower levels (Fingerhut et al, 1991).

Organochlorines. DDT, endrin, toxaphene, Kepone, heptachlor, aldrin, and dieldrin are listed by the International Agency for Research on Cancer (IARC) as probably carcinogenic in humans. In experimental animals these chemicals are carcinogenic at doses far exceeding residue levels in food. Substances like DDT don't show the same effects in all species. DDT has produced liver cancer in rats but not in hamsters or nonhuman primates (Williams and Weisburger, 1986). Based on studies such as these, the IARC stated that the evidence that these substances are carcinogenic in humans is not strong. In fact, several studies have failed to show an association between DDT and several different types of cancer.

Epidemiological studies have failed to show any relationship between cancer incidence and the organochlorines, but the validity of the studies has been questioned because of either inadequate numbers of subjects or inadequate latency periods (Concon, 1988). Similarly, persons involved in the manufacture of heptachlor were not shown to have higher incidence of cancer than the population as a whole (Sharp and Eskenazi, 1986). However, IARC believes that the animal evidence must be considered until conclusive proof of its noncancerous nature in humans is found (IARC, 1982).

Organophosphates. The data on organophosphates indicate that many of them may not be carcinogenic. Malathion did not produce cancer in rats or mice fed the compound for over 2 years using the National Cancer Institute's protocol for testing (Ames, 1989). Similar results were shown with methyl parathion and diazinon. Equivocal data resulted from studies with parathion, and positive results from studies with tetrachlorvinphos. Further, studies did not show any potentiation effects if several pesticides were fed together (Durham, 1987).

Reproductive Hazards and Pesticides

Reproductive hazards from pesticide residues in foods also have been questioned. Certain pesticides at high doses exert adverse effects on fertility. Wives of workers in a dibromochloropropane plant were unable to become pregnant. The male workers were subsequently shown to have low sperm counts; no effects were seen in female workers. Also workers exposed to chlorophenol had higher rates of stillbirths and malformations.

Any association between exposure to DDT and other organochlorine herbicides and documented adverse effects on reproduction has not been found using epidemiological tools, although Kepone and mirex have been shown to have teratogenic effects in experimental animals (Concon, 1988). Women exposed to higher than normal levels of DDT were followed for a 5-year period, and their newborns did not show any changes, even though DDT levels in cord blood were higher than normal. The documentation that organochlorines pass through the placenta and are found in human breast milk certainly creates cause for concern. Furthermore, animal data on pregnant rats receiving very high doses of DDT indicated that offspring showed decreased growth rates and increased mortality. Once again, epidemiological studies leave nagging disquiet about the lack of statistical power to pick up the association, rather than assurance of the lack of effect of current exposure levels on humans.

One possible epidemiological link occurred in 1979 with the herbicide 2,4,5-T. The EPA used its emergency powers to suspend its use for forests, pastures, and right-of-ways because this herbicide was alleged to cause an increased number of miscarriages in female monkeys and women in Alsea, Oregon (Concon, 1988). In 1981 EPA reversed itself when it concluded that there was not enough evidence to support a ban on traditional use of 2,4,5-T, although it is labeled as teratogen for experimental animals (Concon, 1988).

In like manner, an attempt has been made to find epidemiological links between birth defects and Agent Orange, the 2,4,5-T defoliant used during the war in Vietnam. Some studies show no greater rates of reproductive problems among those exposed to Agent Orange than to those who were not exposed. Examination of birth defect records over a 17-year period yielded no association between spraying of the herbicide 2,4,5-T and malformations of the central nervous system, heart, palate, or male genitalia. A contradictory study indicated an increased incidence of neural tube defects and facial clefts among children of Vietnam veterans (Sharp and Eskenazi, 1986). This particular herbicide generates strong emotional responses because accusations have been made that the government has withheld critical data pertaining to the issue.

Even if it is assumed that this defoliant causes birth defects when sprayed in massive quantities and inhaled and absorbed through the skin, these conditions may not have any relationship to the kinds and amounts of pesticides left in food. The amount of dioxin contaminant in Agent Orange was 100–200 times greater than in formulations for agricultural and roadside use. Moreover, Agent Orange was literally dumped on Vietnam, while the agricultural herbicide is used in rigidly controlled amounts.

Other pesticides and herbicides have been reported to have teratogenic effects. Carbaryl is teratogenic to a variety of species when doses are as high as 10 grams per kilogram. Nevertheless, the fact that in some species doses as low as 4 milligrams per kilogram cause an increase in the number of stillbirths is cause for concern, even though this dose is much higher than would be received by humans from residues in food (Concon, 1988). Certain organophosphates such as diazinon, phosmet, and leptophos have also been reported to have taratogenic effects. The more common ones such as parathion and malathion have not (Concon, 1988).

Safety of Pesticides

Consumer fears with regard to pesticide toxicity stem from the logical (but not necessarily correct) argument that if a chemical is lethal to insects, it must be harmful to humans. This logical conclusion is validated for some by knowledge of the harmful effects of pesticides on birds of prey. Some individuals believe that, like birds of prey, the entire population is at risk because of some unmeasured effects of pesticide residues on humans and their environment. Worries about pesticides have increased due to recent scares about alar in apples, EDB in flour, and pesticide chemical spills killing millions of fish in the Rhine River.

These fears cause some consumers to long for zero residues. Canadian consumer activist Linda Pim put it this way: "Zero risk from chemicals is probably unattainable, but that's a poor excuse for not trying to chase it." The question that nags at everyone is: Will these very small traces of pesticide cause harm to humans, or will the age-old dictum "only the dose makes the poison" prove true for pesticide residues?

For some, but certainly not all, fear is lessened by knowledge of the no-effect level or the hours of scientific assessment spent in determining the ADI for humans. In some cases scientists and consumers alike are plagued by doubts. The most consistent result from toxicity studies is their inconsistency. This is partly due to the limitations posed by epidemiological studies, which in many cases assume but don't actually know what the exposure levels were. Results of these studies are further confused by other lifestyle and risk factors such as smoking. To make matters worse, epidemiological studies fail to yield results that can be extrapolated to the miniscule amounts that are encountered through residues in foods. Further, their capability to detect the effects of pesticides is likely to decrease rather than increase, because improved industrial practices and increased environmental awareness continue to reduce occupational and environmental exposure to pesticides and other chemicals.

As discussed in Chapter 3, relying on high-dose animal studies with genetically pure strains raised in controlled environments may under- or overestimate true cancer risks and other risks to humans. Unfortunately, however, we are stuck with data from high-dose animal studies and occupationally exposed persons. Neither is able to accurately appraise risks associated with chronic low-dose exposure, but those are all the data that exist. Nor is there a new method looming on the horizon that will once and for all answer the safety questions. Until such methods exist, false positive results generated by extrapolating from high-exposure data (such as experimental, occupational, or accidental exposure) in humans and animals and unreassuring negative data derived from epidemiological studies will continue to fuel the controversy as to the safety of extremely small amounts of these pesticide chemicals in our food.

WHERE DO WE GO FROM HERE?

First, it is important to look at what we have learned over 50 years of pesticide usage. One of the most important lessons is that reduced pesticide exposure and controlled pesticide use are of critical importance to agriculture, the environment, and human safety. Eras of overuse and abuse resulted in unnecessary damage to the environment, caused crop damage by secondary pests, hastened the development of pesticide resistance, and put at risk both human and animal health (Pimentel, 1987b). Powerful, undiscriminating weapons should never again be applied when small selective ones will do (NAS/NRC, 1986, 1987). The current era is one characterized by the protocols of Integrated Pest Management (IPM). This system uses pesticides on a limited basis together with crop protection methods that were used before the widespread use of chemical pesticides but then were nearly forgotten.

IPM has many aspects (Hegele, 1989). Careful side dressing of plants, rather than broadcasting of pesticides, is one aspect that can substantially reduce not only the quantity of pesticide, but also the amount that becomes airborne and contributes to aerial drift. Drift has been minimized through stringent regulation of aerial spraying and licensure of sprayers. Controlled spraying and side dressing reduce potential damage to nontarget species in surrounding areas.

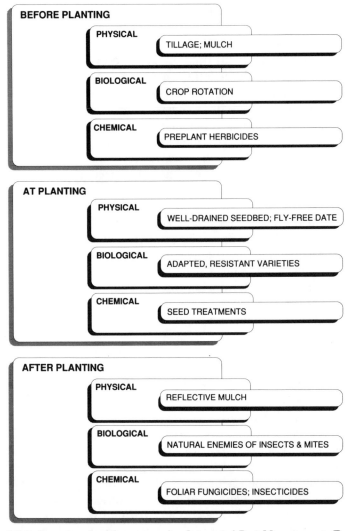

Figure 13.3. Steps to a healthy crop, using Integrated Pest Management. (Reprinted, with permission, from Cook, 1991)

Nonchemical methods are being used, such as the spraying of a light mist of oil to prevent insects from landing on the fruit. Pesticides that decompose quickly are sought. In addition, farmers are using pesticides on a rotating basis to minimize the problems of secondary infestations and pest resistance (CAST, 1983.).

Practices such as fastidious farm and plant sanitation, use of optimal planting times and storage conditions, crop rotation, less monocropping, row planting with mechanical cultivation, and crop "spying" to know when and which particular pests are problems—all have potential for reducing pesticide use by as much as 75% (Mandava, 1985).

Directions for the Future

Multifaceted research has been undertaken by entomologists around the world to accomplish the following objectives:

• Identify, isolate, and manufacture small selective pesticides that will have only a single species as their target.

• Foster the use of biodegradable insecticides and growth regulators that will have minimal environmental damage.

• Determine the optimum time of pesticide application so that application coincides with times when the insect population is on the rise or when the pest is at its most vulnerable phase, such as during the reproductive cycle.

• Explore the use of natural predators, parasites, and pathogens.

• Isolate and identify naturally occurring pesticides in plants, such as pyrethroids, which are highly toxic to insects but have no seeming effects on either rats or dogs. Currently these naturally occurring and frequently target-specific products are very expensive, but their expense can be partially compensated for by the small amounts that would be required (Elliot, 1985; Perkins, 1985).

• Utilize insect and pest traps that contain mating and feeding attractants (pheromones), oil, and other nonchemical means of pest control.

• Continue the search for genetically pest-resistant plants and genetically sterile insects.

• Emphasize control rather than kill of insect populations.

Basic assumptions about pesticides have shifted dramatically in the past 50 years. For the next era, an ethical balance must be struck between the polarized positions of no pesticide use on the one hand and pesticide abuse on the other. Most consumers clearly are not ready to return to weevils, worms, and insects (all of which carry molds, bacteria, parasites, and tapeworms), to fungi (some extremely toxic and carcinogenic), or to vermin vestiges in our food. Neither is the world ready for smaller crop yields on a global basis, nor can it afford such. Yet we must be constantly vigilant so that we can provide the balance necessary to preserve our fragile environment, ensure the safety of farm workers, protect against insectborne disease and epidemics, minimize the risks afforded by long-term low doses of toxic chemicals, and provide adequate food for this hungry planet. Use of these potentially helpful but also destructive chemicals requires Solomon-like decision making. Decisions must be based on the most rigorous possible science and should occur in an environment free from hysteria-producing pronouncements by self-appointed consumer activists.

In terms of residues in foods and the risks they pose, this safety in food must be viewed in terms of other risks and must be separated from other issues about pesticides. As Dr. Bruce Ames and co-workers (1987) pointed out, "We are ingesting in our diet at least 10,000 times more by weight of natural pesticides than of man-made residues." He calculated that the risk from the consumption of a single beer was 10,000 times greater than the risk afforded by the daily dietary ingestion of DDE/DDT. Further, we must be reminded of the statement of the director of the National Cancer Institute's Division of Cancer Etiology, who wrote that he is "unaware of evidence that suggests that regulated and approved pesticide residues" in food contribute to the toll of human cancer in the United States" (Anonymous, 1990).

For the environment, pesticides with a high degree of target specificity and low avian and mammalian toxicity must be sought. In some cases natural pesticides will provide what we are looking for, but we must continually remind ourselves that nature is not always benign. For instance, nicotine is a very good natural pesticide. By using the natural pesticide nicotine in place of carbaryl, we would be moving to a compound with a much higher toxicity—from one with an LD_{50} of 630 milligrams per kilogram to one with an LD_{50} of only 83 milligrams per kilogram.

While society may decide to ban a variety of pesticides for environmental reasons (Myers and Colborn, 1991a) or because of concerns about workers, it should not ban pesticides from the food supply based on the inconsequential risk these chemicals pose to humans ingesting parts-per-million-level residues in food, nor should it decide to enfranchise a pesticide just because it is natural.

Ensuring the safety-in-use of these many chemicals is a dynamic challenge (Campt et al, 1991). We must constantly search for more and better answers, especially where some of the safety data need to be more complete. We must carefully regulate so that chemicals aren't misapplied or illegally used on food, whether domestically or on imported products. We must develop an internally consistent policy toward chemical use in general, whether in highly regulated food systems or in unregulated (except for labeling) use of pesticide chemicals in home and garden products. We must lobby for more money for the FDA to test for the presence of illegal use of pesticides on domestic and imported food.

We may need to broaden our definition of risk to include an array of human and environmental problems (Myers and Colborn, 1991b). We must also prioritize risks so that we spend resources on things that truly create a hazard rather than simply chasing a mythical zero risk (Ames and Gold, 1991). On the other hand, the scientific community must continue to question tolerance levels to see that they ensure human safety. We must continue the trend of less chemical use and the search for safer alternatives. We must continue to balance the risk of using pesticides against the risk of not using them.

REFERENCES

ACSH. 1985. Does Nature Know Best? Natural Carcinogens in Food. American Council on Science and Health, Summit, NJ.

Alvares, A. P. 1984. Environmental influences on drug biotransformations in humans. World Rev. Nutr. Diet. 43:45-49.

Ames, B. N. 1989. Pesticide residues and cancer causation. Pages 223-227 in: Carcinogenicity and Pesticides: Principles, Issues, and Relationships. N. N. Ragsdale and R. E. Menzer, eds. American Chemical Society, Washington, DC.

Ames, B. N., and Gold, L. S. 1991. Cancer prevention strategies greatly exaggerate risks. Chem. Eng. News 69(1):28-32.

Ames, B. N., Magaw, R., and Gold, L. S. 1987. Ranking possible carcinogenic hazards. Science 236:271-280.

Anonymous. 1985. Crop Protection Chemicals Reference. Chemical and Pharmaceutical Publ. Corp., New York.

Anonymous. 1986a. Impotent pesticides. Insight (on the News), June 2.

Anonymous. 1986b. Pesticides may alter brain function. Sci. News 129(6):81-96.

Anonymous. 1987a. Pesticides for breakfast. Minneapolis Star Tribune, May 18.

Anonymous. 1987b. Study links cancer risk to Vietnam war service. Chem. Eng. News 65(Sept. 14):6.

Anonymous. 1990. No evidence of human cancer from pesticide residues. Food Chem News 32(28):39.

Anonymous. 1991. EPA reaffirms de minimus and most positions on NRDC petition. Pestic. Toxic Chem. News 19(15):31-32.

Austin, H., Kehl, J. E., and Cole, P. 1989. A prospective follow-up study of cancer mortality in relation to serum DDT. Am. J. Public Health 79(1):43-36.

Barrons, K. C. 1983. The alleged menace of herbicides. Chem. Eng. News. 61(28):28-29.

Benbrook, C. M. 1991. What we know, don't know, and need to know about pesticide residues in food. Pages 140-150 in: Pesticide Residues and Food Safety: A Harvest of Viewpoints. B. G. Tweedy, H. J. Dishburger, L. G. Ballantine, and J. McCarthy, eds. American Chemical Society, Washington, DC.

Bifani, P. 1987. Socio-economic aspects of technological innovation in food production systems. Pages 177-222 in: Towards a Second Green Revolution. G. B. Marini-Bettòlo, ed. Elsevier, New York.

Black, A. L. 1988. Bioaccumulation in the food chain: Fact or fiction? Proc. Nutr. Soc. Aust. 13:49-54.

Blair, A. 1982. Cancer risks associated with agriculture: Epidemiological evidence. Basic Life Sci. 21:93-111.

Blanc, B. 1981. Biochemical aspects of human milk. World Rev. Nutr. Diet. 36:1-89.

Borlaug, N. E. 1972. Human population, food demands and wildlife needs. Trans. N. Am. Wildlife Nat. Res. Conf., 37th. pp. 19-35.

Brown, W. L. 1986. Current and future environmental issues as seen from the private sector. Cereal Foods World 31:800-801.

Bruhn, C. M. 1991. Pesticide use in California. Food Saf. Notebook 2(2):13-15.

Burmeister, L. F., Van Lier, S. F., and Isacson, P. 1982. Leukemia and farm practices in Iowa. Am. J. Epidemiol. 115:710-718.

Campanini, G., Maggi, E., and Artioli, D. 1980. Present situation of organochlorine pesticide residues in food of animal origin in Italy. World Rev. Nutr. Diet. 35:129-171.

Campt, D. D., Roelofs, J. V., and Richards, J. 1991. Pesticide regulation is sound, protective, and steadily improving. Chem. Eng. News 69(1):44-47.

Carnevale, C. 1987. Food contaminants—FDA's concerns. Cereal Foods World 32:812-813.

Carpenter, W. D. 1991. Insignificant risks must be balanced against great benefits. Chem. Eng. News 69(1):37-39.

Carson, R. 1962. Silent Spring. Houghton Mufflin, Boston.

CAST. 1983. Pesticides and the "superpest problem." Sci. Food Agric. 1(2):22-25.

CAST. 1986. Agriculture and Ground Water. Rep. 103. Council for Agricultural Science and Technology, Ames, IA.

Centers for Disease Control (CDC). 1986. Aldicarb food poisoning from contaminated melons. MMWR 35:254-258.

Chin, H. B. 1991. The effect of processing on residues in foods. Pages 175-181 in: Pesticide Residues and Food Safety: A Harvest of Viewpoints. B. G. Tweedy, H. J. Dishburger, L. G. Ballantine, and J. McCarthy, eds. American Chemical Society, Washington, DC.

Clower, M. 1991. Pesticide residue method development and validation at the Food and Drug Administration. Pages 105-113 in: Pesticide Residues and Food Safety: A Harvest of Viewpoints. B. G. Tweedy, H. J. Dishburger, L. G. Ballantine, and J. McCarthy, eds. American Chemical Society, Washington, DC.

Coffin, D. E., and McKinley, W. P. 1980. Sources of pesticide residues. Pages 678-686 in: Safety of Foods, 2nd ed. H. D. Graham, ed. AVI, Westport, CT.

Concon, J. M. 1988. Pesticides. Pages 1133-1230 in: Food Toxicology. Part B: Contaminants and Additives. Marcel Dekker, New York.

Consumers Association. 1980. Pesticide Residues and Food. The Association, London.

Cook, R. J., and Veseth, R. J. 1991. Wheat Health Management. The American Phytopathological Society, St. Paul, MN.

Donham, K. J., and Mutel, C. F. 1982. Agricultural medicine: The missing component of the rural health movement. J. Fam. Pract. 14:511-520.

Doull, J. 1989. Pesticide carcinogenicity: Introduction and background. Pages 1-5 in: Carcinogenicity and Pesticides: Principles, Issues, and Relationships. N. N. Ragsdale and R. E. Menzer, eds. American Chemical Society, Washington, DC.

Durham, W. F. 1987. Toxicology of insecticides, rodenticides, herbicides, and fungicides. Pages 364-385 in: Handbook of Toxicology. T. J. Haley and W. O. Berndt, eds. Hemisphere, New York.

Eilrich, G. L. 1991. Tracking the fate of residues from the farm gate to the table—A case study. Pages 202-212 in: Pesticide Residues and Food Safety: A Harvest of Viewpoints. B. G. Tweedy, H. J. Dishburger, L. G. Ballantine, and J. McCarthy, eds. American Chemical Society, Washington, DC.

Ekström, G., and Åkerblom, M. 1990. Pesticide management in food and water safety: International contributions and national approaches. Rev. Environ. Contam. Toxicol. 114:23-55.

Elliot, M. 1985. Pyrethroid insecticides in the environment. Pesticide Sci. 16:192-215.

El-Zik, K. M., and Frisbie, R. E. 1985. Integrated crop management systems for pest control. Pages 21-122 in: CRC Handbook of Natural Pesticides: Methods. Vol. 1. CRC Press, Boca Raton, FL.

EPA. 1988. Pesticide Industry Sales and Usage. 1987 Market Estimates, November. Environmental Protection Agency, Washington, DC.

Fagan, K., and Pollak, J. K. 1984. The effect of the phenoxyacetic acid herbicides. Pestic. Rev. 92:30-58.

FAO. 1984. Pesticide Residues In Food—1984. FAO Plant Production and Protection. Paper 62. Food and Agriculture Organization, Rome.

FDA. 1990. Residues in foods—1989. J. Assoc. Off. Anal. Chem. 73:127A-146A.

Fingerhut, M. A., Halperin, W. E., Marlow, D. A., Piacitelli, L. A., Honchai, P. A., Sweeney, M. H., Greife, A. L., Dill, P. A., Steenland, K., and Suruda, A. J. 1991. Cancer mortality in workers exposed to 2,3,7,8-tetrachlorodibenzo-p-dioxin. New Engl. J. Med. 324(4):212-218.

Food Watch Facts. 1991. Agriculture and the Food and Fiber System. ACA Education Foundation, Inc., Washington, DC.

Gallenberg, L. A., and Vodicnik, M. J. 1989. Transfer of persistent chemicals in milk. Drug Metab. Rev. 21:277-317.

GAO. 1986. General Accounting Office suggests revisions to pesticide law. Chem. Eng. News. 64(June 23):24-25.

Gartrell, M. J., Craun, J. C., Podrebarac, D. S., and Gunderson, E. L. 1985a. Pesticides, selected elements, and other chemicals in infant and toddler total diet samples, October 1978–September 1979. J. Assoc. Off. Anal. Chem. 68:842-862.

Gartrell, M. J., Craun, J. C., Podrebarac, D. S., and Gunderson, E. L. 1985b. Pesticides, selected elements, and other chemicals in adult total diet samples, October 1978–1979. J. Assoc. Off. Anal. Chem. 68:862-875.

Georgii, S., Brunn, H., Stojanowic, V., and Muskat, E. 1989. Xenobiotics in foodstuffs—Estimations of a daily intake. 3.3. Chlororganic pesticides in selected foodstuffs, invalid diets and baby foods. Dtsch. Lebensm. Rundsch. 85:385-389.

Gianessi, L. P. 1991. Use of pesticides in the United States. Pages 24-30 in: Pesticide Residues and Food Safety: A Harvest of Viewpoints. B. G. Tweedy, H. J. Dishburger, L. G. Ballantine, and J. McCarthy, eds. American Chemical Society, Washington, DC.

Goldman, L. R., Beller, M., and Jackson, R. J. 1990. Aldicarb food poisonings in California, 1985–1988: Toxicity estimates for humans. Arch. Environ. Health 45(3):141-147.

Gorchev, H. G., and Jelinek, C. F. 1985. A review of the dietary intakes of chemical contaminants. Bull. WHO 63:945-962.

Hanson, D. J. 1987. Science failing to back up veteran concerns about Agent Orange. Chem. Eng. News 65(45):7-14.

Hassall, K. A. 1982. The Chemistry of Pesticides: Their Metabolism, Mode of Action and Uses in Crop Protection. Verlag Chemie, Deerfield Beach, FL.

Hayes, W. J., and Laws, E. R., eds. 1991. Handbook of Pesticide Toxicology. Academic Press, San Diego. 3 vols.

Hegele, F. A. 1989. Integrated pest management—A quality assurance tool? Cereal Foods World 34:296.

Heinonen-Tanski, H., Siltanen, H., Kilpi, S., Simojoki, P., Rosenberg, C., and Makinen, S. 1985. The effect of annual use of some pesticides on barley yields. Pestic. Sci. 16:341-348.

Heinonen-Tanski, H., Siltanen, H., Kilpi, S., Simojoki, P., Rosenberg, C., and Makinen, S. 1986. The effect of the annual use of some pesticides on soil microorganisms, pesticide residues in soils and carrot yields. Pestic. Sci. 17:135-142.

IARC. 1982. Chemicals, Industrial Processes and Industries Associated with Cancer in Humans. International Agency for Research on Cancer.

IFT's Office of Scientific and Public Affairs. 1990. Pesticides in food. Food Technol. (Chicago) 44(2):44 ff.

Jennings, A. L. 1991. Some economic and social aspects of pesticide use. Pages 31-37 in: Pesticide Residues and Food Safety: A Harvest of viewpoints. B. G. Tweedy, H. J. Dishburger, L. G. Ballantine, and J. McCarthy, eds. American Chemical Society, Washington, DC.

Jensen, A. A. 1983. Chemical contaminants in human milk. Residue Rev. 89:1-128.

Kearney, P. C., and Helling, C. S. 1982. Problems caused by pesticides with particular reference to the impact on the agricultural environment. Pages 22-39 in: Agrochemicals: Fate in Food and the Environment. International Atomic Energy Agency, Vienna.

Klingman, G. C., and Nalewaja, J. D. 1984. Hoes and herbicides. Sci. Food Agric. 2:12-15.

Leftwich, R. B., Floro, J. F., Neal, R. A., and Wood, A. J. J. 1982. Dinitrophenol poisoning: A diagnosis to consider in undiagnosed fever. South. Med. J. 75:182-184.

Lever, B. G. 1982. Economics of pest control with emphasis on developing countries. Pages 41-60 in: Agrochemicals: Fate in Food and the Environment. International Atomic Energy Agency, Vienna.

Mandava, N. B. 1985. CRC Handbook of Natural Pesticides. CRC Press, Boca Raton FL.

Marini-Bettòlo, G. B. M., ed. 1987. Towards a Second Green Revolution. Elsevier, New York.

Marosi, G., Iva'n, J., and Nagymigte'nyi, L. 1985. Cardiodepression in organophosphate poisonings. Arch. Toxicol. 8:289-291.

McLoed, H. A., Smith, D. C., and Bluman, N. 1980. Pesticide residues in the total diet of Canada, 1976–1978. J. Food. Saf. 2:141-164.

Medallion Labs. 1986. Daminozide aka "alar". Anal. Prog. 3(1):7-8.

Mehendale, H. M. 1989. Impact of chemical interactions on the development of cancer. Pages 122-141 in: Carcinogenicity and Pesticides: Principles, Issues, and Relationships. N. N. Ragsdale and R. E. Menzer, eds. American Chemical Society, Washington, DC.

Melnikov, N. N. 1971. Chemistry of pesticides. Residue Rev. 36(5):1-480.

Merliss, M. 1975. The pesticide dilemma. Life Health. 90:18-21.

Metcalf, R. L. 1980. Changing role of insecticides in crop protection. Annu. Rev. Entomol. 25:219-256.

Minyard, J. P., and Roberts, W. E. 1991. FOODCONTAM: A state data resource on toxic chemicals in foods. Pages 151-162 in: Pesticide Residues and Food Safety: A Harvest of Viewpoints. B. G. Tweedy, H. J. Dishburger, L. G. Ballantine, and J. McCarthy, eds. American Chemical Society, Washington, DC.

Monheit, B. M., and Luke, B. G. 1990. Pesticides in breast milk—A public health perspective. Community Health Serv. 14:269-273.

Moses, M. 1989. Cancer in humans and potential occupational exposure to pesticides: Selected epidemiological studies and case reports. Am. Assoc. Occup. Health J. 37(3):131-136.

Mott, L., and Snyder, K. 1988. Pesticide alert. Org. Gard. 35(6):70-78.

Moye, H. A. 1991. The Office of Technology Assessment report on pesticide residue methodology for foods. Pages 78-86 in: Pesticide Residues and Food Safety: A Harvest of Viewpoints. B. G. Tweedy, H. J. Dishburger, L. G. Ballantine, and J. McCarthy, eds. American Chemical Society, Washington, DC.

Murkerjee, S. K. 1982. Agrochemicals in India. Pages 3-22 in: Agrochemicals: Fate in Food and the Environment. International Atomic Energy Agency, Vienna.

Murphy, S. D. 1986. Toxic effects of pesticides. Pages 519-581 in: Casarett and Doull's

Toxicology, 3rd ed. C. D. Klaassen, M. O. Amdur, and J. Doull, eds. Macmillan, New York.

Myers, J. P., and Colborn, T. 1991a. Blundering questions, weak answers lead to poor pesticide policies. Chem. Eng. News 69(1):40-43.

Myers, J. P., and Colborn, T. 1991b. Immense unknowns characterize the assessment of pesticide risks. Chem. Eng. News 69(1):53-54.

National Academy of Sciences, National Research Council (NAS/NRC). 1986. Pesticide Resistance: Strategies and Tactics for Management. National Academy Press, Washington, DC.

National Academy of Sciences, National Research Council (NAS/NRC). 1987. Regulating Pesticides in Food: The Delaney Paradox. National Academy Press, Washington, DC.

Parham, W. C., Shen, S., Moye, H. A., Ruby, A., Olson, L. 1988. Pesticide Residues in Food: Technologies for Detection. OTA-F-398. Office of Technology Assessment, Washington, DC.

Pennington, J. A. T., and Gunderson, E. L. 1987. J. Assoc. Off. Anal. Chem. 70:772-782.

Perkins, J. H. 1985. Naturally occurring pesticides and the pesticide crisis, 1945-1980. Pages 297-325 in: Handbook of Natural Pesticides. CRC Press, Boca Raton, FL.

Petersen, B., and Chaisson, C. 1988. Pesticides and residues in food. Food Technol. 44(7):59-64.

Pike, D. R. 1989. 1988 Major Crop Pesticide Use Survey. University of Illinois, Champagne-Urbana.

Pimentel, D. 1985. Pests and their control. Pages 3-19 in: Handbook of Natural Pesticides. N. B. Mandava, ed. CRC Press, Boca Raton, FL.

Pimentel, D. 1987a. Pesticides: Energy use in chemical agriculture. Pages 157-175 in: Towards a Second Green Revolution. G. B. Marini-Bettòlo, ed. Elsevier, New York.

Pimentel, D. 1987b. Is Silent Spring behind us? Pages 175-187 in: Silent Spring Revisited. G. J. Marco, R. M. Hollingworth, and W. Durham, eds. American Chemical Society, Washington, DC.

Pimentel, D. 1990. Introduction. Pages 3-11 in: CRC Handbook of Pest Management in Agriculture, 2nd ed. Vol. 1. D. Pimentel and A. A. Hanson, eds. CRC Press, Boca Raton, FL.

Pimentel, D., and Levitan, L. 1986. Pesticides: Amount applied and amounts reaching pests. Bioscience 36:86-91.

Pimentel, D., Andow, D., Dyson-Hudson, R., Gallahan, D., Jacobson, S., Irish, M., Kroop, S., Moss, A., Schreiner, I., Shephard, M., Thompson, T., and Vinzant, B. 1990. Environmental and social costs of pesticides: A preliminary assessment. Pages 721-743 in: CRC Handbook of Pest Management, 2nd ed. Vol. 1. D. Pimentel and A. A. Hanson, eds. CRC Press, Boca Raton, FL.

Pimentel, D. 1991. Pesticides and world food supply. Chem. Br. 27:646-647 ff.

Reganold, J. P., Papendick, R. T., and Parr, J. F. 1990. Sustainable agriculture. Sci. Am. 262(6):112-119.

Reif, J. S., Pearce, N., and Fraser, J. 1989. Occupational risks for brain cancer: A New Zealand Cancer Registry-based study. J. Occup. Med. 31:863-867.

Sanders, H. J. 1981. Herbicides. Chem. Eng. News 59(Aug. 3):20-24 ff.

Shank, F. R., Carson, K. L., and Willis, C. A. 1989. Evolving food safety. Pages 296-307 in: Carcinogenicity and Pesticides: Principles, Issues, and Relationships. N. N. Ragsdale and R. E. Menzer, eds. American Chemical Society, Washington, DC.

Sharp, D. S., and Eskenazi, B. 1986. Delayed health hazards of pesticide exposure. Annu. Rev. Public Health 7:441-471.

Smith K. 1990. Alar: One Year Later. American Council on Science and Health, New York.

Smith, L. L., and Elcombe, C. R. 1989. Mechanistic studies: Their role in the toxicological evaluation of pesticides. Food Addit. Contam. 6(Suppl. 1):S57-S65.

Stark, A. D., Chang, H.-G., Fitzgerald, E. F, Riccardi, K., and Stone, R. R. 1990. A retrospective cohort study of cancer incidence among New York State Farm Bureau members. Arch. Environ. Health. 45(3):155-162.

Szmedra, P. T. 1990. Pesticide use in agriculture. Pages 649-677 in: CRC Handbook of Pest Management, 2nd ed. Vol. 1. D. Pimentel and A. A. Hanson, eds. CRC Press, Boca Raton, FL.

Taylor, J. R. 1984. Neurotoxicity of certain environmental substances. Clin. Lab. Med. 4:489-497.

Thonney, P. F., and Bisogni, C. A. 1989. Residues of agricultural chemicals on fruits and vegetables: Pesticide use and regulatory issues. Nutr. Today 24(6):6-12.

Tomerlin, J. R., and Engler, R. 1991. Estimation of dietary exposure to pesticides using the dietary risk evaluation system. Pages 192-201 in: Pesticide Residues and Food Safety: A Harvest of Viewpoints. B. G. Tweedy, H. J. Dishburger, L. G. Ballantine, and J. McCarthy, eds. American Chemical Society, Washington, DC.

Tschirley, F. H. 1986. Dioxin. Sci. Am. 254(2):29-45.

Turnball, G. J. 1984. Pesticide residues in food—A toxicological view. J. R. Soc. Med. 77:932-935.

United Fresh Fruit and Vegetable Association. 1990. As Fresh Insights goes to press the Environmental Protection Agency has just proposed further restriction on the use of EBDC Fungicides. Fresh Insights 6(4):1-2.

Upholt, W. M. 1985. The regulation of pesticides. Pages 273-296 in: Handbook of Natural Pesticides, Vol. I. B. N. Mandava, ed. CRC Press, Boca Raton, FL.

Vettorazzi, G. 1980. Handbook of International Food Regulatory Toxicology. Vol. 1. Evaluations. SP Medical and Scientific Books, New York.

Weinstein, K. W. 1991. When pesticides go public: Regulating pesticides by media after Alar. Pages 277-283 in: Pesticide Residues and Food safety: A Harvest of viewpoints. B. G. Tweedy, H. J. Dishburger, L. G. Ballantine, and J. McCarthy, eds. American Chemical Society, Washington, DC.

Wells, J. W., and Fong, W. G. 1991. State pesticide regulatory programs and the food safety controversy. Pages 313-323 in: Pesticide Residues and Food Safety: A Harvest of Viewpoints. B. G. Tweedy, H. J. Dishburger, L. G. Ballantine, and J. McCarthy, eds. American Chemical Society, Washington, DC.

Wilkinson, C. F. 1990. Lessons from Alar. Priorities. Winter:29-30.

Williams, G. M. 1981. Epigenetic mechanisms of action of carcinogenic organochlorine pesticides. Pages 45-56 in: The Pesticide Chemist and Modern Toxicology. American Chemical Society, Washington DC.

Williams, G. M., and Weisburger, J. H. 1986. Chemical carcinogens. Pages 99-173 in: Casarett and Doull's Toxicology, 3rd ed. C. D. Klaassen, M. O. Amdur, and J. Doull, eds. Macmillan, New York.

Woo, Y.-T., and Arcos, J. C. 1991. Role of structure-activity relationship analysis in evaluation of pesticides for potential carcinogenicity. Pages 175-200 in: Carcinogenicity and Pesticides: Principles, Issues, and Relationships. N. N. Ragsdale and R. E. Menzer, eds. American Chemical Society, Washington, DC.

Woodwell, G. M. 1981. Toxic substances: Clear science, foggy politics. Pages 5-19 in: Management of Toxic Substances in Our Environment. Ann Arbor Press, Ann Arbor, MI.

Worthing, C. R., and Hance, R. J., eds. 1991. The Pesticide Manual: A World Compendium. British Crop Protection Council, Farnham, Surrey, U. K.

Worthing, C., and Walker, B. 1983. The Pesticide Manual: A World Compendium, 7th ed. The British Crop Protection Council, Croydon.

Young, A. L. 1987. Minimizing the risk associated with pesticide use: An overview. Pages 1-11 in: Pesticides: Minimizing the Risks. N. N. Ragsdale and R. J. Kuhr, eds. American Chemical Society, Washington, DC.

Incidental Contaminants in Food

CHAPTER 14

CONTAMINANTS ENTERING THE FOOD CHAIN IN AGRICULTURE

In addition to the pesticides discussed in the preceding chapter, other contaminants enter the food supply through their use in agriculture. These include antibiotics, growth promoters, and contaminants and components of fertilizer. Also as part of the issue, organic farming and its produce and sustainable agriculture are discussed in this chapter.

Antibiotics

Antibiotics are used worldwide in animal husbandry. Initially, they were used therapeutically in much the same way as in human medicine. That changed with an experiment in 1950, which showed that a tetracycline antibiotic added to the diet of hogs at 20 ppm or less improved the feed efficiency, i.e., the rate of weight gain per unit of feed (Stokstad and Jukes, 1950).

Based on the data showing their strong positive effect, antibiotics were quickly approved by the FDA for use as feed additives. Prophylactic use of antibiotics to control endemic bacterial disease and improve feed efficiency rapidly gained acceptance. A specific antibiotic combination of chlortetracycline, sulfamethazine, and penicillin was adopted by pork producers and poultry growers. Beef feedlot operators saw immediate benefits from antibiotics with the reduction of bacterial diseases and liver abscesses (Gustafson, 1986).

The many benefits and widespread acceptance of antibiotics as feed additives propelled this segment of the pharmaceutical industry into a multibillion dollar industry. An astonishing 40% of all antibiotics produced in this country are for veterinary use. Of those, feed additive uses far surpass all others. Eighty percent of poultry, 75% of swine, 60% of feedlot cattle, and 75% of dairy calves raised or marketed in the United States receive antibiotics at some time during their life span (Gersema and Helling, 1986). Such wide use is a testament to their benefits to producers, but it also poses some health questions that must be addressed. Do subtherapeutic doses of antibiotics foster development of resistant bacteria? If yes, does this endanger human health? Do the risks outweigh the benefits derived from the use of antibiotics in animal feed? This section explores the answers to these and other questions.

Benefits of antibiotics. Antibiotics are added to animal feed at low doses (less than 200 ppm) for two reasons. First, they increase growth rate and improve feed utilization. Second, they reduce mortality and morbidity from subclinical infections and prevent common animal diseases (ACSH, 1985).

Increased growth rates and improved feed utilization have been documented in hundreds of separate experiments involving up to 20,000 hogs. The daily average gain in weight was 6% greater in antibiotic-treated hogs than in nontreated hogs. In some cases, weight gains were as much as 16% greater in the treated hogs. Further, there was an attendant increase in feed efficiency of up to 7%. The experimental data are thought to underestimate farm experience, where the effects may actually be greater than seen in the controlled conditions of the studies. In addition to better weight gain and feed efficiency, antibiotic-treated animals have lower mortality and significantly better reproductive performance than untreated animals (Hays, 1986).

Precisely how antibiotics promote growth and increase feed efficiency is not known. Postulated mechanisms include one or more of the following:

- a direct growth-promoting effect,
- a metabolic effect resulting in the faster building of tissue,
- a nutrient-sparing effect that reduces the animal's dietary requirements,
- a food uptake effect resulting in improved digestion and absorption of nutrients or increased feed or water intake, or
- an antibiotic effect that eliminates subclinical infections or parasites that would compete for nutrients or otherwise impair growth.

Several mechanisms may be working in tandem. Interestingly, measurements of the growth effect show it to be just as effective in 1981 as in 1956, even in the possible presence of resistant bacteria (Hays, 1985).

Antibiotics have a greater effect in unsanitary environments; in fact, antibiotics do not promote growth in germ-free animals. Their use controls the spread of infectious disease in crowded conditions and is responsible for the success of concentrated feedlots (Jukes, 1986). Diseases controlled by antibiotics include dysentery, mycoplasma pneumonia, and infectious atrophic rhinitis in swine; liver abscesses, shipping fever pneumonia, and footrot in cattle; and mycoplasmosis, pastruellosis, and psittacosis in poultry. Currently, these diseases are limited to a few outbreaks. It is predicted that without antibiotics, the frequency of these diseases would dramatically increase (Addison, 1984). Furthermore, human diseases spread by animals could increase without antibiotic use.

Antibiotics are especially useful in high-risk situations, as with young or stressed animals. Veal is especially stress susceptible and benefits from antibiotics. Stress effects such as those induced by severe weather or from shipping animals to market are also controlled by the use of antibiotics (Gustafson, 1986).

Faster growth and less disease have been estimated to save $2.1 billion in feed costs annually. The net effect is increased amounts of animal protein with less time and input, hence lower production costs (Jukes, 1986). The savings to the producer can be directly seen in prices at the meat counter. It is estimated that, without antibiotics, beef prices would increase 2.7-10.4%; pork prices, 4.5-14.7%; and poultry prices, 10.3-27.6% (Gersema and Helling, 1986). The Council for Agricultural Science and Technology estimated the cost to consumers of the ban in animal feeds of just two antibiotics, tetracycline and penicillin, to be as high as $3.5 billion dollars (Hays and Black, 1989). Supplies of low-cost meat, milk, and eggs are certainly an economic benefit to the American consumer as well as to the producer.

Risks from antibiotics. Since the 1950s, concern has been raised about the subtherapeutic use of antibiotics. Over the years, books such as *Modern Meat* (Schell, 1984) have continued to raise the issue. Some people fear that medicinal uses of these wonder drugs might be compromised. Another fear is of the toxicity of the antibiotic itself. It is quite well documented that all antibiotics are capable of producing inherently toxic effects, depending on the dose, the mode of administration, and the time of exposure. However, most people believe that only medical doses precipitate toxicity and that little danger of acute toxicity arises from subtherapeutic levels in feed.

In an effort to minimize human exposure to antibiotics from feed, animal producers are required to follow prescribed withdrawal periods. These are based on tissue clearance times and are set so that antibiotic levels drop below detectable levels. No measurable residues remain in milk or meat if the required drug withdrawal schedule is followed.

Withdrawal times vary with the particular drug, dosage, duration, and species. Variable schedules are confusing to some operators, and less reputable operators may choose to ignore the specified requirements. In either case, operators who fail to follow prescribed protocols may create a potential hazard. Obviously, the greatest hazard occurs when animals are given therapeutic doses rather than subtherapeutic doses in feed.

A survey by the U.S. Food Safety and Inspection Service found that 5.3% of the carcasses tested positive for antibiotic residues (Livingston, 1986). The violation rates for cattle, chicken, and turkey were 0.2% or less, but they were nearly 10% for swine. Since the biological activity of these antibiotic residues is destroyed by cooking, it is questionable whether this lack of compliance has any biological significance (Jukes, 1986), especially since it is widely known that pork and poultry should be well cooked.

In 1989, drug testing of milk identified the presence of sulfamethazine, an antibiotic not approved for use in lactating animals because it is a carcinogen. Although the assay used for the original article (Anonymous, 1989a) was flawed, the net result was a marked increase in surveillance of milk for antibiotic residues (Place, 1990).

The greatest potential for harm from antibiotic residues in meat occurs in people with an allergic sensitivity to certain antibiotics and from the development of antibiotic resistance.

Antibiotic resistance. Overuse of antibiotics both in agriculture and in medicine may contribute to the development of antibiotic-resistant bacteria. Before the widespread use of antibiotics, resistant bacteria occurred only rarely. Recently there has been a rise in the prevalence of antibiotic-resistant organisms. With this rise comes great fear that therapeutic antibiotics will become ineffective for treating human disease. In fact, some people believe that antibiotic resistance is a time bomb waiting to explode. They believe that at some time in the future, all organisms will gain antibiotic resistance and unstoppable epidemics will be caused by the resistant bacteria.

Evidence that certain pathogens have developed antibiotic resistance has already accrued (Concon, 1988; Frappaolo, 1986). For instance, antibiotic-resistant strains were responsible for one third of 52 outbreaks of salmonellosis investigated by the Centers for Disease Control between 1971 and 1983. Resistant strains were the cause of 13 deaths out of 312 affected persons.

Studies in the United Kingdom of over 4,000 food samples also show that antibiotic resistance is on the rise, despite the strict limitation of antibiotics for use in animal feed. One study included tests for the presence of resistant *Escherichia coli* (Pinegard and Crooke, 1985). Overall, 12% of the samples were contaminated, cakes and confectionery (28%) more frequently than meat products (9%). However, of the contaminated meat products, 14% showed multiple antibiotic resistance, as contrasted with only 1% of the confectionery. Raw milk has been shown to contain many resistant bacteria. Since it is illegal to use antibiotics while animals are lactating, resistance probably was transferred into the milk.

A variety of *Salmonella* species with resistance has been noted in human disease. Data from Friend and Shahani (1983) showed that resistance of *S. typhimurium* to ampicillin, streptomycin, and tetracycline nearly doubled or even tripled in a 5-year period.

While the above data do document some increase in antibiotic resistance, not all data point to this. A 12-year study in 242 U.S. hospitals (Gersema and Helling, 1986) showed that resistance patterns of over 5 million bacterial isolates of *E. coli* and *Staphylococcus aureus* have changed very little over the past decade, although two other organisms did show increased resistance. Overall, this study concluded that resistance patterns have changed very little over the past decade. The authors believe that much of the concern over antibiotic resistance has arisen from the over 7,000 papers written in English stating concern about increased resistance rather than from increased resistance itself.

If there *is* increased resistance, Thomas Jukes, emeritus professor at University of California, Davis, believes it is the result of the therapeutic use of antibiotics. He states that tetracycline resistance is due to overzealous medical use, not subtherapeutic amounts used in food (Jukes, 1986). Several authors cite the useless prescription of

---

Box 14.1

Are Organisms Becoming Antibiotic-Resistant?

Often cited as proof of a relationship between subtherapeutic antibiotic use and human disease is an incident in 1982–1983 involving *Salmonella newport* in ground beef (Holmberg et al, 1984a, 1984b). The accused hamburger in the incident originated from a beef cattle herd fed subtherapeutic chlortetracycline for growth promotion. Eighteen persons from four midwestern states became infected with *S. newport* that was resistant to several antibiotics. Of the 18, 11 were hospitalized and one died. Some of the affected patients had taken antibiotics 24–48 hours before the onset of salmonellosis. Critics of these studies discount meat as the major way that the organism was transmitted. A rigorous sampling of frozen meat from the same farm failed to show the presence of *S. newport* in frozen meat samples. As further evidence, this farm had provided nearly 40,000 pounds of beef that year with no other reported illnesses. It is now thought that a therapeutically treated dairy herd adjacent to the beef cattle location may have been a source of the antibiotic-resistant organism.

antibiotics for viral infections as a major cause of antibiotic resistance. Jukes (1986) states that highly trained doctors abused antibiotics as much as or more than farm workers who were not as well trained. Jukes also cites evidence from England and other countries in which the banning of antibiotics in low doses in animal foods has had little measurable effect on the prevalence of antibiotic resistance.

He also counters a March 1985 *Consumer Reports* article about antibiotic resistance, which claimed that hundreds and thousands of cases exist. He cites a CDC study that showed a total of 55 salmonella outbreaks involving 3,653 people, or an average of 281 cases per year. He then compares those 281 cases per year with the average annual consumption of beef hamburger of 4.2 billion pounds and concludes that the problem is very minor indeed.

He further states that antibiotic-resistant strains are less virulent that other strains. For instance, the Chicago milk salmonella incident in the spring of 1985, the largest outbreak in U.S. history, was of an antibiotic-resistant strain. This incident affected over 14,000 people, yet it resulted in a death rate far below what critics of antibiotics would have predicted. The rate predicted is between 4 and 5%.

Another serious concern is that plasmids may be able to transfer resistance from nonpathogens to pathogens (Levy, 1986; Ziv, 1986). Some data show that routine contact with either antibiotic-treated feed or animals on the farm may allow resistance to be transferred to humans who live and work on the farm. Levy showed that three of eight farm dwellers not taking any antibiotics began excreting tetracycline-resistant bacteria within 4 months after chickens raised on this farm received tetracycline-supplemented feed. There also appear to be other complex patterns of resistance transfer from manure to plants to humans or from carcasses in slaughterhouses to the persons handling them. Thus, the prospect of resistance being transferred to human pathogens is a very real one (Linton, 1986).

Antibiotic use may also result in some minor problems. One is that antibiotics in animal feed may prolong the shedding of organisms such as *Salmonella*, thereby increasing the chance of infection for those working with the animals. Also, if the producer does not adhere to the regulations regarding drug withdrawal, milk used to make cheese and yogurt will have enough residual antibiotic to alter the microbial fermentation needed for a successful product (Shahani and Whalen, 1986). It is also possible that antibiotics in runoff from feedlots could contaminate groundwater (Addison, 1984).

The controversy. Many committees and commissions have evaluated the safety of antibiotic use. In 1960, the British Netherthorpe Committee determined that current practices were not detrimental to the treatment of animals or humans. They, therefore, recommended no change in policy but warned that the use of antibiotics needs continued close observation.

Five years later in England, seven human deaths and thousands of animal deaths were attributed to the use of antibiotics. This incident led to the formation in 1968 of the Swann Committee. After 1 year's study, this committee recommended that antibiotics be divided into two classes—"feed" and "therapeutic." It further recommended that antibiotics used in human medicine be banned for animal feed use.

Two regulations on antibiotics in the United Kingdom resulted. First, therapeutic antibiotics for animal use require the prescription of a licensed veterinarian. Second, only three antibiotics (bacitracin, virginiamycin, and bambermycin) are available without a prescription for feeding of young poultry, pigs, and calves. It is interesting that since the implementation of this law, there has been no change in the amount of resistant bacteria in the United Kingdom.

Worldwide concern about antibiotics also led to the formation of other committees to study their safety, including the U.S. Task Force on the Use of Antibiotics in Animal Feeds. This group issued a statement in 1972 saying that antibiotic resistance in humans had increased and that animals receiving antibiotics might be a source of antibiotic-resistant organisms. The published statement did not reflect the marked dissension that occurred among the members. Only seven of the 15 members endorsed the report (Friend and Shahani, 1983; Livingston, 1986).

The Commissioner of the FDA decided that additional review was necessary and set up another advisory group. It recommended the complete withdrawal of penicillin for promoting growth, increasing feed efficiency, and preventing disease in feed animals. Further, it suggested that tetracycline use be controlled by prescription. Much protest was voiced, and the manufacturers requested a hearing. After several hearings were held, the proposed withdrawal of antibiotics was again recommended. In addition, one hearing concluded that the low-level use of antibiotics posed greater health risks than did the synthetic estrogen feed additive diethylstilbestrol (Schell, 1984).

Congress intervened and rescinded the order to withdraw antibiotics, citing lack of good epidemiological data to indicate that feeding antibiotics to feed animals resulted in the transfer of antibiotic resistance from animals to humans. It should be noted that the budgets of the FDA and USDA are controlled by agriculture committees of Congress and that these committees have many representatives sensitive to the needs of the farm states (Frappaolo, 1986).

In 1980, concern about the safe use of antibiotics had not diminished, so the National Academy of Sciences (NAS) was asked to look at the question of safety. Its report (NAS/Institute of Medicine, 1989) concluded that

> the postulations concerning the hazards to human health from the addition of subtherapeutic antimicrobials to feeds remain neither proven nor disproven. The lack of data linking human illness with subtherapeutic levels of antimicrobials must not be equated with proof that the proposed hazards do not exist. The necessary research has not been done and may not be possible.

Bills have been introduced in Congress that would disallow any antibiotics used therapeutically for humans. Only those antibiotics shown not to increase antibiotic resistance of pathogenic bacteria in humans would be allowed at subtherapeutic doses (Gersema and Helling, 1986). So far, none of these measures has been enacted.

The World Health Organization has also called for a cut in the number of antibiotics cleared for agricultural use. It recommends that countries limit new antibiotics used to treat serious infections to use in medicine. The report (JECFA, 1988) stated that routine use of antibiotics as prophylaxis in the absence of proven infection should never be used as a substitute for good hygiene.

Despite 40 years of extensive use of antibiotics, it is difficult to cite human health problems that can be attributed specifically to meat from animals fed antibiotics or that can be associated with direct or indirect contact with animals fed low levels of antibiotics. The evidence is sparse, indirect, and difficult to evaluate. A risk assessment study done for the FDA by the NAS Institute of Medicine was "unable to find data directly implicating subtherapeutic doses of antibiotics in livestock with illnesses in people" (NAS/Institute of Medicine, 1989). Nevertheless, this seems like an issue on which we ought to err on the conservative side (Levy, 1987). We simply cannot afford to lose the therapeutic effect of the wonder drugs.

It may be judicious for us to adopt the policy of the European Community (EC) and allow in feed only those antibiotics not used in human medicine. In the United States, 21 antibiotics and other antimicrobials including tetracyclines and penicillin are approved for animal feed use; in the EC, only seven are allowed in feed. Less reliance on antibiotics could be achieved by better animal management practices, including purchase of animals from reliable sources, the separation of new and old animal stock, and the construction of animal housing units designed for warmth, ventilation, and lower population density. Further research on methods to minimize antibiotic use would be helpful.

Tighter controls are also needed to ensure that residues are not present in food. We need controls so that meat does not contain antibiotic residues through failure to observe the required time protocols before marketing the animal, through mistakes such as failure to mark treated cows in a large herd, or through lack of information or misunderstanding regarding withdrawal periods. Persons using these drugs must be clearly instructed so that residue levels in food are well below the tolerances.

Adherence to the tolerances is crucial because of hypersensitivities and allergic reactions. Patients sensitized through therapeutic use of an antibiotic can have an allergic reaction with as little as 0.024 milligram of penicillin. For normal individuals, adherence is crucial so that there are no overt toxic effects or disturbance of normal gut flora.

Wonder drugs have a place in modern agriculture because of increased feed efficiency, reduced suffering among animals, and decreased costs at the meat counter. However, their use must be limited and under tighter control. The benefits of eliminating disease-causing microorganisms and parasites from animal products and the other benefits of antibiotics must be balanced against their possible risk (Hays and Black, 1989).

Growth Promoters

The most notorious feed additive is the growth promoter, diethylstilbestrol (DES). This synthetic, estrogenlike hormone was used during and after the 1950s to promote faster weight gain in cattle and sheep. It was used not only in animal feed, but also as an ear implant. Such hormones are estimated to improve weight gain by 5–20%, feed efficiency by 5–12%, and lean meat growth by 15–25% (Egerstrom, 1989). The benefit of DES is that it saves several billion pounds of feed per year, which translates into millions of acres of land.

As with the use of antibiotics, a strict protocol was required for the removal of DES before animals were sent to market. However, DES residues were found in a small number of carcasses, prompting the banning of this substance as a feed additive. It is still used in the United States as an ear implant.

The concern about DES resulted initially not from feed use of DES but from ill effects that developed during its medical use. In the 1940s and 1950s, massive doses of DES were given to pregnant women to prevent miscarriage. About 20 years later, the young adult daughters of these women began to develop vaginal cancers (Herbst et al, 1971, 1972).

Fear about the use as a feed additive resulted even though the residues found in animals where the protocol was not followed were thousands

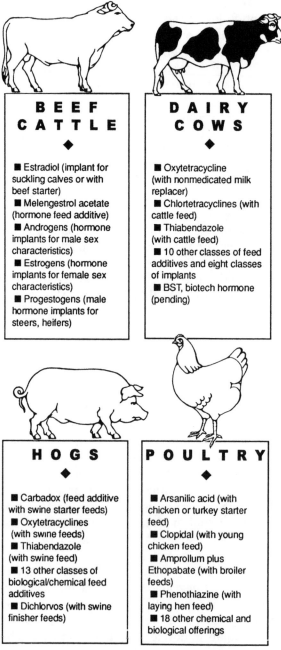

Figure 14.1. A modern menu for farm animals. (Adapted with permission, from Egerstrom, 1989; copyright St. Paul Pioneer Press. Based on data from the University of Minnesota Experiment Station).

of times smaller than the dose taken to prevent miscarriage. For example, even when the protocol is not followed, DES residue is not found in the meat itself. It is found only in the animal's liver at 5–10 ppb. One would have to eat 220 pounds of liver per day to get 1 milligram of DES. This dose is insignificant compared with the daily dose of 65 milligrams given to the women whose daughters have problems (Minnesota Agricultural Extension Service, 1972) and with natural estrogens in plant products such as wheat germ (4,000 ppb), wheat bran (1,500 ppb), soybean oil (2,200 ppb), and peanut oil (1,800 ppb) (Best, 1989). Some people point out that plant estrogens have less potency than DES. Nevertheless, the levels of DES in meat are extrordinarily small in comparison to hormone levels found elsewhere. Levels may be higher in some animals that are consumed, such as bulls or ovulating cows, and they are 5,000 times higher in women taking a typical birth control pill.

Many feel that DES is a case in which the risks from ingestion at the parts-per-billion level were overestimated and confused with risks of ingestion of large amounts, while the benefits were not adequately communicated. Others believe that the benefits are not substantial enough to justify any risk, especially in countries where protein amounts are adequate to excessive (Quattrucci, 1987). The WHO/FAO Joint Expert Committee on Food Additives (JECFA, 1988) has reviewed a variety of growth-promoting hormones, but despite its finding that there was little risk, the EC has banned the import of meat from hormone-treated animals (Anonymous, 1989c).

Fertilizers

All plants require fertile soil to grow. Soil fertility may be decreased after several years of growing, but fertility must be maintained if crop yields and produce size are to be maintained or optimized.

Sixteen nutrients are essential for plant life, and 13 of them come from the soil (Box 14.2). Nutrients are classed as primary or secondary nutrients or as micronutrients. Among the essential elements, those that are primary nutrients are required in relatively large amounts and are often present in inadequate amounts. Most commercial fertilizers contain one or more of these. Elements that are secondary nutrients are essential but seldom limit plant growth, especially in limed soils. Micronutrient elements are required in very small amounts.

Fertility can be restored and maintained by the addition of commercial fertilizers or through the use of "organic" fertilizers such as compost, humus, and manure. With either class of fertilizer, basic free elements such as nitrogen and phosphorus are absorbed by the plant. With compost, the free elements may have to be released by the decaying matter in order for the plant to absorb them. The advantage of "organic" fertilizers is that they help build good soil characteristics such as tilth, aeration, and holding capacity for water and nutrients. These advantages are especially important in very sandy or heavy clay soils.

"Green manure" is used in certain cases where crops are specifically grown to be plowed back into the soil. Crop rotation is another way to help restore fertility to the soil, especially if the plants in the rotation are those that fix nitrogen, such as legumes. Intercropping helps prevent erosion (Hileman, 1990)

Despite many claims as to the superiority of organic produce, it has been repeatedly shown that there is no difference in either the macro- or micronutrient content between plants and vegetables grown organically or with fertilizer. For instance, a 25-year study conducted at the U.S. Soil and Nutrition Laboratory in Ithaca, New York, shows that the vitamin and mineral contents of rye and potatoes were the same whether fertilized with manure or chemical fertilizer (Anonymous, 1983). Similar results were seen in a 34-year study at an experimental farm in England (Alther, 1972), a 10-year study by Michigan State on its experimental farm, and also a study by the French Ministry of Health (Newsome, 1990).

These results are not surprising. The nutrient content of a plant is more the product of its genetics, the amount of sunlight it receives and other climatic effects, its maturity when picked, and the conditions of shipping and storage than the effect of the type of fertilizer, assuming that the correct level of nutrients was present in either case. A letter to the *Journal of the American Dietetic Association* cited nutritional differences in organically grown crops such as less total protein but protein of a higher quality. However, the author was quick to point out that the differences would not be important nutritionally (Lafler, 1990).

Nevertheless, some measurable differences occur in produce raised

---

Box 14.2

Soil-Derived Elements Necessary for Plant Growth

Primary nutrients: nitrogen, phosphorous, potassium

Secondary nutrients: calcium, magnesium, sulfur

Micronutrients: boron, chlorine, copper, manganese, molybdenum, iron, zinc

by organic rather than regular methods. Nitrate levels in produce were less if vegetables were raised with compost rather than with chemical fertilizer or blood meal. For example, compost-raised leeks absorbed only half the nitrate of chemically fertilized plants. Vegetables fertilized with wood chips absorbed the fewest nitrates. For example, wood-chip-fertilized turnips had only one tenth the nitrate content of plants raised on blood meal. Nitrate loads on the vegetable are related to the rate of nitrogen release. Since compost must be broken down by bacteria before the nitrogen is available to the plant's root system, the uptake of nitrate is slowed.

In one study, nitrogen fertilizer applied to forest blueberries caused the formation of aluminum nitrate, which was taken up by the berries (Nilsson and Lindahl, 1986). Three hens fed these berries died, and some hens laid eggs with defective shells. Although autopsy did not show any signs of macroscopic organ injury, the authors suggested that aluminum from the solubilized aluminum nitrate was the cause of their findings. They pointed out that the level of berries the hens received would be the equivalent to a human eating 35 liters of berries. Further studies are needed to substantiate these results and to determine the effects of smaller quantities of aluminum nitrate.

Concerns about the use of nitrate fertilizer stem not only from their effects on food but also from their effects on workers and on groundwater, although men occupationally exposed to nitrates at a fertilizer plant in England did not have any higher gastric or other cancer incidence than those not so exposed. Furthermore, death rates from other causes were not higher despite high exposures to nitrate, and there were no differences in fertility rates (Al-Dabbagh et al, 1986; Pocock, 1985).

All nitrogen fertilizer, whether "organic" or chemical, increases nitrate in groundwater. Nitrate levels are both an environmental concern and a concern to human health. As discussed in Chapter 10, nitrates can form nitrosamines, which are powerful gastric carcinogens. However, data assessing groundwater nitrate levels do not help clarify the risk posed by nitrate in the groundwater. In parts of the United Kingdom, groundwater nitrate levels were negatively associated with gastric cancer, whereas in Chile they were positively associated. The nitrate problem needs to be addressed in a way that systematically looks at the effects of all nitrate sources on human health and the environment (CAST, 1986).

Another concern about commercial fertilizers is their level of cadmium. Cadmium is a natural component of phosphate rock. It is truly a matter of concern because it belongs to the family of toxic heavy metals. Like other heavy metals, it is not readily eliminated and therefore accumulates in organisms over time. Cadmium raises blood pressure; interferes with the metabolism of other divalent metals such as calcium, copper, and zinc; and has been related to increased risks of hypertension, cancer, and coronary heart disease.

The amount of cadmium in phosphate fertilizers is dependent on the amount naturally present in the phosphate deposits. Phosphates from Finland, the Soviet Union, and South Africa have low cadmium levels (less than 3 milligrams per kilogram). In the United States, phosphate rock has 20–120 milligrams of cadmium per kilogram. In some countries in Africa, the levels are up to 280 milligrams per kilogram.

There is fear that levels of cadmium in some European countries are already above the limit recommended by WHO (Box 14.3). Although cadmium comes from many environmental sources, one way to reduce the cadmium load is to remove it from phosphate rock, increasing the price of the fertilizer (Layman, 1988). Scandinavian countries and Switzerland have set limits on the cadmium levels allowed in fertilizer. For instance, the Finnish government has set the cadmium level at 30 ppm. The EC may also limit cadmium. Although such reduction appears commendable, there is a question whether it will actually reduce the cadmium contamination of plant foods, as the phosphate in fertilizer already retards absorption of cadmium. Cadmium contamination of foodstuffs appears to correlate well with fields being in close proximity to well-traveled highways and with the plant's capacity to accumulate cadmium (Concon, 1988).

The risks from organically grown produce rest in three areas. First is the increased risk of biological and microbial contamination. Second is the increased risk of contamination by pollutants that are bioconcentrated. The third risk is the difficulty of achieving adequate yields on a widespread basis and without increased cost if organic methods totally replace chemical fertilizers.

In terms of microbial problems, we have already discussed the *Listeria* contamination of cabbage fertilized with sheep manure. Salmonellosis, roundworms, and other microbial diseases passed through manure or raw sewage have also been reported. Raw sewage is a potential carrier of pathogenic organisms, and its use is never recommended.

Processed sewage sludge does not cause any microbial problems because the processes kill the pathogens. Digested sewage sludge is sludge that has been anaerobically decomposed, in which process most pathogenic organisms are destroyed. Activated sludge is sewage that has been aerobically decomposed, dried, and ground. It has a relatively high nitrogen content and is free of pathogenic organisms.

Even though sludges are are free of pathogens, they are not recommended for root crops and vegetables that are in contact with the soil.

Sludge can be a source of heavy metals and other toxicants that are bioconcentrated and, in fact, sludge has been found to be a source of copper, zinc, lead, nickel, cadmium, chromium, and mercury (Komsta-

---

Box 14.3

WHO/FAO Limits for Trace and Toxic Metal Components and Contaminants of Food

Levels of safe intake, called provisional tolerable weekly intake (PTWI) levels, have been set for food components or contaminants with cumulative properties. These include heavy metals like cadmium, lead, and mercury.

	PTWI	
Metal	(milligrams per person)	(milligrams per kilogram of body weight)
Cadmium	0.5	0.008
Mercury	0.3	0.005
Lead	3.0	0.003

Components and contaminants without cumulative properties are divided into three classifications. Those that are essential but also toxic (e.g., iron, arsenic, and tin) are given provisional maximum tolerable daily intake (PMTDI) levels. Since they are essential, a minimum level may also be set if data are adequate for the determination. Those that are not essential and are toxic also have a PMTDI. A third class is for those that are essential and for which the maximum dose is known. These have a maximum tolerated daily intake (MTDI) level. An example in this class is phosphorus; it is an essential nutrient, and large amounts of it might enter the food supply. The values given are for nutritionally adequate diets; the levels might be more toxic if, for instance, the body is deficient in calcium.

Metal	PMTDI (milligrams per kilogram of body weight)
Arsenic	0.002
Copper	0.5
Tin	2.0
Zinc	1.0

(Adapted from Vettorazzi, 1987)

Szumska, 1985). Cadmium and mercury are toxic at all levels, and the other trace minerals are toxic at high levels. Toxicants like polychlorinated biphenyls (PCBs), which also might be bioconcentrated in sludge, are not taken up by the plant root system, so they are not found in plants grown with sludge containing PCBs. However, PCBs and other chlorinated hydrocarbons in the soil can sometimes be ingested along with the plant and thereby increase animal exposure (Biehl and Buck, 1987). Other lesser concerns about the use of sludge stem from its odor, which may not make its use desirable at all locations, and its slow nitrogen release.

Thus, organic and inorganic fertilizers can be sources of heavy metals. In both instances, the human hazards must be managed. Economically feasible ways should be sought to remove heavy metals and other contaminants from both sludge and chemical fertilizer.

ORGANIC PRODUCE

Yields with Organic Methods

The picture of what happens to yields and profitability with totally organic methods is not consistent. CAST (1980) reported the yield of most major grain crops to be reduced 50% with organic methods. A small sample of Australian farmers (36) was asked about their experience with organic methods (Conacher and Conacher, 1982). Most felt that the health of animals had declined. Fourteen said it was too early to tell whether there were any effects on yield or on profitability with respect to yields; seven said they were higher, 10 lower, three the same. Those who had a positive experience listed as reasons little or no soil erosion, ecologically balanced agriculture, satisfactions with a safer environment, and/or that they were happy in the belief in that they were producing safer food.

These farmers listed their major problems as weeds. Interestingly, the neighbors of the farmers who switched to organic farming had fairly negative or mixed attitudes in regard to these practices.

In a British study, farmers using organic or mixed methods reported easier and greater soil tillability and milk yields (Alther, 1972). For cereal crops, yields were greater with chemically fertilized or mixed farming methods, but no clear pattern emerged for other crops. An Iowa study showed 50% less erosion using organic methods (Jukes, 1977).

Reasons given for a switch to sustainable methods included less loss of top soil, fewer energy-intensive methods, reduced reliance on pesticides and fertilizer, less pesticide resistance from years of pesticide use, and reduced need for expensive machinery (Hileman, 1990).

An article in *Scientific American* in 1990 showed that sustainable methods resulted in lower yields but about the same profitability, as the input costs were less (Reganold et al, 1990). Furthermore, it should be pointed out that, although the yields of the organic farms were less than those of other fields in the study, they were not less than the average for the region studied. CAST (1990) and NAS/NRC (1989) have published comprehensive reviews of so-called alternative agriculture.

The Consumer and Organic Produce

After many scares—real or media-produced—some consumers desire organic food, but the purchase of organic food introduces several consumer concerns. One is price. Many consumers said they were willing to pay more for organic produce (Hay, 1989), but the question is, how much more? Organic food can cost up to 300% more and averages 20–50% more than its nonorganic counterpart (Anonymous, 1989b). However, higher price is no guarantee of higher quality. In some cases, the produce is of lower quality and has cosmetic damage (Anonymous, 1991b).

Even more disheartening is the fact that paying a premium price for organic produce does not guarantee that the food is pesticide-free. Some unscrupulous dealers will state that their produce is pesticide-free in order to exact a higher price. Several states have instituted legislation to define the use of the term "organic." The difficulty is that there is no good way to determine whether the produce is organic, which is why the U.S. government has been slow to define the term legally. For instance, an item stated to be organically grown may have more pesticide residues than its regular supermarket counterpart. This may be due to pesticide drift and pesticide left in the soil, or it may be due to sellers cashing in on higher margins in the "organic" section by selling regular produce as organic (Simco and Jarosz, 1990).

Some associations of organic food growers have tried to deal with the problem by setting up a certification program to inspect farms and fields to ascertain that the produce so grown is organic.

The trend toward "organic" seems to be growing. Nearly one quarter of the respondents in a California survey indicated that they would look for organic foods when shopping, and slightly over half considered them better than other foods (Jolly et al, 1989). Organic produce enthusiasts claim that organically grown produce tastes better; however, studies by the Good Housekeeping Institute failed to substantiate this notion (Sadler, 1990).

Each person must decide whether organic foods pose a truly lower risk or whether they simply introduce some different risks, such as increased microbial or heavy metal contamination from the manure. Then the decision must be made as to whether the additional cost is worth the reduced risk of pesticide contamination (if present) for both the food and the environment.

POLLUTANTS AS INCIDENTAL ADDITIVES

In addition to unintentional additives that enter the food supply as the direct or indirect result of agriculture, industrial pollutants have made their way into the food supply. These include substances like heavy metals, PCBs, and other contaminants from packaging. This last section deals with some of these.

Polychlorinated Biphenyls

PCBs were discovered around the turn of the century. Their property of extreme stability and nonflammability made them valuable in electrical transformers and capacitors and other electronic parts (JECFA, 1990). They were also used as vehicles for pesticide and paints, suspension agents for carbonless paper, and flame retardants. About 8 million pounds of PCBs were produced in the United States before their production was curtailed in 1974. Much of this was discarded with no thought about environmental impact. In fact, the extreme chemical stability of these compounds caused many people to assume that they must be of low toxicity and of little consequence to the environment.

This assumption was shown to be wrong by several incidents in the 1960s. In 1966, PCB contamination of wildlife was first discovered. Two years later in Japan, a poisoning outbreak affected 1,000 persons and caused stillbirths and problems to infants whose mothers ate PCB-contaminated rice oil. The poisoning was given the name Yusho disease. During the search for a cause, it was learned that PCBs leaked out of the machinery during the heating of rice oil and contaminated it (Kuratsune, 1989). Since that initial incident, many investigators have found PCB residues in human populations, even those not occupationally exposed (Anderson, 1989; Jensen, 1989; Kimbrough and Grandjean, 1989).

Concern about the toxicity of PCBs caused their production in the United States to cease in 1974. Even though no new PCBs are produced, many are still in use in old transformers (Kimbrough, 1987). However, the extreme stability of these compounds means that, even though no new sources are being added, the supply is not being degraded. In fact, this compound has a half-life approaching infinity! They are not degraded in the environment nor are they completely metabolized or excreted. Instead, they come to a steady state, bioconcentrate in the food chain, and produce low-level contamination of humans (Jensen, 1989; Rogan et al, 1986). However, there may be a way to rid the environment of PCBs with quicklime (Anonymous, 1991a).

PCBs and food. The tragic rice oil incident awakened the food industry and consumers to the idea that PCBs could contaminate food. Several sources of PCB contamination of food have been located. One was through the use of recycled paper (which included PCB-containing carbonless paper) to make cartons for food packaging. PCBs can migrate from the packaging into the food. Another was through PCB-containing coatings for silo interiors. Silage fed to cattle from these coated silos caused PCB contamination of the milk and meat. Both these sources of contamination have been virtually eliminated (Kreiss, 1985).

In the U.S. population without occupational exposure to PCBs, the average blood levels are 4–8 milligrams per milliliter. In fact, 95% of the population has levels less than 20 milligrams per milliliter. Current human exposure to PCBs in food is primarily through fish (Anderson, 1989; Belton et al, 1986) and breast milk (Jensen, 1983). Thus, sports fishermen and others consuming fish from Lake Michigan, parts of the Mississippi and Hudson Rivers, Newark Bay, and certain lakes in Alabama had serum PCB levels much greater than the average.

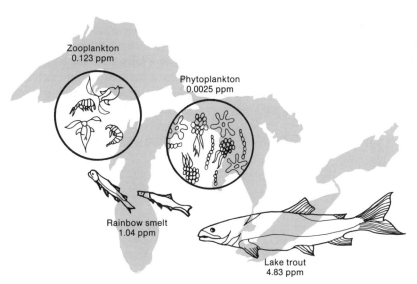

Figure 14.2. Biomagnification of polychlorinated biphenyls (PCBs) in the aquatic food chain of the Great Lakes. As phytoplankton take nutrients necessary for plant growth from the water, they also collect minute amounts of PCBs. The zooplankton that feed on them and the small fish that feed on zooplankton concentrate the chemicals. Large fish may accumulate quantities high enough to cause their death. (Adapted, with permission, from Hileman, 1988. ©1988 American Chemical Society)

PCBs in fish appear to be related to the size of the fish, with larger fish having higher levels. The PCB content of fish and other food is reduced by cooking (Table 14.1).

Although it is clear that PCBs in game fish can be a concern, the PCB levels of fish that are of greatest concern are those in the edible portion. Many media reports give values for the fish that including parts not customarily eaten, such as the skin and internal organs. While the values are accurate, they overestimate the amounts that would be ingested (Armbruster et al, 1987), especially because cooking may also decrease PCBs.

Health effects of PCBs. Acute exposure to PCBs causes chloracne with dark brown pigmentation; itching; skin rash; burning of eyes, nose, and throat; upper respiratory irritation; nausea; weakness; and dizziness. In the Yusho incident, the fetus was very adversely affected. There were many stillbirths, and many pregnancies were difficult to carry to term. Babies that were born had the PCB syndrome, which causes dark brown pigmentation of the skin.

One study followed 117 children born between 1978 and 1985 to women exposed to contaminated rice oil in Taiwan in 1978. In addition to having some of the signs of PCB poisoning, including lower birth weight, swollen eyelids, and deformed fingernails, all the children showed some degree of developmental delay (Rogan et al, 1986).

Breast feeding poses a particular risk because PCBs stored in the mother's fat may be transferred to fat in the breast milk. Infants who nursed on breast milk contaminated with PCBs had problems with decayed nails and teeth. Other toxicities have been reported in children exposed to high levels of PCBs in breast milk, including motor immaturity in infants and adverse effects on visual memory (Foran et al, 1989). However, the effects of PCBs cannot be separated from those of other contaminants to which these children were exposed.

Table 14.1. Loss of Polychlorinated Biphenyls (PCBs) During Cooking of Carp by Various Methods[a]

Cooking Method	PCB (micrograms)	Percent change in PCB
Charbroiled		
Raw	330	−93
Cooked	23.7	
Poached		
Raw	346	−94
Cooked	20.2	
Deep fried		
Raw	304	+21
Cooked	414	

[a] Data from Zabik et al (1982).

The FAO/WHO Joint Expert Committee on Food Additives has stated that the benefits of breast feeding outweigh any risks (JECFA, 1990).

Chronic exposure of various experimental animals to PCBs may increase their cancer risk. The actual numbers of tumors and specific effects are dependent on the isomer of PCB fed and the animal species studied. Subhuman primates, the guinea pig, and the mink are more sensitive to PCBs than the rat or the mouse. Humans also appear to be less sensitive to the toxic effects of PCBs than some species. The species difference is thought to be related to ability to store adipose tissue. Subhuman primates, guinea pigs, and mink have less adipose tissue than humans. The ability to store vitamin A may also have an effect. Higher storage levels of vitamin A protect the animal from adverse effects of PCBs.

Epidemiological data seem to bear out the theory that humans are less susceptible to cancers from PCBs than are other species. Studies on workers exposed to PCBs at two capacitor plants in the United States (2,567 persons) showed an all-cause mortality and cancer mortality lower than expected (Kreiss, 1985). From data such as these and additional data from the National Cancer Institute, the FDA concluded that lifetime consumption of fish contaminated at the FDA maximum-allowable level of 2 ppm would result in 7.2 cancers per 100,000 persons. The Environmental Protection Agency also calculated this risk, using data from the Centers for Disease Control and the assumption of a lifetime daily consumption of 38.6 grams (approximately 1½ ounces) of fish contaminated with 2 ppm PCBs, and predicted a larger number of cancers, 450 per 100,000.

Despite the discrepancy in the predicted number of cancers, a review by Kimbrough (1987) from the Centers for Disease Control in Atlanta ended with the following statement:

> When humans were exposed only to PCBs or PBBs [polybrominated biphenyls], the only observed acute effects have generally been minor. So far, no significant chronic effects have been causally associated with exposure to PCBs or PBBs.

The potential of this chemical to cause great harm together with statements that the effects may be minor has resulted in a great deal of controversy about some public policy issues with respect to PCBs. For example, many people are bitter about fishing bans in the Northeast for fish such as striped bass (Belton et al, 1986; Flynn, 1987). Some insist that the risks have been overstated and that the cost to fishermen and their families (in fact, to a whole way of life) is great. Others argue that strict regulation and control have reduced human dietary exposure to PCBs (Biehl and Buck, 1987), which may be important in light of the toxic and possibly latent carcinogenic effects of high levels of exposure (Mussalo-Rauhamaa et al, 1990).

Tolerances for PCBs have been established in the United States, and if they are exceeded, the food is subject to seizure (Farley, 1988).

Polybrominated Biphenyls

While PCB contamination is a problem in many states, polybrominated biphenyl (PBB) contamination of the food chain primarily occurred in the early 1970s in Michigan. It was the result of a mistake when flame retardant was inadvertently mixed with livestock feed. This resulted in the contamination of meat, milk, and eggs sold in the Michigan area. The mistake was indeed costly, as 30,000 cattle, 6,000 swine, 15,000 sheep, and 1.5 million chickens were destroyed, along with 5 million eggs (Biehl and Buck, 1987).

PBB health effects are similar to those from PCBs. Like PCBs, they produce changes in the liver, cause tumors, affect reproduction, and promote biochemical change in experimental animals (Biehl and Buck, 1987; Wasito and Sleight, 1989).

Persons from quarantined farms and other groups with high exposures have been followed since the early 1970s. In all these studies, there were no findings that indicated that exposure to PBBs in any way affected the health of those exposed (Kimbrough, 1987). Although PBBs cause tumors in experimental animals, the levels used are over 10 times the highest measured human exposure. Thus, PBBs at the exposure levels encountered so far show no measurable chronic health effects. Like PCBs, PBBs also decrease with cooking (Zabik, 1987).

Indirect Additives from Packaging

Packaging of food introduces another risk-benefit problem. It is true that materials used in the packaging can migrate or leach into the food (Risch, 1988), but it is also true that packaging reduces deterioration and contamination. Furthermore, consumers tend to be schizophrenic in their behavior toward packaging. According to a survey by the Food Marketing Institute, consumers complain about too much packaging while at the same time complaining about lack of tamper-evident packaging. While consumers worry about no tamper-evident packaging, they are increasing their purchases in bulk—the ultimate vehicle for contamination and tampering. The concern of this discussion, however, is indirect additives from packaging that may be toxic.

Materials from paperboard, cans, and plastic can migrate into the food. Contamination by PCBs from paperboard was discussed in this chapter. Since PCBs are no longer used in carbonless paper, food cartons are not a source of this contamination. People also used to be concerned about lead solder from metal cans, which could leach into acid foods. Lead solder is no longer being used in cans.

There is now some concern about the safety of certain plastics, especially those that may be used and subsequently heated (Koelsch and Labuza, 1991a, 1991b). Much research is now being done on the safety of the microwave susceptors used for microwave foods, which need high temperatures in order to brown or to be crisp. These packaging materials are approved for food use, but the high temperatures reached in the microwave may allow some components to be changed and possibly migrate into the food. Certain components such as polyvinyl chloride and acrylonitrile, even though found at the parts-per-million level, should not be used in food-grade plastics.

Nontoxic packaging materials are being sought to replace the nontoxic but environmentally dangerous chlorofluorocarbons used to make foam packages for fast food service and as trays for meat and poultry. Testing must ensure that these new materials are totally inert when in contact with the food.

SUMMARY

Indirect additives pose unique problems. Their risks are diverse. They arrive in our food but weren't intended to be there. We must be ever vigilant in looking for and monitoring these contaminants. The real question then relates to the benefit they provide versus the risk they pose. We must try to sort out, as with direct additives, whether these components actually pose a risk when ingested in very small quantities. None of these answers will come easily, but the problems must be considered in an unbiased manner with maximum information.

REFERENCES

ACSH. 1985. Antibiotics in Animal Feed: A Threat to Human Health? American Council on Science and Health, Summit, NJ.

Addison, J. B. 1984. Antibiotics in sediments and run-off waters from feedlots. Residue Rev. 92:1-28.

Al-Dabbagh, S., Forman, D., Bryson, D., Stratton, I., and Doll, R. 1986. Mortality of nitrate fertilizer workers. Br. J. Ind. Med. 43:507-518.

Alther, L. 1972. Organic farming on trial. Nat. Hist. 81(9):16-24.

Anderson, H. A. 1989. General exposure to environmental concentrations of halogenated biphenyls. Pages 325-344 in: Halogenated Biphenyls, Terphenyls, Naphthalenes, Dibenzodioxins and Related Products, 2nd ed. R. D. Kimbrough and A. A. Jensen, eds. Elsevier, New York.

Anonymous. 1983. Organically grown food. In: Foods and Nutrition Encyclopedia, Vol. 2. E. H. Ensminger, M. E. Ensminger, J. E. Konlande, and J. R. K. Robson, eds. Pegus Press, Clovis, CA. pp. 1693-1696.

Anonymous. 1989a. Dairy dilemma—Milk is found tainted with a range of drugs farmers give cattle. Wall Street J., Dec. 29, p. 1.

Anonymous. 1989b. Just what is organic food—And is it good for you? Bus. Week, Sept. 25, p. 232.

Anonymous. 1989c. Brie and hormones. The Economist, Jan. 7, pp. 21-22.

Anonymous. 1991a. Quicklime may be safe, cheap way to rid environment of PCBs. St. Paul Pioneer Press, Mar. 12.

Anonymous. 1991b. Perspectives on pesticides and organic foods. Food Saf. Noteb. 2(5):37-38.

Armbruster, G., Gerow, K. G., Gutenmann, W. H., Littman, C. B., and Lisk, D. J. 1987. The effects of several methods of fish preparation on residues of polychlorinated biphenyls and sensory characteristics in striped bass. J. Food Saf. 8:235-243.

Belton, T., Roundy, R., and Weinstein, N. 1986. Managing the risks of toxic exposure. Environment 28(9):19-37.

Best. D. 1989. Hormones in meat: What are the real issues? Prep. Foods (3):116-117.

Biehl, M. L., and Buck, W. B. 1987. Chemical contaminants: Their metabolism and their residues. J. Food Prot. 50:1058-1073.

CAST. 1980. Organic and Conventional Farming Compared. Report 84. Council for Agricultural Science and Technology, Ames, IA.

CAST. 1986. Agriculture and Ground Water. Report 103. Council for Agricultural Science and Technology, Ames, IA.

CAST. 1990. Alternative Agriculture: Scientists' Review. Special Publ. 16. Council for Agricultural Science and Technology, Ames, IA.

Conacher, A., and Conacher, J. 1982. Organic Farming in Australia. Geowest No. 18, Nedlands, Australia.

Concon, J. M. 1988. Food Toxicology: Contaminants and Additives. Part B. Marcel Dekker, New York. pp. 1172-1229.

Egerstrom, L. 1989. The fight over food. St. Paul Pioneer Press, Sept. 10., pp. 1H, 3H.

Farley, D. 1988. Chemicals we'd rather dine without. FDA Consumer 22(6):10-15.

Flynn, B. T. 1987. PCBs and the fishing ban—Cost without benefit. ACSH News Views 8(3):1-2.

Foran, J. A., Cox, M., and Croxton, D. 1989. Sport fish advisories and projected cancer risks in the Great Lakes Basin. Am. J. Public Health 79:322-325.

Frappaolo, P. J. 1986. Risks to human health from the use of antibiotics in animal feeds. Pages 100-111 in: Agricultural Uses of Antibiotics. W. A. Moats, ed. American Chemical Society, Washington, DC.

Friend, B. A., and Shahani, K. M. 1983. Antibiotics in foods. Pages 47-61 in: Xenobiotics in Foods and Feeds. American Chemical Society, Washington, DC.

Gersema, L. M., and Helling, D. K. 1986. The use of subtherapeutic antibiotics in animal feed and its implications on human health. Drug Intell. Clin. Pharm. 20(3):214-218.

Gustafson, R. H. 1986. Antibiotics use in agriculture: An overview. Pages 1-6 in: Agricultural Uses of Antibiotics. W. A. Moats, ed. American Chemical Society, Washington DC.

Hay, J. 1989. The consumer's perspective on organic foods. Can. Inst. Food Sci. Technol. J. 22(2):95-99.

Hays, V. W. 1985. Antibiotics for animals. Sci. Food Agric. 3(2):17-26.

Hays, V. W. 1986. Benefits and risks of antibiotics use in animals. Pages 74-86 in: Agricultural Uses of Antibiotics. W. A. Moats, ed. American Chemical Society, Washington DC.

Hays, V. W., and Black, C. A. 1989. Antibiotics for Animals: The Antibiotic Resistance Issue. Council for Agricultural Science and Technology, Ames, IA.

Herbst, A. L., Ulfelder, H., and Poskanzer, D. C. 1971. Adenocarcinoma of the vagina: Association of maternal stilbesterol therapy with tumor appearance in young women.

New Engl. J. Med. 284:878-888.

Herbst, A. L., Kurman, R. J., Scully, R. E., and Poskanzer, D. C. 1972. Clear cell adenocardinoma of the genital tract in young females. New Engl. J. Med. 187:1259-1264.

Hileman, B. 1988. The Great Lakes cleanup effort. Chem. Eng. News 66(6):22-39.

Hileman, B. 1990. Alternative agriculture. Chem. Eng. News 68(13):26-40.

Holmberg, S. D., Osterholm, M. T., Senger, K. A., and Cohen, M. L. 1984b. Drug-resistant *Salmonella* from farm animals fed antimicrobials. New Engl. J. Med. 311:617-622.

Holmberg, S. D., Wells, J. G., and Cohen, M. L. 1984a. Animal-to-man transmission of antimicrobial-resistant *Salmonella*: Investigations of U.S. outbreaks, 1971–1983. Science 225:833-835.

Jensen, A. A. 1983. Chemical contaminants in human milk. Pestic. Rev. 89:1-128.

Jensen, A. A. 1989. Background levels in humans. Pages 345-380 in: Halogenated Biphenyls, Terphenyls, Naphthalenes, Dibenzodioxins and Related Products, 2nd ed. R. D. Kimbrough and A. A. Jensen, eds. Elsevier, New York.

Joint Expert Committee on Food Additives (JECFA). 1988. Evaluation of certain veterinary drug residues in food, 32nd Rep. WHO Tech. Rep. Series 763. World Health Organization, Geneva.

Joint Expert Committee on Food Additives and Contaminants (JECFA). 1990. Evaluation of Certain Food Additives and Contaminants. WHO Tech. Rep. Series 789. World Health Organization, Geneva. pp. 30-33, 36-37.

Jolly, D. A., Schutz, H. G., Diaz-Knauf, K. V., and Johal, J. 1989. Organic foods: Consumer attitudes and use. Food Technol. (Chicago) 43(11):60-66.

Jukes, T. H. 1977. Organic food. CRC Crit. Rev. Food Sci. Nutr. 9(4):395-418.

Jukes, T. H. 1986. Effects of low levels of antibiotics in livestock feeds. Pages 112-127 in: Agricultural Uses of Antibiotics. W. A. Moats, ed. American Chemical Society, Washington, DC.

Kimbrough, R. D. 1987. Human health effects of polychlorinated biphenyls (PCBs) and polybrominated biphenyls (PBBs). Annu. Rev. Pharmacol. Toxicol. 27:87-111.

Kimbrough, R. D., and Grandjean, P. 1989. Occupational exposure. Pages 485-508 in: Halogenated Biphenyls, Terphenyls, Naphthalenes, Dibenzodioxins and Related Products, 2nd ed. R. D. Kimbrough and A. A. Jensen, eds. Elsevier, New York.

Koelsch, C. M., and Labuza, T. P. 1991a. Packaging, waste disposal, and food safety. I: Landfilling, source reduction, and recycling of plastics. Cereal Foods World 36:44.

Koelsch, C. M., and Labuza, T. P. 1991b. Packaging, waste disposal, and food safety. II: Incineration or degradation of plastics, and a possible integration solution. Cereal Foods World 36:284.

Komsta-Szumska, E. 1985. Environmental impact of sewage sludge on livestock: A review. Vet. Hum. Toxicol. 28(1):31-37.

Kreiss, K. 1985. Studies on populations exposed to polychlorinated biphenyls. Environ. Health Perspect. 60:193-199.

Kuratsune, M. 1989. Yusho, with reference to Yu-Cheng. Pages 381-400 in: Halogenated Biphenyls, Terphenyls, Naphthalenes, Dibenzodioxins and Related Products, 2nd ed. R. D. Kimbrough and A. A. Jensen, eds. Elsevier, New York.

Lafler, J. M. 1990. More on organic foods. J. Am. Diet. Assoc. 90:920-921.

Layman, P. L. 1988. Changes in European farm policies trouble fertilizer industry. Chem. Eng. News 66(11):7-12.

Levy, S. 1986. Antibiotics—Risk assessment. IFT seminar, June 16, Dallas, TX.

Levy, S. 1987. Antibiotic use for growth promotion in animals: Ecological and public health consequences. J. Food Prot. 50:616-619.

Linton, A. H. 1986. Flow of resistance genes in the environment from animals to man. J. Antimicrob. Chemother. 18(Suppl. C):189-197.

Livingston, R. C. 1986. Antibiotic residues in food: Regulatory aspects. Pages 128-136 in: Agricultural Uses of Antibiotics. W. A. Moats, ed. American Chemical Society, Washington, DC.

Minnesota Agricultural Extension Service. 1972. Drug residues in meat "unnecessary and inexcusable." Dep. of Information and Agric. Journalism, University of Minnesota, December 29.

Mussalo-Rauhamaa, H., Häsänen, E., Pyysalo, H., Antervo, K., Kauppila, R., and Pantzar, P. 1990. Occurrence of beta-hexachlorocyclohexane in breast cancer patients. Cancer 66:2124-2128.

NAS/Institute of Medicine. 1989. Human Health Risks with the Subtherapeutic Use of Penicillin or Tetracyclines in Animal Feed. FDA's Center for Veterinary Medicine, Rockville, MD.

National Academy of Sciences/National Research Council (NAS/NRC). 1989. Alternative Agriculture. National Academy Press, Washington, DC.

Newsome, R. L. 1990. Organically grown foods. Food Technol. (Chicago) 44(6):26, 29.

Nilsson, N., and Lindahl, O. 1986. The effect of nitrogen fertilization on forest blueberries. Nutr. Health 4:151-153.

Pinegar, J. A., and Crooke, E. M. 1985. *Escherichia coli* in retail processed foods. J. Hyg. 95:39-46.

Place, H. R. 1990. Animal drug residues in milk: The problem and the response. Dairy Food Environ. Sanit. 10:662-664.

Pocock, S. J. 1985. Nitrates and gastric cancer. Hum. Toxicol. 4:471-474.

Quattrucci, E. 1987. Hormones in food: Occurrence and hazards. Pages 103-138 in: Toxicological Aspects of Food. K. Miller, ed. Elsevier Applied Science, New York.

Reganold, J. P., Papendick, R. I., and Parr, J. F. 1990. Sustainable agriculture. Sci. Am. 261(6):112-120.

Risch, S. J. 1988. Migration of toxicants, flavors, and odor-active substances from flexible packaging materials to food. Food Technol. (Chicago) 42(7):96-102.

Rogan, W. J., Gladen, B. C., McKinney, J. D., Carreras, N., Hardy, P., Thullen, J., Tingelstad, J., and Tully, M. 1986. Polychlorinated biphenyls (PCBs) and dichlorodiphenyl dichloroethene (DDD) in human milk: Effects of maternal factors and previous lactation. Am. J. Public Health 76(2):172-177.

Sadler, M. 1990. The greening of consumer demand. Food Policy 15:448-450.

Schell, O. 1984. Modern Meat: Antibiotics, Hormones, and the Pharmaceutical Farm. Random House, New York.

Shahani, K. M., and Whalen, P. J. 1986. Significance of antibiotics in foods and feeds. Pages 88-99 in: Agricultural Uses of Antibiotics. W. A. Moats, ed. American Chemical Society, Washington, DC.

Simco, M. D., and Jarosz, L. 1990. Organic foods: Are they better? J. Am. Diet. Assoc. 90:367-371.

Stokstad, E. L., and Jukes, T. H. 1950. Further observations on the "animal protein factor." Proc. Soc. Exp. Biol. Med. 73:523-528.

Vettorazzi, G. 1987. The importance of toxicology in food science. Pages 1-16 in: Toxicological Aspects of Food. K. Miller, ed. Elsevier Applied Science, New York.

Wasito, and Sleight, S. D. 1989. Promoting effect of polybrominated biphenyls on tracheal papillomas in Syrian golden hamster. J. Toxicol. Environ. Health 27(2):173-187.

Zabik, M. E., Merrill, C., and Zabik, M. J. 1982. PCBs and other xenobiotics in raw and cooked carp. Bull. Environ. Contam. Toxicol. 28:710-715.

Ziv, G. 1986. Therapeutic use of antibiotics in farm animals. Pages 8-22 in: Agricultural Uses of Antibiotics. W. A. Moats, ed. American Chemical Society, Washington, DC.

Radionuclides in Food

CHAPTER 15

The 1986 Chernobyl incident in the USSR, the most serious nuclear accident in the 45-year history of nuclear reactor operations, reawakened public awareness of the possibility of radionuclides in foods. Since the banning of above-ground testing in 1963, public concern about radionuclides in food had waned but now increased again.

Regardless of the level of public concern, the FDA and other scientists study and monitor radionuclides in food and the environment on a regular basis. According to the National Council on Radiation Protection and Measurement, one finding is that radiation from natural sources accounts for more than 80% of the exposure received by an average member of the population. The total effective dose equivalent rate has been estimated to be about 3,600 micro-Sieverts(μSv) per year, with 3,000 μSv (80%) from natural sources. Box 15.1 gives a breakdown

---

Box 15.1

Involuntary Sources of Radiation

Source	**Percent of Total**
Radon	54
Natural radiation	
Inside body	11
Cosmic radiation	8
Rocks and soil (other than radon)	8
Medical X-rays	11
Nuclear medicine	4
Consumer products	3
Nuclear power production	0.1
Other	<1

(From Anonymous, 1989, and ACSH, 1988)

---

of some background radiation levels, showing that the largest natural portion comes from inhaled radon decay products. Only 400 μSv per year are found in the body, and this level must be from food and water. Thus, food comprises roughly 10% of the average annual dose (Moeller, 1988; Silini, 1988; Stroube et al, 1985).

Box 15.2 lists some types of background radiation and compares them with other sources of radiation, including ones that have a voluntary component such as cigarette smoking. Note that the irradiation of foods for preservation, as discussed in Chapter 12, does not appear even minimally, as it does not result in radioactivity.

Box 15.2

Radiation Exposures

Source	Dose (mrem/year)
Natural background	
Airborne radon	200
Cosmic radiation and body potassium-40	100
At sea level	26
In Leadville, Colorado (12,000 ft)	100
At Three Mile Island	100
Granite walls	
Grand Central Station	200
U.S. Capitol	20
Cigarettes, two packs per day	1,300
Medical procedures	
X-ray diagnosis	39
Nuclear medicine	14
Consumer products, excluding watches,	
eyeglasses	5–13
Travel	
Over the ocean	20
In the air, 0.5 millirem per hour	
Passenger, 10 hours per year	5
Airline personnel, 1,500 hours per year	750
On a freshwater lake	0
All other sources: weapons testing fallout,	
nuclear power, etc.	1

(From Moeller, 1988, and Yallow, 1987)

RADIOACTIVITY IN FOODS

Natural Radioactivity

Radionuclides have existed in foodstuffs from the beginning of the universe. Their existence in food is readily accounted for, as many elements in food have radioactive counterparts that can replace them. The radioactive ones have the same chemical properties, with one difference: they decompose, with the liberation of ionizing radiation.

Natural radionuclides enter the food chain from the soil. The level in the plant depends first on the amount in the soil and second on how well a particular plant takes up the elements through the root system. Thus, the soil is a prime source of geographic variability.

Isotopes found naturally in food include radium-226, radium-228, carbon-14, rubidium-87, polonium-210, thorium-228, and potassium-40. Potassium-40 is the most prevalent, contributing about 0.01% of total body potassium and providing 25% of the background radiation received by cells (Miller and Miller, 1986).

Most foods contain 1–10 micro-Curies (μCi) per gram of K-40. For example, rice contains a very small amount, less than 1 μCi per gram; coffee and tea contain 35 μCi per gram.

Radioactivity as a Result of Human Activity

Table 15.1 gives the three sources of radioactivity in foods that are products of human activity or fallout. Fallout is the result of the fission (splitting) of uranium or plutonium. During this splitting, a tremendous amount of energy is produced, along with a large number of possible radionuclides (about 200). Some of these compounds escape as gases and are diluted by the atmosphere. They may be inhaled; however, inhalation is not a major route of entry of radioactivity into the body. Most particles attach to dust and debris that settle onto the earth's surface. Rain can wash the particles out of the atmosphere and cause

Table 15.1. Sources of Radiactivity in Foods[a]

Type	Source	Year
Industrial	Effluent from nuclear power plants	Continuing
	Fossil fuel power plants	Continuing
	Laboratories	Continuing
	Medical facilities	Continuing
Weapons testing	Debris from above-ground tests	1950s and 1960s
	Residual activity	Some continuing
Nuclear accidents	Windscale, Great Britain	1957
	Chernobyl, USSR	1986

[a] Data from Anonymous (1989) and ACHS (1988).

their deposition in soil and on vegetation.

"Hot" soil has the potential of irradiating the whole body with pene-trating gamma-rays, affecting the skin with nonpenetrating beta-particles, or contaminating plants used for human and animal food. Once deposited on soil or leaves, radioactive isotopes can be taken up by the plant root system. The material on the plants can be transferred to animals that eat the vegetation. As with persistent pesticides, the element can concentrate as it moves through the food chain.

Nuclear testing has greatly diminished since 1962, and the levels of radionuclides in the environment have also continuously decreased. In general, only low levels of several long-lived radionuclides are pre-sently being measured in the environment and in the food chain (Carter and Hanley, 1988).

PHYSIOLOGICAL EFFECTS OF IONIZING RADIATION

Physiological data carefully gathered following the bombing of Japan and from the days of above-ground nuclear testing until its banning in 1963 give a large body of information on the biological effects of ionizing radiation. It affects human tissues by depositing energy as it passes. This extra energy from the radiation strips an electron from an atom. Thus the atom, which previously was neutral (carried no charge), is now positively charged because the negatively charged electron has been stripped away. Both the unpaired electron and the ionized atom are very unstable and highly reactive, so they start a complex chain of reactions that results in the production of free radicals. Many different products can be formed in just a few millionths of a second. The biological result can be the immediate death of the cell or changes that, over the course of time, lead to changes in the genetic material of the cell or cause cancer or destruction of the immune system. (For a review, see Silini, 1988, or Upton and Linsalata, 1988.) High levels of ionizing radiation can be lethal or give rise to effects that severely hamper many cell functions. The systems most likely to be affected are the bone marrow, skin, gastrointestinal lining, lens of the eye, ovary and testes, lung, kidney, and central nervous system. In general, functional parenchymal cells are lost and are replaced with connective tissue, giving rise to both loss of function and increase in tissue fibrosis. Lower levels can give rise to mutations and cancer if the radiation alters the DNA of cells that may later divide.

The extensive knowledge gained has been useful in predicting the effects of ionizing radiation for use in medicine and in production of nuclear power. It has even enabled scientists to predict the effect of the 4-5 millirems per year of radioactivity we receive from spurious sources such as luminous dial watches, airport luggage inspections, television sets, radon, air travel, and the naturally occurring background radiation.

Factors Affecting Toxicity

Radiation may be absorbed by the body in two ways. In the first, the radiation source is outside the body, as in an X-ray. This is called external radiation. The second, internal radiation, occurs when the radiation occurs within the body, as when radionuclides are either inhaled or ingested.

With food we are concerned with internal radiation that results when contaminated agricultural products or water are ingested. The first effects are on the gut. Luckily, most of the particles are insoluble; they simply pass through the gastrointestinal tract, causing minor irradiation to the stomach and the intestine during their short transit. Obviously, the shorter the transit time before excretion, the less the time available in which to inflict any cell damage.

Other factors that affect the toxicity are the energy of the radiation, the chemical half-life, and the biological half-life. Since most of the products have a short half-life and decay rapidly, the damage to humans and animals is minimal. Others are highly insoluble; they do not enter the metabolic scheme and therefore have short biological half-lives and little health consequence.

Characteristics of Specific Radionuclides

Of the various types of radiation, the high-energy α-emitter can cause great damage because of its high ionizing capability. Luckily, its large size prevents deep penetration into tissues.

Those radionuclides that are the most destructive are those that can penetrate the soft tissue and become part of active metabolism. Cesium-137 has these noxious properties. Its chemical similarity to potassium means that it is rapidly adsorbed by the bloodstream and can be distributed in all cells of the body. Its half-life is 27 years.

Carbon-14 can also irradiate the whole body since all organic molecules contain carbon. Carbon-14 is a soft beta-emitter with a half-life of 5,600 years.

Other radionuclides that wreak physiological havoc include strontium-90 and iodine-131. Strontium-90, with a half-life of 28 years, is produced in abundance from both the detonation and the fuel cycle of nuclear weapons. The toxicity of this alkaline earth radionuclide has been studied extensively in mice, rats, dogs, and pigs. Since strontium is an analog of calcium, it is readily absorbed either from the gastrointestinal tract or the lung and is deposited in the bone. The actual amount absorbed by the gut is very dependent on the body's trace mineral status. Adequate calcium and phosphate intakes substantially decrease its absorption.

A single brief oral intake of strontium-90 results in a high incidence of cancers of the bone and leukemias. Age at exposure to the ionizing radiation determines the susceptibility to leukemias. Children under 10 are at greatest risk. The Federal Radiation Council has stated that strontium-90 should not exceed 1,500 millirems above background levels. The same recommendation applies to iodine-131.

Radioiodine is produced in abundance in operating nuclear reactors and during nuclear weapons fallout. High levels of this radionuclide result in the almost complete destruction of the thyroid, with an attendant decrease in thyroid hormone production. Levels of iodine-131 that damage the thyroid but leave the tissue capable of proliferation lead to hyperplastic cancers. Children appear to be twice as susceptible as adults. As a preventive in situations where exposures may be greater than 10 rads, potassium iodide should be administered to minimize the dose to the thyroid. Luckily, this isotope has a short half-life of only 8.1 days (Upton and Linsalata, 1988).

Since internal irradiation by radionuclides ingested in food and water is only a very small part of the average total dose received by most tissues, the carcinogenic effect attributed to such irradiation is indeed small. It is estimated that, on average, 0.3% of fatal cancers can be attributed to this cause. However, in areas where the natural background level of radiation is high, as in certain parts of India, China, or Brazil, internal radiation doses are higher. Since a dose relationship appears to exist between level of radiation and cancer risk, in areas where the background radiation is high the cancer rates attributable to internal irradiation are also likely to be relatively high (Upton and Linsalata, 1988).

MONITORING THE LEVELS OF RADIOACTIVITY IN THE FOOD CHAIN

Since 1955, U.S. federal agencies, as part of their total Market Basket Survey, have kept track of the numbers of radionuclides in foods. Radioactive iodine in milk is the primary indicator of whether or not a public health risk exists. Iodine in milk is used for several reasons. First, radioactive iodine has high toxicity. Second, it is found in greater abundance after nuclear reactor accidents such as Chernobyl. Third, iodine-131 appears in milk within 3–4 days after release of nuclear material into the environment, and milk reaches consumer markets faster than most other foods with potential for contamination from fallout. Finally, milk is frequently a major source of calories and nutrients for infants, the group most sensitive to radioactive iodine (Blanc, 1981; Hile, 1986).

In addition to the milk- and market-basket sampling program, the FDA samples imported foods and the milk, fish, and food crops near 40 nuclear power plants in the United States (Gartrell et al, 1985a, 1985b). Most radionuclides are below the limits of detection even in food crops raised near nuclear power plants. Occasionally, some samples show strontium-90 and cesium-137. Overall, levels of strontium-90 showed a decreasing trend for the 5-year period from 1978 to 1982 (Stroube et al, 1985; Yang and Nelson, 1986). Earlier studies showed the rise and decay of radionuclides following the now-banned above-ground nuclear tests. After several above-ground tests in the late 1950s, significant surface contamination of strontium-90 was noted. The deposition was heaviest after heavy rains in the spring and summer. Market basket studies also showed an increase in strontium-90 levels in foods after the above-ground tests. These levels decreased by half in 18 months.

Strontium-90 is still monitored. Certain food supplements such as bone meal are not recommended as a calcium source because of strontium-90 levels. Ironically, this health food is marketed for building bones as the substance most like natural bone.

Data gathered on powdered milk (Dunning, 1961) show 41 μCi of cesium-137 in 1956, 57.1 in 1959, and 16.0 in 1961, thus reflecting the increase in nuclear testing and then the ban on above-ground tests.

Cesium-137 is also monitored in soil. Soil samples from Europe showed more residual cesium-137 in the soils from weapons test fallout than was added by Chernobyl. However, while much can be learned from samples collected during nuclear weapons testing, what happened at Chernobyl is indeed different. In weapons testing, radioactive debris is launched high into the stratosphere and is deposited widely and continuously. In nuclear accidents, the fallout reaches only the troposphere and is deposited mostly with rain.

Hazards from Nuclear Accidents

After the nuclear accident at Chernobyl on April 26, 1986, there was great concern in the affected areas. In the immediate vicinity, radiation doses reached 1,000 mega-Curies (see Table 12.2 for units used to measure radiation.) In the Eastern Europe areas with the greatest contamination—northern Poland, Czechoslovakia, and Hungary, radiation doses reached 30–50 mega-Curies. Persons in these areas were told to avoid milk and fresh fruits and vegetables (Palca, 1986). To give some perspective, the Three Mile Island incident in the United

States in 1979 yielded a radiation dose of only 15 Curies—a dose many million times less than that from Chernobyl.

Ground-level deposits after rainfall at locations close to Chernobyl and in Poland may have reached levels 200–400 times greater than background levels. At distances as far as 1,000 miles from Chernobyl, when there was heavy rainfall, values reached 30–40 times the background level. In areas without rainfall, the background level was increased by a factor only somewhat greater than 2.

Over the next 50 years, 75 million people of western Russia will receive an average dose of 29 million rems. It is assumed that one additional cancer will occur for every 1,000 rems, or that there will be 3,000 cancer deaths above what would otherwise be expected in those areas of Russia (Levi, 1986).

After the Chernobyl accident, monitoring of food, air, and water was increased in the United States through an increase in the number of market-basket samples. Measurements were also taken at U.S. embas-

Figure 15.1. This photograph, shown on Soviet TV on April 30, 1986, is of damage to the Chernobyl nuclear plant from the accident on April 26. The right side of the photograph was enhanced to clarify the damage. (Printed by permission of AP/Wide World Photos)

sies throughout the affected regions. Other countries also stepped up their testing programs.

Even in countries like Austria, where the fallout was high, the hazard to adults was reported as extremely small, although there was concern about possible hazards for fetuses and young infants. Luckily, the concentration of iodine-131 in breast milk was found to be one tenth of that in cow's milk. The highest levels from cow's milk were reported in June at distances up to 1,000 miles (approximately 1,500 kilometers) from Chernobyl, with varying concentrations of 0–610 Becquerels per liter. The readings show a great deal of variability, depending on the rainfall in the region and on whether cows were permitted to graze on open pasture. After 4 months, levels were reduced to 40 Becquerels (Haschke et al, 1987).

In Japan, milk was measured as having an iodine-131 content four to five times higher than that of rainwater, although milk processing decreased the levels (Nishizawa et al, 1986).

There were reports that some contaminated foods such as dry milk were unscrupulously shipped to other countries. A Brazilian federal court disclosed that for 8 months its citizens had been eating food from Western European countries with cesium levels 10 times greater than allowed. In Sao Paulo alone, 70 metric tons of food were pulled off the shelves (Egger, 1987).

In addition to iodine-131 contamination of milk and dairy products, meat and meat products from animals and wild game grazing on contaminated plants showed increased radioactivity (Vasilenko, 1986). Consumption of game birds and other wild game was limited in some countries. Italy considered a ban on bird hunting, as Chernobyl was located in a major migratory flyway where birds traveled north from Africa and through the Ukraine (Davis, 1986).

Wales, Umbria, and Scotland instituted a ban on the slaughter of lambs with abnormal amounts of cesium-137. In some cases, the cesium activity was measured as being tens or even hundreds of Becquerels per kilogram. The silver isotope, Ag-110, was found in livers of beef and lamb but not in the meat (Jones et al, 1986; Twomey, 1987).

Perhaps the most devastating effects of this accident outside the immediate area were those on the reindeer in Lapland (1,300 miles from Chernobyl). The major food of the reindeer, lichen, soaked up much radiation and may not be safe to eat for years. Reindeer meat was measured in November 1986 as having 40,000 Becquerels per kilogram of meat. By March of 1987, it was down to 5,000–6,000 Becquerels per kilogram. The Swedish government bought all the contaminated reindeer meat, but the whole way of life of the Lapps, which is dependent on the reindeer, was threatened (Knight, 1987).

Other nuclear accidents had far less impact than Chernobyl. One was at the Windscale reactor on the northwest coast of England, another was at Three Mile Island, Pennsylvania. After the Windscale accident, the level of radioactive iodine in the milk supply increased for up to a 200-mile radius. No other radionuclides were found above background levels in milk or other foodstuffs (Concon, 1988). After the Three Mile Island incident, about 25% of the milk samples from surrounding areas were positive for radioactive iodine. However, concentrations measured were not considered hazardous even for infants drinking a quart of milk a day. In the 31 days following the accident, no reactor-produced radionuclides were found in the nearly 400 samples of food tested. Even if a person received the maximum cumulative dose from the accident, the risk of fatal cancer was calculated to be 1 in 50,000. In terms of excess fatal risks, the exposure would amount to an insignificant additional risk (Concon, 1988). Using the estimated average loss of life expectancy, the American Council on Science and Health calculated that, from the 1-millirem exposure, the average resident of Harrisburg, PA, lost 1.2 minutes of life (ACSH, 1988).

RISKS FROM FOOD

Although the radiation contamination of food and the environment as discussed in this chapter is a concern and every effort should be mounted to prevent it, it should not in any way be confused with the irradiation of food for preservation as discussed in Chapter 12. As

---

Box 15.3

How to Avoid Radiation

Dr. Dade Moeller, a member of the National Council on Radiation Protection and Measurement, suggests that to avoid radiation one would need to make the following changes in lifestyle:

1. Live at sea level around the equator, in a well-ventilated frame house or in a wooden houseboat on a freshwater lake.

2. Avoid mountain climbing, air travel, cigarette smoking, well waters, and foods high in naturally occurring nuclides.

Food *is* a source of radionuclides. One should take every possible precaution while avoiding undue fear and paranoia.

(Adapted from Moeller, 1988)

pointed out there, subjection of a food product or ingredient to gamma-radiation from cobalt-60 or to electron beam radiation cannot produce radioactive isotopes or radiation, nor has intensive investigation for over 40 years been able to demonstrate any health hazard from radiolytic products that could result.

On the other hand, concern about contamination by radioactive isotopes is warranted because in toxicological terms these compounds must be viewed as having no threshold, and they may be carcinogenic, mutagenic, and teratogenic (Concon, 1988). Furthermore some radionuclides have affinities for an organ that makes the dose in the organ several times higher than the absorbed dose and may allow accumulation over time as more compound is absorbed. Since several radionuclides have long half-lives, the isotopes may emit radiation throughout a significant portion of a person's life.

Fear of increased cancer from radiation is real. However, actual figures are usually far less than anticipated. Dr. Norman Rasmussen, a nuclear engineer at Massachusetts Institute of Technology, stated that there was a lot of evidence that low levels of radiation do not cause harm and may even do some good. He pointed out that humans and animals evolved in a world of natural low-level radiation (Anonymous, 1989).

The number of cancers predicted for the 80,000 Japanese survivors of the bombs dropped during World War II was 4,500. Forty years later there were 100 cancers and 91 leukemias among this group, clearly many less than expected (Yallow, 1987). Other studies on radiation exposure also showed rates to be less than expected. For example, X-ray technicians in the 1950s were exposed to 5–15 rem per year (now the level has been reduced to under 1 rem per year). A study comparing X-ray technicians with pair-matched other laboratory technicians showed no difference in cancer incidence or death rates for the two groups (Yallow, 1987). Surprisingly, cancer death rates in the seven states with the highest levels of natural radiation are 10–15% lower than the national average.

It is very difficult to predict the effect of low-level radiation. For instance, equations predict that cancer death rates in Denver should be greater than in New Orleans because Denverites receive twice the background level of radon, but the cancer death rate in New Orleans is actually higher than in Denver. Especially when the level of radiation is below 10 millirems, other confounding factors such as smoking apparently have an impact on cancer rates.

Based on her studies, Nobel laureate Dr. Rosalyn Yallow (1987) questioned the doomsday predictions that the Chernobyl accident would result in 9,000 extra leukemia deaths in Finland by 1991. She felt that the levels would be much lower, which so far has turned out to be the case.

REFERENCES

ACSH. 1988. Health Effects of Low Level Radiation. American Council on Science and Health, Summit, NJ.

Anonymous. 1989. Living with radiation. Natl. Geogr. 175 (4):403-437.

Blanc, B. 1981. Biochemical aspects of human milk. World Rev. Nutr. Diet. 36:1-89.

Carter, M. W., and Hanley, L. 1988. Food-chain contamination from testing nuclear devices. Pages 172-194 in: Radionuclides in the Food Chain. J. H. Harley, G. D. Schmidt, and G. Silini, eds. Springer-Verlag, New York.

Concon, J. J. 1988. Radionuclides in foods. Pages 1231-1247 in: Food Toxicology. Pt. B. Additives and Contaminants. Marcel Dekker, New York.

Davis, L. 1986. Concern over Chernobyl-tainted birds. Sci. News 130(1):54.

Dunning, G. M. 1961. Foods and fallout. Borden's Rev. Nutr. Res. 23(1):1-20.

Egger, D. 1987. West Germany pours hot milk; Chernobyl's cup runneth over. Nation 244:392.

Gartrell, M. J., Craun, J. C., Podrebarac, D. S., and Gunderson, E. L. 1985a. Pesticides, selected elements and other chemicals in infant and toddler total diet samples, October 1978–September 1979. J. Assoc. Off. Anal. Chem. 68:842-861.

Gartrell, M. J., Craun, J. C., Podrebarac, D. S., and Gunderson, E. L. 1985b. Pesticides, selected elements and other chemicals in adult total diet samples, October 1978–September 1979. J. Assoc. Off. Anal. Chem. 68:862-875.

Haschke, F., Pietshnig, B., Karg, V., Vanura, V., and Schuster, E. 1987. Radioactivity in Austrian milk after the Chernobyl accident. New Engl. J. Med. 316:409-410.

Hile, J. P. 1986. Radionuclides in imported foods. FDA Compliance Policy Guide 7119.14. U.S. Food and Drug Admin., Washington, DC.

Jones, G. D., Forsyth, P. D., and Appleby, P. G. 1986. Observation of 110m-Ag in Chernobyl fallout. Nature (London) 322(6077):313.

Knight, R. 1987. The legacy of Chernobyl: Disaster for the Lapps. U.S. News World Rep. 102(Mar. 23):36.

Levi, B. G. 1986. Soviets assess cause of Chernobyl nuclear accident. Phys. Today 39(Dec.):17.

Miller, E. C., and Miller, J. A. 1986. Carcinogens and mutagens that may occur in foods. Cancer 58:1795-1803.

Moeller, D. W. 1988. Radiation exposures from natural background. ACSH News Views 9(3):7-8 ff.

Nishizawa, K., Hamada, N., Kayama, Y., Kojima, S., Ogata, Y., Ohshimo, M. 1986. I-131 in milk and rain after Chernobyl. Nature (London) 324(6095):308.

Palca, J. 1986. British embassies turn to monitors. Nature (London) 321(6069):458.

Silini, G. 1988. Biological effects of ionizing radiation. Pages 35-44 in: Radionuclides in the Food Chain. J. H. Harley, G. D. Schmidt, and G. Silini, eds. ILSI Monographs, Springer-Verlag, New York.

Stroube, W. B., Jelinek, C. F., and Baratta, E. J. 1985. Survey of radionuclides in foods, 1978-1982. Health Phys. 49:731-735.

Twomey, T. 1987. Radioactivity and its measurement in foodstuffs. Dairy Food Sanit. 7:452-457.

Upton, A. C., and Linsalata, P. 1988. Long-term health effects of radionuclides in food and water supplies. Pages 218-234 in: Radionuclides in the Food Chain. J. H. Harley, G. D. Schmidt, and G. Silini, eds. ILSI Monographs, Springer-Verlag, New York.

Vasilenko, I. Y. 1986. A radiation-hygienic appraisal of biosphere contamination with I-129. J. Hyg. Epidemiol. Microbiol. Immunol. 30:243-248.

Yallow, R. 1987. Radioactivity. Speech at Macalester College, St. Paul, MN. Sept. 10.

Yang, Y.-Y., and Nelson, C. B. 1986. An estimation of daily food usage factors for assessing radionuclide intake of the U.S. population. Health Phys. 50:245-257.

Epilogue—
Where Do We Go from Here?

One pervading theme of the preceding chapters is that there are many perceptions of truth, stemming from a variety of positions and ways of gathering data. Some of the positions are well thought out and based on plausible assumptions; others are less well drawn but may be equally plausible. Some seem to take into account the weight of evidence, others to ignore the evidence for a gut feeling or intuitive position. Some are held and proselytized with religious fervor. Some are based on fear, others on concern for the environment. Some are based on greed, others on misinformation. Some are based on a one-sided or biased screening of factual information, while others were formed before the facts were fully in place. While decisions made on the basis of partial information are not desirable, information about complicated issues is nearly always incomplete. When such is the case and the possibility of making the wrong choice can have bad or dire consequences, fear often results.

Another pervading theme of this book is that food safety issues are incredibly complex. A myriad of possible interrelationships exist among chemicals in our food (both those put there by nature and those added somewhere during the process of getting the foods from the fields, woods, and streams into our bodies). Other complicating factors include the compositions of diets, other interacting constituents of the environment, and personal factors such as age, state of health, level of activity, and genetic makeup.

This mind-boggling complexity makes it nearly impossible to obtain all the data desirable for public health recommendations. At least some of the recommendations inevitably will be wrong or subject to subsequent substantial amendment. The trick is to make recommendations that follow the spirit of the Hippocratic oath in medicine, which is to do some good at best but to certainly do no harm. But here we find a Catch 22: various groups are unable to agree about either the good or the harm. The concept of risk-benefit is often not applied, and so neither risk nor benefit is assessed with accuracy. As a result, various factions can make sincere but diametrically opposed recommendations about what is to be done.

Incredible complexity coupled with ignorance of scientific issues leads at best to consumer confusion and at worst to poor consumer choices. A Johnny Carson monologue about the food label was indicative of the fact that consumers don't understand the chemical terms found there. The monologue asked whether the calcium stearate additive in a particular food was calcified pieces of steer. While this is funny, it is also scary because most consumers do not know what calcium stearate is, why it is in the food, or whether it is good or bad for them. Inadequate knowledge of chemistry, especially in a generation that doesn't know much about food preparation and food ingredients, results in very confused consumers.

This confusion contributes to the inability of some consumers to make moderate responses. For example, at the height of the oat bran craze in 1989, oats and oat bran were in such demand that the United States became a net importer of oats—a previously unheard-of situation. Once some consumers hear that a food component is beneficial, they think that if some is good, more is better. What saves most of these consumers from the adverse effects of too much of the component is human nature itself. That is, unless a person is totally compulsive, the tendency toward overconsumption is soon compensated for by the lack of resolve or a shift to another more-recent fad.

If the food component is reported to be harmful, the response is often equally overreactive. After a 1989 report on alar in apples, the apple industry lost $325 million dollars in sales (Smith, 1990), and apples rotted on the shelves. A consumer query from one women showed not only the depth of her misunderstanding but also the depth of her fear. She asked not whether her apple juice was safe to drink, but whether it was safe to dump it down the drain or if she needed to take it to a toxic waste dump. For many consumers, the news that any amount of a substance is harmful is translated into the idea that they should take in none at all, even if the amount tested could never be consumed under even very abnormal eating patterns. Furthermore, once a constituent's safety has been questioned, it tends to remain on a consumer blacklist despite any subsequent findings that might exonerate it.

In like manner, consumers understand what is present naturally in a food even less well than what was added to it. Thus, my purpose in exploring naturally occurring toxic hazards in food and plants (Chapter 5) was not to persuade readers to avoid these common foods. Rather, I thought it important to put some perspective on the natural chemicals in our food and to show clearly that their toxicology (usually unknown) needs to be better understood. The risks and benefits of chemicals naturally present in food must be weighed, just as we weigh the risks and benefits of chemicals that are added to food.

Another problem facing us in the food safety arena is that of inadequate knowledge of farms and food production. The average consumer does not have a clear picture of what the life of a well-cared-for farm animal is like, does not know how oranges get from a warm climate to the supermarket or the many steps that wheat goes through before it becomes bread, and does not have a picture of the potential for loss in both the amount of available food and the quality of food at any point along the chain.

Many are unaware that the leisure time and lifestyle that exist in the United States and other Western countries today are in large measure made possible by a very efficient agribusiness complex. The fact that around 15% of the disposable income in the United States is devoted to food is taken for granted and not appreciated. In countries where the systems are not as efficient, as much as 50% of people's disposable income goes for food, and a very much larger percentage of the population is involved directly in agriculture and food production.

Another theme that may have been gleaned from this book is that getting the facts is often as difficult as agreeing on them. There are many problems with animal test methods. We are forced to use high-dose extrapolation techniques and also species that may or may not behave or metabolize the way we do. Even if we use accidental exposures that occur industrially or by other means, the high doses may not be extrapolatable to the chronic low-dose exposure to toxins that may occur through ingestion of food. Any of these methods can over- or underestimate the actual risk to humans.

A further problem for consumers and scientists alike is that people have little interest in a news story proving that a certain food or food chemical is safe; thus the flow of information is lopsided. For instance, a 1990 study showing that FD&C Green No. 3 fed over the lifetime of a rat at high doses caused neither cancer nor any other measurable effect (Borzelleca, 1990) remains buried in a technical journal and receives no press at all. One can only wonder how differently the report would have been treated if the data had showed any adverse effects. Making matters worse is the fact that news programs and programs with news formats must be as concerned about television ratings as are other programs on the air. A story about your child's safe apples would hardly excite even the most profoundly interested.

A still further problem has to do with trust. There is much skepticism about information from manufacturers or research funded by industry in terms of possible bias. However, an equivalent skepticism about reports from consumer interest groups is often lacking because consumers believe that these interest groups have their needs at heart. For instance, a public relations firm created a highly successful media campaign for its client, the Natural Resources Defense Council (NRDC). It was highly successful for the NRDC because having its name in the news helped to sell many copies of the NRDC book on pesticides and to increase markedly both the number and dollar amount of contributions to the organization (Smith, 1990). While such selling might not seem as overtly motivated by profit as the selling of food products themselves, it *is* selling. Consumers should be aware that not all campaigns are designed solely with their better interests in mind, especially when only one side of the story is aired.

Lack of trust is not limited to the food industry. Many consumers no longer trust the government or academics either. There have been too many scandals for them to view the government as a credible source. Even the FDA is looked upon by some people as a noncredible source of information, and some feel that the FDA has sold out to industry interests and no longer protects the consumer. Many consumers have similar thoughts about academics selling out to industry because of grants from that source, rather than having consumer interests at heart. However, independent university researchers or health groups such as the American Cancer Society are still regarded by many consumers as credible sources of information.

These are the problems. Steps toward solving them should be addressed. Many people realize the need to address the issue of science illiteracy. In addition, we need to address the issue of food illiteracy. We need more and better education about what happens on a farm, in a processing plant, and in the food marketing chain. Getting credible information out about the food supply and its risks and benefits should be a priority goal. A strong campaign should be launched to help consumers keep and prepare food safely so that they are not at risk for foodborne disease. Risks should be dealt with in an honest manner so that credibility can return.

Dynamic ways to get out consumer information are crucial. The information should be stated in a way that is both interesting and clear. The message should be stated as simply as possible without being fuzzy. Consider the simple directive to eat less red meat. Some consumers followed the directive literally but did not change their level of fat consumption, which was the reason for the message. They simply chose meat that was pink rather than red.

Marketers and advertisers for food companies should not be allowed to tout a product for not having any preservatives while selling another product that does have such additives, expecting that the consumers will not be concerned about them. Nor should they tout the benefit of one attribute of a food without also declaring that the product may contain a detractor. The new U.S. labeling law adopted in 1990 to take effect in 1993 may curb some of these practices.

We need to seek improvements in agricultural production methods with the objective of sustainability and enhanced yield, as well as to seek improvements in food processing and preservation that enhance quality, availability, and variety in the food supply. We must ban things that might put us at risk without needlessly eliminating those that pose marginal risk and provide abundant benefits in maintaining a cheap, readily available food supply. We also must somehow avoid the chemophobia trap, whereby we utilize limited regulatory resources simply to chase a shrinking zero rather than to increase our health and well-being.

Consumers, activists, industry, government, and academia must devise better ways of collaborating so that food safety issues are dealt with from all perspectives and all available resources are utilized. In this way, decisions about the food supply will be made using all available evidence rather than basing them on half truths with the motivating factor of fear.

Finally, as beneficiaries of an abundant food supply and users of the delivery system that other countries are trying to emulate, U.S. consumers should worry less and enjoy more, should be cognizant of real risks but separate them from trivial and irrational ones. Food is one of the great pleasures that humankind has, so in addition to being wary, we should make wise choices and then enjoy them.

REFERENCES

Borzelleca, J. F. 1990. Lifetime toxicity/carcinogenicity studies of FD&C Green No. 3 in rats. Food Chem. Toxicol. 28:813-818.

Smith, K. 1990. Alar: One Year Later—A Media Analysis of a Hypothetical Health Risk. American Council on Science and Health, New York.

Glossary

ACSH—American Council on Science and Health, a group that publishes scientific opinions about many controversial issues

acute exposure—doses of a substance sufficient to cause an immediate adverse reaction

ADI—acceptable daily intake, the amount of a specific food additive thought to be the maximum level that should be ingested on a daily basis, as determined by experts convened for WHO/FAO. There is a margin of safety between the ADI and the lowest level shown to have any physiological effect.

aerobic organism—an organism that grows in air

aflatoxin—toxins produced by molds, commonly from *Aspergillus flavus*

agglutinate—to cause the clumping of cells with an antibody-antigen reaction

alkaloid—a basic substance found in parts of plants. They usually taste bitter and are often used as active principles in drugs.

AMA—American Medical Association, the professional organization for physicians in the United States

anaerobic organism—an organism that grows in the absence of air

anaphylactic shock—an extreme hypersensitive reaction to a small dose of a compound. Shock is thought to result from a reaction between an antigen and a specific antibody fixed in the cell.

anticarcinogens—compounds that inhibit cancer growth

antinutrients—compounds that block the absorption, utilization, or functioning of a nutrient

antibiotic resistance—a property acquired by some microorganisms that renders them unharmed by antibiotics

antioxidant—a substance that inhibits oxidation and thus prevents the breakdown of fats and other compounds that are damaged by oxygen and other forms of activated oxygen such as peroxides

bacteremia—the presence of viable bacteria in the blood stream

bacterial spore—a reproductive form that is produced by certain bacteria and is often very resistant to destruction by common means such as heat, acid, and salt concentration

beta particle—an electron or positron emitted from a nucleus during beta decay

BHA—butylated hydroxyanisole, an antioxidant that works especially well with fat

BHT—butylated hydroxytoluene, an antioxidant that works especially well with fat

calories—a unit indicating the amount of energy in a food. In this book, the word is used in the way that consumers use it, to mean the number of calories in food; technically, it should be kilocalories or Calories.

cancer initiator—a substance that begins the growth of a cancer

cancer promoter—a substance that enhances a tumorous growth

carditis—inflammation of the heart

carrier—a person who has no apparent disease but who harbors a pathogenic organism that can be transferred to another person and that is capable of producing a disease in the other person

CAST—Council for Agricultural Science and Technology, an organization of professors and extension personnel from the Land Grant Universities dedicated to assessing and informing the public about issues concerning food production and agriculture

CCMA—Certified Color Manufacturers Association, the association of companies that manufacture FD&C colors

CDC—Centers for Disease Control, Atlanta; a part of the U.S. Department of Health and Human Services that tracks disease incidence in the United States

cellulitis—inflammation of cellular or connective tissue

cestodes—internally parasitic flatworms comprising the tapeworms

CFR—Code of Federal Regulations. This contains the federal laws. Title 21 of the CFR deals with food laws.

chelator—a substance that can strongly bind divalent metals

chloracne—an acne caused by contact with chlorinated compounds

chronic exposure—a long-term low dose of a chemical ingested on a regular basis

Codex Alimentarius—a document containing international regulations about food

colohemorrhagic—capable of causing hemorrhage of the colon

competitive organism—an organism the growth of which helps to prevent the growth of another (often pathogenic) organism

cross contamination—the spread of pathogenic microorganisms from one food to another via another vehicle, such as an unclean knife or cutting board, that has been in contact with a food containing the microorganism

cumulative toxicity—a toxic reaction that results because a chemical is held in the body and increases in concentration

cyanogenic—capable of producing cyanide; said of plants such as sorghum and wild cherry

cytotoxin—a substance that inhibits the normal functioning of cells or causes the destruction of the cells, or both

2,4-D—2,4-dichlorophenoxyacetic acid, a common herbicide

DAL—defect action level, the number of defects allowed in a food product before the product is considered unwholesome or contaminated

danger zone—the temperature range that enables rapid proliferation of microorganisms

depuration—a process to remove waste or foul products

DDT—dichlorodiphenyltrichloroethane, a persistent pesticide in the organochlorine family

deiodinate—to remove the iodine from a compound

de minimus—an amount so small that it is thought to be inconsequential. The phrase is Latin and is used in the law to say that the law does not concern itself with trifles.

DES—diethylstilbestrol, a synthetic estrogen used as a growth promoter in animals

dinoflagellate—a pigmented protozoa having some plantlike characteristics and using two or more flagella for movement

DNA alkylation—a change caused in DNA by the addition of another compound to the DNA

dose response—a response that is graded in proportion to the amount of material administered

DRV—daily reference value, the value established for the 1990 Nutrition Labeling and Education Act for substances that do not have RDAs and that could cause problems if taken in excessive amounts

dyspepsia—indigestion

EC—European Economic Community

EDB—ethylene dibromide, a banned fumigant once used for grains and produce

emetic—an agent that causes vomiting

encephalitis—inflammation of the brain

enteric—relating to the intestines

enterocolitis—inflammation of the mucous membrane of the small and large intestines

enteroinvasive organism—an organism that invades the intestinal tract

enteropathogenic—pathogenic for the gastrointestinal tract

enterotoxigenic organism—an organism or compound toxic to the mucous membrane cells of the small intestine

EPA—Environmental Protection Agency

epidemiological data—data derived from study of population groups

facultative anaerobe—an organism able to live with or without oxygen

FAO—Food and Agriculture Organization

FASEB—Federation of the American Societies of Experimental Biologists. Groups of scientists from this federation are often asked to summarize the scientific literature or to conduct studies that help to evaluate the safety of various food components.

FDA—U.S. Food and Drug Administration

FIFRA—Federal Insecticide, Firearm and Rodenticide Act, the federal law regulating pesticide application

FSC—Food Safety Council

FSIS—Food Safety and Inspection Service of the USDA, a body that inspects meat and poultry

gamma ray—radiation with a very short wavelength, emitted by elements such as cobalt or cesium as they disintegrate spontaneously

GAO—General Accounting Office

gastroenteritis—inflammation of the mucous membranes of the stomach and intestines

goitrogen—a compound that causes a goiter (an enlargement of the thyroid gland that is not due to a tumor)

GRAS—generally recognized as safe; refers to those food additives in existence at the time of the passage of the Miller Food Additive Amendment. These compounds were deemed safe because they were already in the food supply and had therefore established their safety through use.

Guillain-Barré syndrome—a neurological syndrome marked by muscle weakness or flaccid paralysis

HACCP—hazard analysis and critical control points, a system designed to eliminate contamination of food in the processing chain by carefully controlling and monitoring crucial points in the process

half-life—the time required to dissipate half of a radioactive material; the biological half life is the time required by a radioactive isotope within the body to lose half of its activity

hallucinogen—a compound that can result in optical or auditory hallucinations or disturbances in perception or thought processes

hemolytic anemia—an anemia resulting from the destruction of the red blood cells

HERP—hazard exposure risk potential, a scheme to evaluate carcinogens that was developed by Dr. Bruce Ames, based on the cacinogenic potency of the compound and the level of exposure in the diet and the environment

hyperestrogenic syndrome—a group of symptoms resulting from too much estrogen in the body

hypotension—low blood pressure

IARC—International Agency for Research on Cancer

IFT—Institute of Food Technologists, a professional association for food professionals

immunocompromised—describing the decreased ability of the immune system to combat disease due to the taking of an immunosuppressive drug or due to an underlying disease such as AIDS or cancer

immunosuppressive—describing an agent that impairs the functioning of the immune system

incidental additive—an additive that makes its way into food at some place between the farm and the table but that was not added intentionally to perform a function

indoles—ring compounds derived from cruciferous vegetables (cabbage, broccoli, etc.) and thought to have anticarcinogenic properties

in vitro—occurring in the test tube

in vivo—occurring in the body

IPM—integrated pest management, a system using a wide variety of methods to minimize contamination by pests

JECFA—Joint Expert Committee on Food Additives, an international committee of experts convened to assess the safety of food constituents used worldwide.

It is under the auspices of FAO and WHO.

JECFI—Joint Expert Committee on Food Irradiation. An international committee of experts convened to assess the safety of food irradiation. It is under the auspices of FAO and WHO.

kwashiorkor—extreme malnutrition caused by inadequate food, especially inadequate protein

LD$_{50}$—the administered dose at which half of the test animals die; this number is used as a way to asses acute toxicity

lupus erythematosus—a disease of the skin and mucous membranes

MAP—modified atmosphere packaging, a process in which air is removed from a food package and replaced with other gases, usually carbon dioxide or a mixture of gases such as carbon dioxide and nitrogen, to extend the shelf life of a product

meningitis—inflammation of membranes of the brain and spinal cord

methemoglobinemia—a condition in which the blood can no longer deliver oxygen to and pick up oxygen from the tissue

MMWR—*Morbidity and Mortality Weekly Report*, a weekly publication of the Centers for Disease Control in Atlanta

MRL—maximum residue level, the maximum dose of a pesticide that can be found in a particular food; this is a level far below the ADI

MTD—maximum tolerated dose, the dose determined in prechronic studies to be the highest that will not impair longevity of animals other than by possibly inducing neoplasms. This is a dose used for long-term and carcinogenicity studies.

MTDI—Maximum tolerated daily intake, a level set for compounds that are essential and for which the maximum safe dose is known, as for phosphorus

mutagen—a compound capable of producing a mutational change in genetic material

mycotoxin—a toxin produced by molds and capable of causing cancer, altering estrus cycles, and disturbing other body functions

NAS—National Academy of Sciences

necrotic—pertaining to the death of cells or tissue

neoplasm—literally, any new tissue growth, but used in referring to uncontrolled growth such as occurs in tumors

nematode—a common name for a parasitic worm

nephrosis—degeneration of the lining of the renal (kidney) tubules

neurotoxin—a compound poisonous to the nervous system

NIH—National Institutes of Health

NLEA—Nutrition Labeling and Education Act of 1990. This act established new standards for the nutrition label that must be in place on all food packages by May 1993.

NOEL—no observable effect level, the highest level of a substance that can be administered to an experimental animal without any evidence that the substance has been taken in

nonphysiological dose—a dose so high that it would not be encountered in normal metabolism or eating, even if the eating pattern were quite distorted

NRC—National Research Council of the National Academy of Sciences. The Food and Nutrition Board of this group evaluates and sets RDAs and makes other dietary recommendations.

OTA—Office of Technology Assessment

organic farming—agriculture practiced without chemical fertilizers, pesticides, growth promoters, or subtherapeutic drugs as feed additives

organic food—Although this term has no legal definition, it is considered to be food grown through organic farming methods without any chemical additives added during processing; for some people it also means that the food has been minimally processed.

PAH—polycyclic aromatic hydrocarbons, compounds that can be mutagenic and carcinogenic and that can be produced during heating

parasite—an organism that lives on or in another and uses it for food

pathogen—an organism capable of producing disease

PBB—polybrominated biphenyl, a chlorinated hydrocarbon that finds its way into the food chain through contaminated animal feed

PCB—polychlorinated biphenyl, a chlorinated hydrocarbon that finds its way into food inadvertently through packaging made from recycled carbonless paper or through industrial pollution

photocarcinogen—a compound that causes cancer under the influence of light

plasmid—an extrachromosomal hereditary determinant of bacteria that replicates and is transferred independently of the chromosomme. It may carry special traits such as antibiotic resistance.

PMTDI—provisional maximum tolerable daily intake, a level set for food components that don't have cumulative properties but are toxic at high levels. Some of these may be essential at low levels but toxic at higher levels, as determined by committees of WHO/FAO.

polyploidy—a condition in which cells contain more chromosomes than customary; the condition of containing more than three haploid chromosomes

psychrotroph—an organism that can grow at cold temperatures

PTWI—provisional tolerable weekly intake, a level set for food contaminants such as cadmium that accumulate in the body, as determined by committees of WHO/FAO

radionuclide—a nuclide that exhibits radioactivity. A nuclide is a species of atom characterized by the number of its protons and neutrons and by the energy content of the nucleus.

RDA—Recommended dietary allowance, a value established by the National Academy of Sciences, which sets required nutrient values for all healthy people in the United States. This value is not an average or a minimum required for a particular population group; rather it takes into account the needs of all individuals, and thus the word *allowance* is used. The U.S. RDA stands for U.S. recommended daily allowance. This uses the values for teenaged males, the group with the highest nutrient requirements of all groups except those with the special needs of pregnancy and lactation. This value was used on nutritional labels from 1971 until it was phased out by the NLEA.

RDI—recommended dietary intake. This value replaced the U.S. RDA for nutritionally labeled products. It is based on the average of the requirements for all age groups over 4 years old (excluding pregnant and lactating women).

Reiter's syndrome—a set of symptoms comprising a fever that lasts over 2 weeks and is accompanied by pains in the joints, conjunctivitis, inflammation of the urethra, and an enlarged spleen

septicemia—a systemic disease caused by the presence of microorganisms or their toxins in the bloodstream

sustainable agriculture—agriculture that enables the production of adequate food while maintaining the resources needed to continue producing adequate food

2,4,5-T—2,4,5-trichlorophenoxyacetic acid, a herbicide

TCDD—2,3,7,8-tetrachlorodibenzo-*p*-dioxin, a herbicide

teratogen—a substance capable of producing abnormalities; literally, capable of producing a monster

toxemia—the presence of a noxious substance circulating in the blood as a result of toxic materials being either absorbed from food or produced by microorganisms

trematode—any of the parasitic flatworms, including flukes

UFW—United Farm Workers, a labor union that represents people who work as farm laborers

URP—unique radiolytic product, a product believed to be formed during food irradiation. However, its uniqueness is controversial.

USDA—United States Department of Agriculture

vegetative cell—the cell phase of a microorganism that is actively growing

virulent strain—a strain of a microorganism capable of producing disease

viruses—living organisms that cause disease in plants, animals, and humans. They are incapable of growth and reproduction outside of a living cell.

VSL—virtually safe level, a level of substance that, when ingested, is thought to be safe

WHO—World Health Organization

xenobiotic—not occurring in nature; used especially for synthetic organic compounds that may be resistant to biodegradation

Index

AAF. *See* 2-Acetylaminofluorene
AAP. *See* American Academy of Pediatrics (AAP)
Abbott Laboratories, and cyclamate, 215, 216
Absolute safety. *See* Safety
Academia, as source of information, 422
Acceptable daily intake (ADI), 40
  of pesticides, 348-349, 351, 367-368
  of residues, 347
  of sulfite, 237
Accum, Frederick. *See* Marcus, Frederick C.
Acesulfame-K (sweetener), 224-225
Acetic acid, 245
2-Acetylaminofluorene, 43, 44
Acetylcholine, 76
Acidulants, 245
Acquired immunodeficiency syndrome (AIDS), 60, 160, 311
ADA. *See* American Diabetes Association (ADA)
ADD (attention deficit disorder). *See* Hyperactivity
Additives, 203-249. *See also* Colors; Flavors; Sweeteners; and additives by name
  banned, 212-214
  carcinogenicity of, 8, 21, 212-214
  chemical, 1, 172, 420-421
  classification of, 205-206, 207-212
  consumer concern about, 8, 69-70, 203, 211-212
  consumption of, 203, 204-205, 230
  defined, 204-206
  direct, 207
  in fast foods, 175
  feed, 8, 379-387
  function of, 205-206
  history of, 203
  indirect, 207
  intake of, 230
  intentional, 207
  laws on, 12, 203, 207, 212-214
  and mycotoxin reduction, 146
  need for use, 248-249

and processing, 172, 174, 175
  for quality, 174
  reasons for use of, 206
  regulating, 21
  safety of, 8, 21, 207-210, 215
  unintentional, 207
  uses of, 206
Additives, indirect. *See* Contaminants
ADI. *See* Acceptable daily intake (ADI)
Adulteration, 203
  by adding colorants, 259
  descriptions of, 19
  laws on, 15-20, 203
Advertising, regulation of, 28
Aeromonas
  hydrophilia, 135
  sobria, 135
AFB. *See* Aflatoxins, blue (AFB)
AFB1. *See* Aflatoxins, B1 (AFB1)
AFG. *See* Aflatoxins, green (AFG)
Aflatoxicosis, 143-144. *See also* Aflatoxins
  in humans, 143-144, 145, 147, 148
  in poultry, 143, 147
  symptoms, 144
Aflatoxins, 147-149. *See also* Aflatoxicosis; Carcinogens; Toxicants, natural
  adverse effects of, 147, 149
  B1 (AFB1), 147-148
  blue (AFB), 143
  carcinogenicity of, 144, 147, 148
  in corn, 144-145, 148
  in dairy products, 141
  defined, 147
  in feed and grains, 143-144, 145
  green (AFG), 143
  in milk, 141
  in peanuts, 141, 145, 148
  in poultry, 143
  in selected foods, 144
  toxicity of, 147-148
  in wheat, 145
Africa
  aflatoxicosis in, 143-144
  monellin (sweetener) use in, 227

433